METHODS IN MOLECULAR BIOLOGY

Series Editor
John M. Walker
School of Life and Medical Sciences,
University of Hertfordshire, Hatfield,
Hertfordshire, UK

For further volumes:
http://www.springer.com/series/7651

For over 35 years, biological scientists have come to rely on the research protocols and methodologies in the critically acclaimed Methods in Molecular Biology series. The series was the first to introduce the step-by-step protocols approach that has become the standard in all biomedical protocol publishing. Each protocol is provided in readily-reproducible step-by step fashion, opening with an introductory overview, a list of the materials and reagents needed to complete the experiment, and followed by a detailed procedure that is supported with a helpful notes section offering tips and tricks of the trade as well as troubleshooting advice. These hallmark features were introduced by series editor Dr. John Walker and constitute the key ingredient in each and every volume of the Methods in Molecular Biology series. Tested and trusted, comprehensive and reliable, all protocols from the series are indexed in PubMed.

Fibrous Proteins

Design, Synthesis, and Assembly

Edited by

Shengjie Ling

School of Physical Science and Technology, ShanghaiTech University, Shanghai, China

 Humana Press

Editor
Shengjie Ling
School of Physical Science and
Technology
ShanghaiTech University
Shanghai, China

ISSN 1064-3745   ISSN 1940-6029   (electronic)
Methods in Molecular Biology
ISBN 978-1-0716-1576-8   ISBN 978-1-0716-1574-4   (eBook)
https://doi.org/10.1007/978-1-0716-1574-4

This Humana imprint is published by the registered company Springer Science+Business Media, LLC, part of Springer Nature.
The registered company address is: 1 New York Plaza, New York, NY 10004, U.S.A.

Preface

Fibrous proteins, such as silk, collagen, elastin, resilin, and keratin, are widely present in different biological tissues or biological structural materials. The significant difference between fibrous proteins and other proteins is their fibrillar structures, which usually are constructed by the highly ordered secondary structures and repetitive amino acid motifs. Unlike globular proteins, most fibrous proteins do not have complex tertiary and quaternary structures but tend to form highly ordered one-dimensional nanoarchitectures, such as nanofibrils, nanofibril bundles, and microfibrils. These nanoarchitectures are further assembled into elaborate one-dimensional, two-dimensional, or three-dimensional ordered structures in biological systems, matching the versatile bio-functional needs. Accordingly, the first part of this volume introduces the structure of representative fibrous proteins, including animal silks, collagen, elastin, resilin, and keratin. The structural hierarchy and structure-property relationship are highlighted for each kind of fibrous proteins so as to provide fundamental guidance for the understanding and utilization of these proteinic nanomaterials.

Some fibrous proteins, such as cocoon silks, collagen, and keratin, have a shining commercial past due to their wide availability and sustainability. For example, silk and animal hair have been used as textile fibers for thousands of years. More remarkably even, these fibrous proteins, especially their nanobuilding blocks, are proven to have a bright future in these years because of their unique combination of biological features (e.g., outstanding biocompatibility and biodegradability) and nano effects (e.g., ultrahigh specific surface area and aspect ratio). Fibrous proteins can be utilized in their native forms, but, importantly, they can be chemically modified and can even be disassembled in preparation and re-assembled under artificial conditions. Lately, a variety of bottom-up and top-down approaches have been developed to isolate or re-assemble fibrous proteins from their natural products. Additionally, the fibrous proteins can be synthesized using gene recombination technologies or chemosynthetic strategies (e.g., solid-phase synthesis and liquid-phase synthesis). These technologies are particularly important for practical applications of fibrous proteins with limited natural sources, such as spider silks, elastin, and resilin. Therefore, Parts II to IV aim to provide detailed experimental protocols for the synthesis, assembly, and characterization of natural, regenerated, and recombinant fibrous proteins. The methods and strategies introduced in this volume are expected to help researchers master the fibrous protein preparations and provide a broad and interdisciplinary perspective to understand structure-property-function relationships of natural and reconstituted fibrous proteins.

The volume was supported by the National Natural Science Foundation of China (nos. 51973116, U1832109, 21935002), the Users with Excellence Program of Hefei Science Center CAS (2019HSC-UE003), the starting grant of ShanghaiTech University, and State Key Laboratory for Modification of Chemical Fibers and Polymer Materials.

Shanghai, China   *Shengjie Ling*

Contents

Contributors

LEITAO CAO • *School of Physical Science and Technology, ShanghaiTech University, Shanghai, China*

NING FAN • *Biomass Molecular Engineering Center and Department of Materials Science and Engineering, School of Forestry and Landscape Architecture, Anhui Agricultural University, Hefei, Anhui, China*

YIMIN FAN • *Jiangsu Co-Innovation Center of Efficient Processing and Utilization of Forest Resources, International Innovation Center for Forest Chemicals and Materials, College of Chemical Engineering, Nanjing Forestry University, Nanjing, China*

WENLI GAO • *Key Lab of State Forest and Grassland Administration on Wood Quality Improvement & High Efficient Utilization, School of Forestry and Landscape Architecture, Anhui Agricultural University, Hefei, Anhui, China*

HONGCHONG GUO • *School of Physical Science and Technology, ShanghaiTech University, Shanghai, China*

NA KONG • *School of Physical Science and Technology, ShanghaiTech University, Shanghai, China*

SHENGJIE LING • *School of Physical Science and Technology, ShanghaiTech University, Shanghai, China*

QIANG LIU • *School of Physical Science and Technology, ShanghaiTech University, Shanghai, China*

YAWEN LIU • *School of Physical Science and Technology, ShanghaiTech University, Shanghai, China*

ZHUOCHEN LV • *School of Physical Science and Technology, ShanghaiTech University, Shanghai, China*

YING PEI • *School of Materials Science and Engineering, Zhengzhou University, Zhengzhou, China*

PING QI • *School of Physical Science and Technology, ShanghaiTech University, Shanghai, China*

HUANHUAN QIAO • *Biomass Molecular Engineering Center and Department of Materials Science and Engineering, School of Forestry and Landscape Architecture, Anhui Agricultural University, Hefei, Anhui, China*

JING REN • *School of Physical Science and Technology, ShanghaiTech University, Shanghai, China*

TING SHU • *School of Physical Science and Technology, ShanghaiTech University, Shanghai, China*

YUZHAO TANG • *National Facility for Protein Science in Shanghai, Shanghai Advanced Research Institute, Chinese Academy of Sciences, Shanghai, China*

YUELONG XIAO • *School of Materials Science and Engineering, Zhengzhou University, Zhengzhou, China*

CHAO YE • *School of Physical Science and Technology, ShanghaiTech University, Shanghai, China*

WENJIE YU • *National Facility for Protein Science in Shanghai, Shanghai Advanced Research Institute, Chinese Academy of Sciences, Shanghai, China*

JICONG ZHANG • *School of Physical Science and Technology, ShanghaiTech University, Shanghai, China*

WENWEN ZHANG • *Jiangsu Co-Innovation Center of Efficient Processing and Utilization of Forest Resources, International Innovation Center for Forest Chemicals and Materials, College of Chemical Engineering, Nanjing Forestry University, Nanjing, China*

CHENXI ZHAO • *School of Physical Science and Technology, ShanghaiTech University, Shanghai, China*

KE ZHENG • *Biomass Molecular Engineering Center and Department of Materials Science and Engineering, School of Forestry and Landscape Architecture, Anhui Agricultural University, Hefei, Anhui, China*

JIAJIA ZHONG • *National Facility for Protein Science in Shanghai, Shanghai Advanced Research Institute, Chinese Academy of Sciences, Shanghai, China*

LIANG ZHOU • *Key Lab of State Forest and Grassland Administration on Wood Quality Improvement & High Efficient Utilization, School of Forestry and Landscape Architecture, Anhui Agricultural University, Hefei, Anhui, China*

XIAOJIE ZHOU • *National Facility for Protein Science in Shanghai, Shanghai Advanced Research Institute, Chinese Academy of Sciences, Shanghai, China*

Part I

Introduction to Fibrous Proteins

Chapter 1

Structure of Animal Silks

Wenwen Zhang and Yimin Fan

Abstract

As an abundant fibrous protein, animal silks have received a variety of interests in both traditional and high-tech industries, such as textiles, decoration, and biomedicine, due to their unique advantages in mechanical performance, sustainability, biocompatibility, and biodegradability. While developing applications of animal silks, the structure of animal silks has also received more and more attention in these decades. Briefly, most animal silks can be considered as semicrystalline fibers, which are composed of β-sheet nanocrystals and amorphous regions. However, different animal silks have similarities and also have obvious differences at different structural levels. In this chapter, we will introduce the structures of the three most representative animal silks, that is, spider dragline silk, tussah silk, and mulberry silk. The similarities and differences in their structures will be highlighted, so as to provide fundamental guidance for the research and use of these animal silks.

Key words Animal silks, Secondary structure, Primary structure, Nanofibril organization

1 Introduction

Animal silks (including mulberry silk, tussah silk, and spider silks) and their associated silk proteins have received extensive attention in these decades owing to their unique advantages of sustainability, mechanical performance, and good biocompatibility [1–3]. Among these animal silks, the silkworm cocoon silks, such as mulberry and tussah silks, are relatively easy to obtain. Therefore, their-related research, especially their application research, is also more extensive. The development of their applications has expanded from the traditional textile industry to various emerging fields, such as regenerative medical tissues, drug carriers, photonic crystal materials, separation membranes, and smart devices [4–10].

Compared with silkworm cocoon silks, the application research on spider silk protein is relatively lagging because it is difficult to obtain it in large quantities from nature. However, with the rapid development of synthetic biology in recent years, recombinant proteins with spider silk protein-like sequences have been successfully expressed in different host systems, such as bacteria, yeast, and

Shengjie Ling (ed.), *Fibrous Proteins: Design, Synthesis, and Assembly*, Methods in Molecular Biology, vol. 2347,
https://doi.org/10.1007/978-1-0716-1574-4_1, © Springer Science+Business Media, LLC, part of Springer Nature 2021

goats [11–14]. This progress makes it possible to use spider silk-like proteins. For example, in these years, a series of recombinant spider silk materials have been developed, such as fibers, hydrogels, and tissue engineering scaffolds, and have shown promising applications in biomedicine and fiber derivatives [15–20].

A detailed comparison of the properties of silk protein materials and animal silk reveals that the mechanical properties of most silk protein materials are inferior to those of animal silk. One of the critical reasons is that so far, the understanding of animal silk is still insufficient, although the research on the structure of animal silk has spanned half a century. For instance, Pauling et al. began to use X-ray diffraction to analyze the crystal structure of silkworm cocoon silks as early as the 1950s. With the development of characterization techniques, the understanding of animal silk has significantly deepened. However, even so, there are still some debates in the understanding of the structure of animal silk. For example, there is still no consensus on the types and relative content of the secondary structure of animal silk. Take mulberry silkworm silk as an example; its measured crystal content varies in a wide range, roughly from 20% to 50%.

Beyond secondary structure, the mesostructure of animal silks, such as nanofibril, nanofibril bundle, and nanofibrillar interface, has also received attention in the most recent years, since these mesostructures have been proven to play a vital role in the determination of fracture resistance and ultralow temperature toughness of animal silks [21–25]. Therefore, in this chapter, we will take spider silk, mulberry silk, and tussah silk as examples to systematically introduce the hierarchical structure of animal silk, including the amino acid sequence, secondary structure, and the organization of nanofibrils. As mentioned above, there are still some differences in the understanding of the structure of animal silk, this chapter therefore also summarizes these differences in more detail, so as to provide objective guidance for the follow-up study of animal silk or silk protein materials.

2   Primary Structure of Animal Silk

Owing to the diversity of silk species, different silks show considerable complexity in their construction. At the molecular scale, the constituent of proteins varies with the variety of silks (Fig. 1). The amino acid residues of the spider silk, mulberry silk, and tussah silk are mainly alanine (A) and glycine (G), and the ratio of the two residues in the three different kinds of silk is very similar. However, the other amino acid residues vary widely [26–29]. Spider dragline silk mainly contains charged amino acid residues, such as glutamic acid (E) and arginine (R). In contrast, the main amino acid residue in mulberry silk and tussah silk is serine (S).

Mulberry silk :

GAGAGSGAGAGSGAGAGSGAGAGSGAGAGSGAGAGSGAGAGYGAGVGVGYGAGYGAGAGAGYGAGAGS
GAASGAGAGSGAGAGSGAGAGSGAGAGSGAGAGSGAGAGSGAGAGSGAGAGSGAGAGSGAGAGSGAGA
GSGAGVGSGAGAGSGAGAGVGYGAGAGVGYGAGAGSGAASGAGAGSGAGAGSGAGAGSGAGA
GSGAGAGSGAGAGSGAGAGSGAGAGSG

Tussah silk :

AAAAAAAAAAAAAAGSGAGGSGGYGGYGGYGSDSAAAAAAAAAAAAAAGSSAGGAGGGYGWGDG
GYGSDSAAAAAAAAAAAAAA

Nephila clavipes (N.clavipes) MaSp1 :

AAAAAAGGAGQGGYGGLGSQGAGRGGLGGQGAGAAAAAAGGAGQGGYGGLGGQGAGQGGYG
GLGSQGAGRGGLGGQGAGAAAAAAAA

Nephila clavipes (N.clavipes) MaSp2 :

AAAAAAAASGPGQQGPGGYGPGQQGPGGYGPGQQGPSGPGSAAAAAAAASGPGQQGPGGYGP
GQQGPGGYGPGQQGPGQQGPGGYGPGQQGPGQLSGPGSAAAAAAAA

Latrodectus Hesperus(L.hesperus) MaSp1 :

AAAAAAAAAGGAGQGGQGGYGQGGYGQGGAGQGGAGSAAAAAAAAAGGAGQGGYGRGGAGQ
GGAGAAAAAAAAA

Latrodectus Hesperus (L.hesperus) MaSp2 :

AAAAAAAAAAGGAGPGRQQGYGPGSSGAAAAAAAAGGPGYGGQQGYGPGGAGAAAAAAAAAA

Fig. 1 The primary structure of animal silk. Typical amino acid sequences of repetitive core of Mulberry silk H-chain, Tussah silk, and the major ampullate spidroin of *Nephila clavipes* and *Latrodectus hesperus*. The GAGAGS and A_n which refers to the β-sheet forming segments are highlighted

2.1 Mulberry Silk   In the primary structure, mulberry silk fibroin consists of three major proteins: a heavy chain (H-chain), a light chain (L-chain), and a glycoprotein (P25) [30]. The heavy and light chains are linked by disulfide bonds and then bound to P25 by noncovalent action [31, 32]. In addition, the H-chain, L-chain, and P25 are assembled in a ratio of 6:6:1 to form a high molecular weight structural unit [33]. The H-chain is 5263 amino acid residues long, with a molecular weight of ~391 kDa. It is composed of 45.9% glycine (G), 30.3% alanine (A), 12.1% serine (S), 5.3% tyrosine (Y), 1.8% valine (V), and 4.7% of the other 15 amino acid types. The primary structure of the heavy chain consists of two less repetitive ends (N- and C-termini) and a highly repetitive and glycine-rich center which is divided into four typical domains: GAGAGS, tyrosine-containing (Y) fragments, GAAS and nonrepetitive fragments [34] (Fig. 1). The highly repetitive GAGAGS mainly make up the crystalline regions, and tyrosine-containing fragments mainly constitute the semicrystalline regions [29]. The L-chain is 266 amino acid residues long, with a molecular weight of ~26 kDa and has a less differentiated amino acid composition and nonrepetitive sequence. The glycoprotein P25 has a molecular weight of 30 kDa [35].

2.2 Tussah Silk

In terms of the amino acid sequence of the main molecular chain of tussah silk fibroin, it is composed of 2639 amino acid residues, and the molecular weight is about 216 kDa [36]. In addition to the 155 amino acid residues at the N-terminal, the tussah silk fibroin is divided into 80 polyalanine-containing structural units, which are connected end to end. Each structural unit consists of a repetitive sequence fragment of A_n (polyalanine, $10 < n < 15$) and a G-rich fragment. A_n, similar to the GAGAGS in mulberry silk, is the basic unit of β-sheet structure, but more hydrophobic than the GAGAGS in mulberry silk. Among G rich fragments, tussah silk is mainly composed of some nonrepeating fragments, such as GGYG, GGAG, GSGA, and RGD [26] (Fig. 1). It is worth noting that because of many polyalanine fragments with a degree of polymerization over 12 in tussah silk, the dominant conformation in the silk fibril is still β-sheet, but there should also contain a fraction of α-helix structure.

2.3 Spider Dragline Silk

Similar to the amino acid sequence of tussah silk fibroin, the spider dragline silk generally made by two different spidroins with molecular weight up to ~350 kDa [37]: the major ampullate spidroin 1 (MaSp1) [38] and the major ampullate spidroin 2 (MaSp2) [35]. Except for a few spider silk such as *Latrodectus hesperus* (*L. hesperus*), the complete amino acid sequence of most spider silk is still unclear, so the spider silk protein sequences collected by the protein library are mostly small fragments. In addition to C- and N-terminal fragments, spidroins are generally divided into three domains, including A_n (GA_n), GGX, or GPGXX, where X stands for tyrosine, leucine, and glutamine [39, 40] (Fig. 1). The function of A_n fragment is similar to the GAGAGS of mulberry silk fiber, which is the unit that forms β-sheet structure [41], and the $(GA)_n$ fragment plays the same role in spider dragline silk [42]. The GGX fragment may form a 3₁ helix structure [43, 44]. GPGXX fragment is considered to form a β-turn spiral structure and provides good elongation for the dragline silk and capture silk of spider [45].

3 Secondary Structure of Silk

Animal silk proteins differ significantly in both amino acid composition and sequence of amino acid residues in the primary structure. However, due to the specific regularity cycle of amino acid sequences in silk proteins and spidroin, almost all animals contain well-defined and similar secondary structures, such as β-sheet, β-turn, α-helix, 3₁ helix, and random coil [46] (Fig. 2). The β-sheet conformation can be divided into parallel and antiparallel

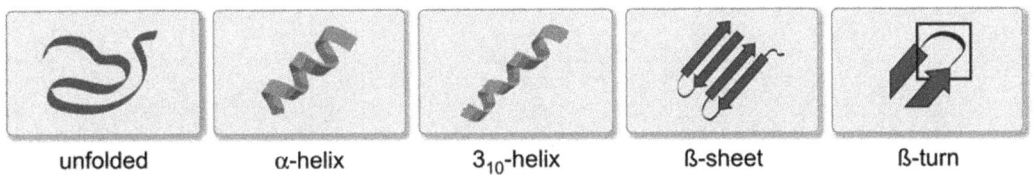

unfolded   α-helix 3_{10}-helix   ß-sheet   ß-turn

Fig. 2 Common secondary structure of animal silk. Adapted by permission from Elsevier [46], Copyright 2008

β-sheet conformation [47]. Among them, because of the better arrangement of the molecular chain segments, the antiparallel β-sheet conformation is more stable.

3.1 Mulberry Silk

As a dominant component of repetitive core, the repeated GAGAGS fragments are considered as the constituent units of the β-sheet structure in mulberry silk. The tyrosine-containing (Y) fragments may contain constituent other β-sheet crystals with different cell parameters from those of the GAGAGS fragments [48]. GAAS, on the other hand, is thought to form a four-residual β-turn structure, which plays a role in perturbing the repeated segments of crystalline $(GX)_n$ (X represents nonglycine residue) [49]. The nonrepetitive fragments are hydrophilic, which can form a loop structure and promote the formation of antiparallel β-sheet structure [29].

3.2 Tussah Silk

The A_n fragments, like the hexapeptide in mulberry silk fibroin, function as the β-sheet structure former in tussah silk. Therefore, the main crystal structure of tussah silk was found to be a β-sheet structure. While compared with mulberry silk fiber, the sequence of amino acids has many "polyalanine" fragments with alanine residue in 10–15, in consequence, the tussah silk contains a small amount of α-helix structure [50].

3.3 Spider Dragline Silk

While the secondary structure of spider silk is more complex, the alanine residual of "polyalanine" fragments in spidroin is between 4 and 7, which can be the subject of β-sheet conformation, but also has the possibility to form α-helix. In addition, there is evidence for the existence of 31 helix conformation in spider silk [43]. A series of simulation results show that the higher stiffness, strength, and toughness can be achieved when the β-sheet crystallite size is limited to a certain range (a few nanometers), and the uniform shear deformation and molecular "stick-slip" deformation owing to the limitation of the size can also enhance their performance significantly [24, 51, 52]. In fact, due to the existence of nonperiodic lattice structure, the crystalline sizes obtained by different testing methods are different, for example, the crystallite size in spider dragline silks is calculated to be ~2 × 5 × 7 nm from the X-ray

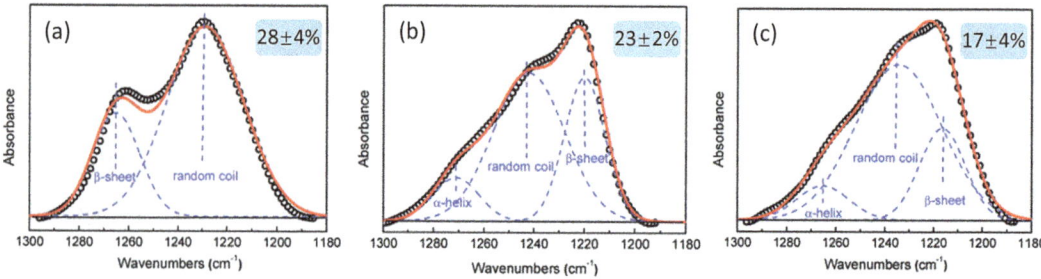

Fig. 3 Deconvolution results of amide III band and the β-sheet content of single animal silk fibers: Mulberry silk, Tussah silk, and *Nephila clavipes*. Adapted by permission from American Chemical Society [62], Copyright 2011

diffraction patterns [53] while 70–500 nm by the analytical transmission electron microscopy (TEM) and electron diffraction experiments [54, 55]. The amorphous structure is more complicated than the crystalline region. The wide-angle X-ray scattering (WAXS) verified the presence of oriented non-crystalline phase [56] (a third phase) [57]. The small-angle X-ray scattering (SAXS) results suggest that the third phase combined with the crystalline phase to form the nanofibrillar structure, and then embedded into semicrystalline matrix [56].

Generally, natural animal silks are considered to be semicrystalline biopolymers with highly organized antiparallel β-sheet nanocrystals embedded into amorphous regions constituted by the random coil and/or helix structure [58]. The crystalline region is highly oriented along the fiber axis and determines the strength and stiffness of animal silk [59], while the amorphous region defines the elasticity [60]. Because of more crystallizable GAGAGS fragments in mulberry silk fibroin, the crystallinity of mulberry silk is higher than that of tussah silk and spider dragline silk. The results of synchrotron FTIR show that the content of β-sheet in the spider dragline silk, mulberry silk, and tussah silk is $17 \pm 4\%$, $28 \pm 4\%$, and $23 \pm 2\%$, respectively [61, 62] (Fig. 3).

4 Nanofibril Organization in Animal Silks

Commonly, both the spider and silkworm cocoon silks feature a similar hierarchical structure (Fig. 4). At the nanoscale, all single silk nanofibrils show a beaded topological structure, which is composed of uniform nanoscale globules with a diameter of 3–5 nm. At the micron scale, these nanofibrils are bundled into 20–200 nm thick microfibrils along the longitudinal fiber axis and further organized into silk fibroin filaments [63]. The researches on nanofibril structure of animal silks mainly depend on the dissolving of

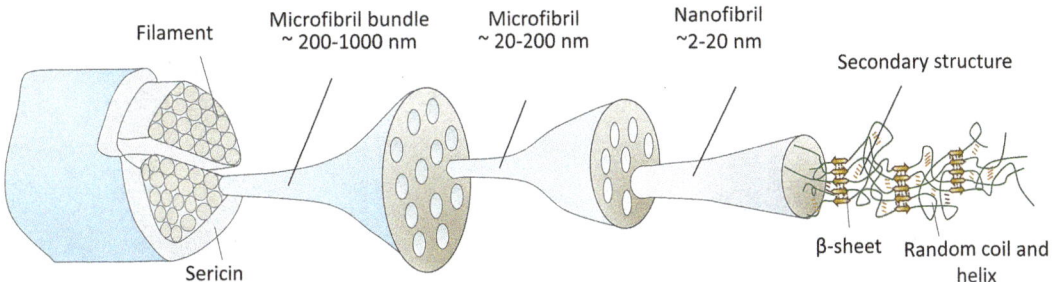

Fig. 4 Schematic of the hierarchical structure of animal silk fiber

silks or the investigating of the cross section of silks by electron microscopy, atomic force microscope (AFM), optical microscope, and scattering techniques.

4.1 Mulberry Silk

The morphology of the nanofibrils in mulberry silk has been widely investigated. The nanofibrils with diameters of 20–30 nm can be observed on the surface of single mulberry silk after degumming (remove the outermost layer of sericin) [64] (Fig. 5a). The internal structure of single mulberry silk has been revealed by degumming and peeling or slicing. Two irregularly shaped central protein brins with numerous nanofibrils have been uncovered by the AFM and SEM observation. These nanofibrils with diameters of 60–200 nm are arranged parallel to the fiber direction [65–68] (Fig. 5b, c). In addition, the layers of nanofibrils aligned at various angles featured a "pseudolayered" structure [69]. Extracting nanofibrils from mulberry silk by chemical treatment is another effective method. Single nanofibril with a diameter of 20 nm and length of 500 nm has been exfoliated from mulberry silk by HFIP treatment [9]. Niu et al. obtained fibrils with a diameter of 30 nm and further isolated silk nanoribbons with a width of 25 nm and a thickness of 0.4 nm [70]. In general, the nanofibrils obtained by chemical dissolution affected by chemical regents inevitably, and the diameter of nanofibril will systematically be underestimated.

4.2 Tussah Silk

Similar to the investigation of nanofibrils in mulberry silk, the nanofibrils of tussah silk could also be enquired by the methods of direct observation and chemical extraction. Through the polarized light microscopy, the highly aligned fibrillar bundles with high orientation can be observed under cross-polarized light on the surface of degummed *A. yamamai* silk (Fig. 5d). Furthermore, through exfoliating the surface layer of the degummed *A. yamamai* silk, the ribbon-like microfibrils with a width of 500–1000 nm and numerous porous with a width of ~200 nm can be detected by the SEM images [5] (Fig. 5e). Meanwhile, the fibrillar organization could prevent crack propagation and enhance the mechanical properties of silk fibers [21] (Fig. 5f). Moreover, the researches by Fu et al. showed that the highly aligned and relatively

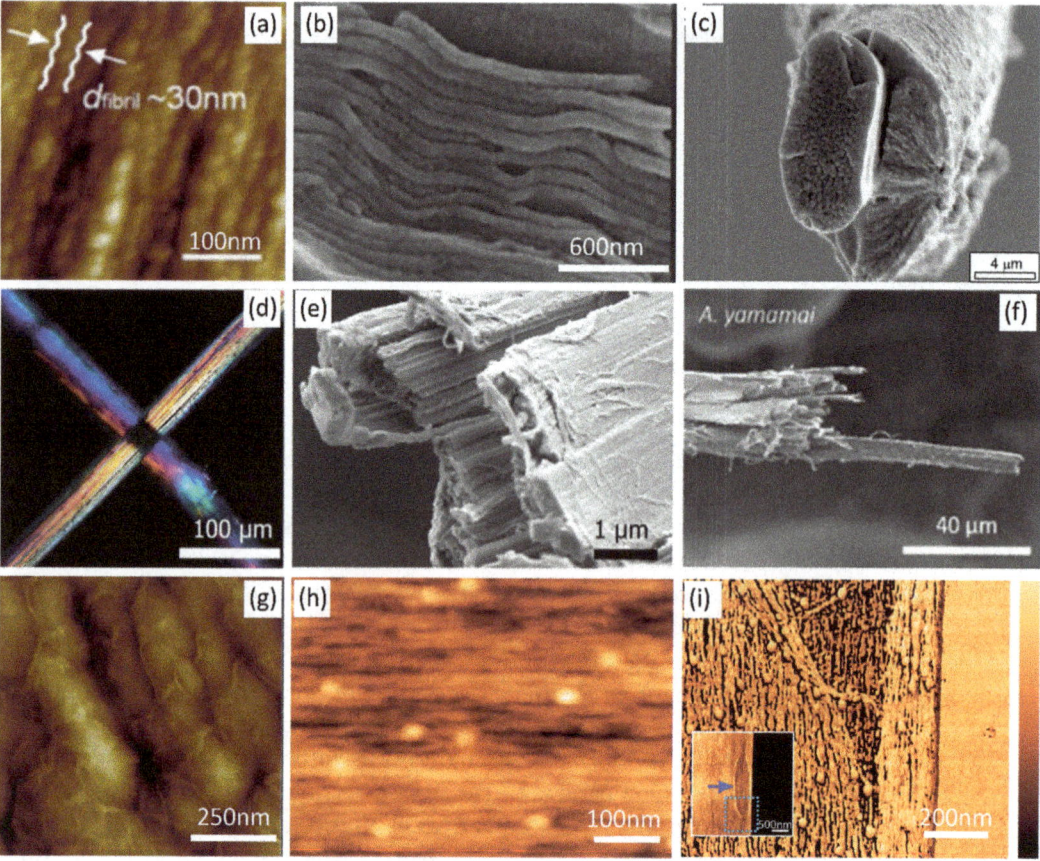

Fig. 5 The morphology of the nanofibrils in animal silk fibers. (**a**) AFM image of fibril structure observed on the degummed fiber surface. Adapted by permission from Royal Society of Chemistry [66], Copyright 2014. SEM image of peeled Mulberry silk fibers (**b**) [67] and fracture surface of *B. mori* silk (**c**) [68]. Adapted by permission from Elsevier [67], Copyright 2000 and Elsevier [68], Copyright 2002. (**d**)Polarized light microscopy image of *A. yamamai* silk. (**e**) Cross-sectional SEM image of *A. yamamai* silk shows the structure of microfibrils. (**f**) SEM image of *A. yamamai* silk shows the microfibrillar slipping and pulling after tensile fracture. (**g**) AFM image of fibril structure in the core region spider silk fibers. Adapted by permission from American Chemical Society [73], Copyright 2012. (**h**) Surface morphology of draglines. Adapted by permission from American Chemical Society [69], Copyright 2018. (**i**) AFM image of nanofibrils from a defect surface of *Loxosceles laeta* spider. Adapted by permission from American Chemical Society [69], Copyright 2018

independent nanofibrillar structure of tussah silk could be active the partly molecular chain to induce crack blunt and then keep the ductile failure of the whole fiber at cryogenic temperature [25]. Through the combination of partial dissolution and mechanical isolation, the single nanofibril with a diameter of 5 nm can be obtained, which is similar to the diameter of single mulberry nanofibrils [5]. By modulating parameters of the mechanical treatments, the full-sized mesosilks can be obtained, including silk microfibrils (1–2 μm in diameter), nanofibrils (200–300 nm in diameter), nanorods (11 ± 4 nm in diameter and 150–300 nm in length), and nanoparticles (2–7 nm) [71].

4.3 Spider Dragline Silk

A skin-core organization has been identified by studying the micro-tomed or fractured cross section of *Nephila* spider dragline silk. And the diameter of the fibrils has been observed to be 100–150 nm by confocal microscopy [72]. However, the nanofibrils with a diameter of 200 nm and the globular protrusions distributed along the axes with a width of 100 nm and length of 150 nm have been identified by AFM [73] (Fig. 5g–i). The corresponding simulations show that the convex morphology of nanofibers allows controllable local slippage to disperse the energy and could improve the strength of silk fibrils effectively [73, 74]. Giesa et al. used the coarse-graining mesoporous model to investigate the size effects on nanofibrils: the geometric confinement of silk fibrils to diameters of 20–80 nm is critical to map the nanoscale properties to the macroscale and further provide enhanced strength, extensibility, and toughness to the silk fibers [21], which is consistent with the size of the fibrils observed in the experiment. In general, the morphology and diameter of the nanofibrils vary with the spices of spiders. For example, the nanofibrils with a diameter of 60 ± 30 nm and aligned angle of $69° \pm 3°$ have been revealed by the AFM topography images of *Latrodectus hesperus* spider dragline silk [75]. Whereas, the diameter of nanofibrils in the skin layer and core region of *Argiope bruennichi* spider dragline silk is 100 nm and 6.3 nm, respectively [76]. Different from the cylindrical structure of dragline silk in most spiders, the *Loxosceles* silk features a thin ribbon morphology with a width of 6–8 nm and thickness of ~50 nm [69, 77]. It is found that different kinds of silk fibers possess different structural characteristics, such as wrinkled structure [78], helically wound nanofiber tube [79], helical nanofiber bundles, and layered structure with cross angles between different layers of nanofibers [67].

At a larger scale, animal silks share the same core–shell structure [80]. The mulberry silk and tussah silk fibroin filaments are coated with a layer of sericin. Silk fibroin protein is the main component of silk and accounting for ~70% in mulberry silk and ~85% in tussah silk [81]. Sericin plays a binding and protective role in the surface layer of silk fiber, accounting for about 25% and 15% in mulberry and tussah silk, respectively. In addition, it is a kind of water-soluble protein abundant in serine, which can be removed by degumming. While the spider dragline silk has a more complex multilayer structure, and the composition of each layer is different, mainly including protein, lipid, and glycoprotein [82] (Fig. 6).

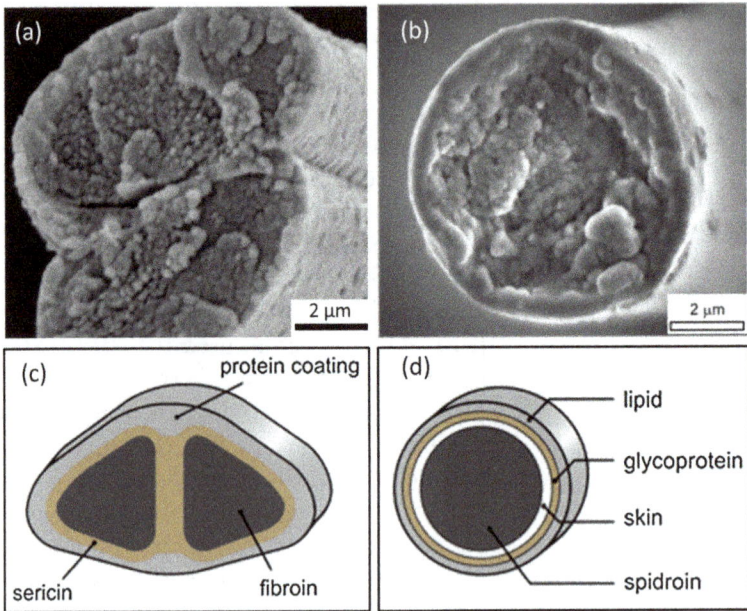

Fig. 6 Core–shell structure of animal silk. SEM image of native silkworm silk (**a**) and cross-sectional of *A. trifasciata* fibers (**b**) [68]. Adapted by permission from Elsevier [68], Copyright 2002. Schematic of the core–shell structure of silkworm (**c**) and spider silk (**d**). Adapted by permission from Elsevier [46], Copyright 2008

5   Summary

Taking mulberry silk, tussah silk, and spider silk as examples, this chapter summarizes the hierarchical structure of animal silk, including primary structure, secondary structure, and nanofibril structure. These structures dominate the performance of animal silk and associated silk protein materials. For example, the crystalline structure in animal silk is generally believed to determine the strength and modulus of the silk, while the amorphous structure is thought to control the toughness. At the higher structural scale, the nanofibril structure is considered to control the breakage resistance and ultralow temperature toughness of animal silk. Of course, the impact of each structural level on material properties is not isolated but synergistic. Understanding these synergistic can promote our understanding of the structure–function relationship of animal silk and also provide a blueprint on the design of regenerated functional fibers following silk fiber design strategies.

References

1. Yarger JL, Cherry BR, Arjan VDV (2018) Uncovering the structure–function relationship in spider silk. Nat Rev Mater 3:18008

2. Ling S, Kaplan DL, Buehler MJ (2018) Nanofibrils in nature and materials engineering. Nat Rev Mater 3:18016

3. Meyers MA, Chen P-Y, Lin AY-M, Seki Y (2008) Biological materials: structure and mechanical properties. Prog Mater Sci 53:1–206

4. Lin S, Ye C, Zhang W, Xu A, Chen S, Ren J, Ling S (2019) Nanofibril organization in silk fiber as inspiration for ductile and damage-tolerant fiber design. Adv Fiber Mater 1:231–240

5. Zhang W, Ye C, Zheng K, Zhong J, Tang Y, Fan Y, Buehler MJ, Ling S, Kaplan DL (2018) Tensan silk-inspired hierarchical fibers for smart textile applications. ACS Nano 12:6968–6977

6. Ling S, Jin K, Kaplan DL, Buehler MJ (2016) Ultrathin free-standing Bombyx mori silk nanofibril membranes. Nano Lett 16:3795–3800

7. Shengjie L, Zhao Q, Wenwen H, Sufeng C, David L, Kaplan (2017) Design and function of biomimetic multilayer water purification membranes. Sci Adv 3:e1601939

8. Ling S, Wang Q, Zhang D, Zhang Y, Mu X, Kaplan DL, Buehler MJ (2018) Integration of stiff graphene and tough silk for the design and fabrication of versatile electronic materials. Adv Funct Mater 28:1705291

9. Ling S, Li C, Jin K, Kaplan DL, Buehler MJ (2016) Liquid exfoliated natural silk nanofibrils: applications in optical and electrical devices. Adv Mater 28:7783–7790

10. Guo J, Li C, Ling S, Huang W, Chen Y, Kaplan DL (2017) Multiscale design and synthesis of biomimetic gradient protein/biosilica composites for interfacial tissue engineering. Biomaterials 145:44–55

11. Zheng K, Ling S (2019) De novo design of recombinant spider silk proteins for material applications. Biotechnol J 14:e1700753

12. Arcidiacono S, Mello C, Kaplan D, Cheley S, Bayley H (1998) Purification and characterization of recombinant spider silk expressed in Escherichia coli. Appl Microbiol Biotechnol 49:31–38

13. Jansson R, Lau CH, Ishida T, Ramström M, Sandgren M, Hedhammar M (2016) Functionalized silk assembled from a recombinant spider silk fusion protein (Z-4RepCT) produced in the methylotrophic yeast Pichia pastoris. Biotechnol J 11:687–699

14. Teulé F, Cooper AR, Furin WA, Bittencourt D, Rech EL, Brooks A, Lewis RV (2009) A protocol for the production of recombinant spider silk-like proteins for artificial fiber spinning. Nat Protoc 4:341–355

15. Scheller J, Henggeler D, Viviani A, Conrad U (2004) Purification of spider silk-elastin from transgenic plants and application for human chondrocyte proliferation. Transgenic Res 13:51–57

16. Wang Y, Kim H-J, Vunjak-Novakovic G, Kaplan DL (2006) Stem cell-based tissue engineering with silk biomaterials. Biomaterials 27:6064–6082

17. Rammensee S, Huemmerich D, Hermanson KD, Scheibel T, Bausch AR (2005) Rheological characterization of hydrogels formed by recombinantly produced spider silk. Appl Phys A 82:261

18. Huemmerich D, Slotta U, Scheibel T (2006) Processing and modification of films made from recombinant spider silk proteins. Appl Phys A 82:219–222

19. Yoshimoto H, Shin YM, Terai H, Vacanti JP (2003) A biodegradable nanofiber scaffold by electrospinning and its potential for bone tissue engineering. Biomaterials 24:2077–2082

20. Gustafsson L, Jansson R, Hedhammar M, van der Wijngaart W (2018) Structuring of functional spider silk wires, coatings, and sheets by self-assembly on superhydrophobic pillar surfaces. Adv Mater 3:1704325

21. Giesa T, Arslan M, Pugno NM, Buehler MJ (2011) Nanoconfinement of spider silk fibrils begets superior strength, extensibility, and toughness. Nano Lett 11:5038–5046

22. Du N, Liu XY, Narayanan J, Li L, Lim ML, Li D (2006) Design of superior spider silk: from nanostructure to mechanical properties. Biophys J 91:4528–4535

23. Giesa T, Buehler MJ (2013) Nanoconfinement and the strength of biopolymers. Annu Rev Biophys 42:651–673

24. Keten S, Xu Z, Ihle B, Buehler MJ (2010) Nanoconfinement controls stiffness, strength and mechanical toughness of beta-sheet crystals in silk. Nat Mater 9:359–367

25. Fu C, Wang Y, Guan J, Chen X, Vollrath F, Shao Z (2019) Cryogenic toughness of natural silk and a proposed structure–function relationship. Mater Chem Front 3:2507–2513

26. And YN, Asakura T (2002) High-resolution 13C CP/MAS NMR study on structure and structural transition of antheraea pernyi silk fibroin containing Poly(l-alanine) and Gly-rich regions. Macromolecules 35:2393–2400

27. Work RW, Young CT (1987) The amino acid compositions of major and minor ampullate silks of certain orb-web-building spiders (Araneae, Araneidae). J Arachnol 15:65–80

28. Shao Z, Vollrath F, Yang Y, Thogersen HC (2003) Structure and behavior of regenerated spider silk. Macromolecules 36:1157–1161

29. Ha SW, Gracz HS, Tonelli AE, Hudson SM (2005) Structural study of iIrregular amino acid sequences in the heavy chain of Bombyx mori silk fibroin. Biomacromolecules 6:2563

30. Takei F, Kikuchi Y, Kikuchi A, Mizuno S, Shimura K (1987) Further evidence for importance of the subunit combination of silk fibroin in its efficient secretion from the posterior silk gland cells. J Cell Biol 105:175–180

31. Tanaka K, Kajiyama N, Ishikura K, Shou W, Mizuno S (1999) Determination of the site of disulfide linkage between heavy and light chains of silk fibroin produced by Bombyx mori. Biochim Biophys Acta 1432:92–103

32. Tanaka K, Mori K, Mizuno S (1993) Immunological identification of the major disulfide-linked light component of silk fibroin. J Biochem 114:1–4

33. Inoue S, Tanaka K, Arisaka F, Kimura S, Ohtomo K, Mizuno S (2000) Silk fibroin of Bombyx mori is secreted, assembling a high molecular mass elementary unit consisting of H-chain, L-chain, and P25, with a 6:6:1 molar ratio. J Biol Chem 275:40517–40528

34. Cong-Zhao Z, Fabrice C, Nadine M, Yvan Z, Catherine E, Yang T, Michel J, Joel J, Michel D, Roland P (2000) Fine organization of Bombyx mori fibroin heavy chain gene. Nucleic Acids Res 12:2413–2419

35. Lewis RV (1992) Spider silk: the unraveling of a mystery. Acc Chem Res 25:392–398

36. Sezutsu H, Yukuhiro K (2000) Dynamic rearrangement within the Antheraea pernyi silk fibroin gene is associated with four types of repetitive units. J Mol Evol 51:329–338

37. Xu M, Lewis RV (1990) Structure of a protein superfiber: spider dragline silk. Proc Natl Acad Sci U S A 87:7120–7124

38. Thamm C, Scheibel T (2017) Recombinant production, characterization, and fiber spinning of an engineered short Major Ampullate Spidroin (MaSp1s). Biomacromolecules 18:1365–1372

39. Lewis RV (2006) Spider silk: ancient ideas for new biomaterials. Chem Rev 106:3762–3774

40. Rising A, Nimmervoll H, Grip S, Fernandez-Arias A, Storckenfeldt E, Knight DP, Vollrath F, Engström W (2005) Spider silk proteins—mechanical property and gene sequence. Zool Sci 22:273–281

41. Holland GP, Lewis RV, Yarger JL (2004) WISE NMR characterization of nanoscale heterogeneity and mobility in supercontracted nephila clavipes spider dragline silk. J Am Chem Soc 126:5867–5872

42. Holland GP, Jenkins JE, Creager MS, Lewis RV, Yarger JL (2008) Solid-state NMR investigation of major and minor ampullate spider silk in the native and hydrated states. Biomacromolecules 9:651–657

43. van Beek JD, Hess S, Vollrath F, Meier BH (2002) The molecular structure of spider dragline silk: folding and orientation of the protein backbone. Proc Natl Acad Sci 99:10266–10271

44. Lefevre T, Rousseau ME, Pezolet M (2007) Protein secondary structure and orientation in silk as revealed by Raman spectromicroscopy. Biophys J 92:2885–2895

45. Brooks AE, Stricker SM, Joshi SB, Kamerzell TJ, Middaugh CR, Lewis RV (2008) Properties of synthetic spider silk fibers based on argiope aurantia MaSp2. Biomacromolecules 9:1506–1510

46. Hardy JG, Römer LM, Scheibel TR (2008) Polymeric materials based on silk proteins. Polymer 49:4309–4327

47. Asakura T, Okonogi M, Nakazawa Y, Yamauchi K (2006) Structural analysis of alanine tripeptide with antiparallel and parallel beta-sheet structures in relation to the analysis of mixed beta-sheet structures in samia cynthia ricini silk protein fiber using solid-state NMR spectroscopy. J Am Chem Soc 128:6231–6238

48. Ha S-W, Gracz HS, Tonelli AE, Hudson SM (2005) Structural study of irregular amino acid sequences in the heavy chain of Bombyx mori silk fibroin. Biomacromolecules 6:2563–2569

49. Drummy LF, Farmer BL, Naik RR (2007) Correlation of the β-sheet crystal size in silk fibers with the protein amino acid sequence. Soft Matter 3:877–882

50. Hallmark V, Rabolt JF (1989) Fourier-transform Raman studies of secondary structure in synthetic polypeptides. Macromolecules 22:500–502

51. Termonia Y (1994) Molecular modeling of spider silk elasticity. Macromolecules 27:7378–7381

52. Blackledge TA (2012) Spider silk: a brief review and prospectus on research linking biomechanics and ecology in draglines and orb webs. J Arachnol 40:1–12

53. Hakimi O, Knight DP, Vollrath F, Vadgama P (2007) Spider and mulberry silkworm silks as compatible biomaterials. Compos B Eng 38:324–337

54. Thiel BL, Kunkel DD, Viney C (1994) Physical and chemical microstructure of spider dragline: a study by analytical transmission electron microscopy. Pept Sci 34:1089–1097

55. Thiel BL, Guess KB, Viney C (2015) Non-periodic lattice crystals in the hierarchical microstructure of spider (major ampullate) silk. Biopolymers 41:703–719

56. Riekel C, Vollrath F (2001) Spider silk fibre extrusion: combined wide- and small-angle X-ray microdiffraction experiments. Int J Biol Macromol 29:203–210

57. Grubb DT, Jelinski LW (2010) Fiber morphology of spider silk: the effects of tensile deformation. Macromolecules 30:2860–2867

58. Fu C, Shao Z, Fritz V (2009) Animal silks: their structures, properties and artificial production. Chem Commun (Camb):6515–6529

59. Müller M (2007) Silkworm silk under tensile strain investigated by neutron spectroscopy and synchrotron X-ray diffraction. Macromolecules 40:1035–1042

60. Krasnov I, Diddens I, Hauptmann N, Helms G, Ogurreck M, Seydel T, Funari SS, Muller M (2008) Mechanical properties of silk: interplay of deformation on macroscopic and molecular length scales. Phys Rev Lett 100:048104

61. Ling S, Qi Z, Knight DP, Huang Y, Huang L, Zhou H, Shao Z, Chen X (2013) Insight into the structure of single Antheraea pernyi silkworm fibers using synchrotron FTIR microspectroscopy. Biomacromolecules 14:1885–1892

62. Ling S, Qi Z, Knight DP, Shao Z, Chen X (2011) Synchrotron FTIR microspectroscopy of single natural silk fibers. Biomacromolecules 12:3344–3349

63. Nguyen AT, Huang QL, Yang Z, Lin N, Xu G, Liu XY (2015) Crystal networks in silk fibrous materials: from hierarchical structure to ultra performance. Small 11:1039–1054

64. Shen Y, Johnson MA, Martin DC (1998) Microstructural characterization of Bombyx mori silk fibers. Macromolecules 31:8857–8864

65. Miller LD, Putthanarat S, Eby RK, Adams WW (1999) Investigation of the nanofibrillar morphology in silk fibers by small angle X-ray scattering and atomic force microscopy. Int J Biol Macromol 24:159–165

66. Xu G, Gong L, Yang Z, Liu XY (2014) What makes spider silk fibers so strong? From molecular-crystallite network to hierarchical network structures. Soft Matter 10:2116–2123

67. Putthanarat S, Stribeck N, Fossey SA, Eby RK, Adams WW (2000) Investigation of the nanofibrils of silk fibers. Polymer 41:7735–7747

68. Poza P, Pérez-Rigueiro J, Elices M, Llorca J (2002) Fractographic analysis of silkworm and spider silk. Eng Fract Mech 69:1035–1048

69. Wang Q, Schniepp HC (2018) Strength of recluse spider's silk originates from nanofibrils. ACS Macro Lett 7:1364–1370

70. Niu Q, Peng Q, Lu L, Fan S, Shao H, Zhang H, Wu R, Hsiao BS, Zhang Y (2018) Single molecular layer of silk nanoribbon as potential basic building block of silk materials. ACS Nano 12:11860–11870

71. Zheng K, Zhong J, Qi Z, Ling S, Kaplan DL (2018) Isolation of silk mesostructures for electronic and environmental applications. Adv Funct Mater 28:1806380

72. KITAGAWA M, KITAYAMA T (1997) Mechanical properties of dragline and capture thread for the spider Nephila clavata. J Mater Sci 32:2005–2012

73. Brown CP, Harnagea C, Gill HS, Price AJ, Traversa E, Licoccia S, Rosei F (2012) Rough fibrils provide a toughening mechanism in biological fibers. ACS Nano 6:1961–1969

74. Cranford SW (2013) Increasing silk fibre strength through heterogeneity of bundled fibrils. J R Soc Interface 10:20130148

75. Riekel C, Burghammer M, Dane TG, Ferrero C, Rosenthal M (2017) Nanoscale structural features in major ampullate spider silk. Biomacromolecules 18:231–241

76. Gould SAC, Tran KT, Spagna JC, Moore AMF, Shulman JB (1999) Short and long range order of the morphology of silk from Latrodectus hesperus (Black Widow) as characterized by atomic force microscopy. Int J Biol Macromol 24:151–157

77. Koebley SR, Vollrath F, Schniepp HC (2017) Toughness-enhancing metastructure in the recluse spider's looped ribbon silk. Mater Horiz 4:377–382

78. Li SF, McGhie AJ, Tang SL (1994) New internal structure of spider dragline silk revealed by atomic force microscopy. Biophys J 66:1209–1212

79. Vollrath F, Holtet T, Thøgersen HC, Frische S (1996) Structural organization of spider silk. Proc R Soc London Ser B 263:147–151

80. Shao Z, Vollrath F (2002) Surprising strength of silkworm silk. Nature 418:741

81. Kundu SC, Kundu B, Talukdar S, Bano S, Nayak S, Kundu J, Mandal BB, Bhardwaj N, Botlagunta M, Dash BC, Acharya C, Ghosh AK (2012) Invited review nonmulberry silk biopolymers. Biopolymers 97:455–467

82. Sponner A, Vater W, Monajembashi S, Unger E, Grosse F, Weisshart K (2007) Composition and hierarchical organisation of a spider silk. PLoS One 2:e998

Chapter 2

Structure of Collagen

Chenxi Zhao, Yuelong Xiao, Shengjie Ling, Ying Pei, and Jing Ren

Abstract

Collagen is the most abundant fibrous protein in nature and widely exists in tissues such as connective tissue, tendon, skin, bone, and cartilage. On the one hand, collagen provides mechanical support in tissues, and on the other hand, plays an important role in controlling cell adhesion, cell migration, and tissue repair. A systematic understanding of the structure of collagen can promote the understanding of the biological functions of collagen scaffolds, and also provide theoretical guidance for applications of these natural fibrous protein materials. Therefore, this chapter centers on introducing the structure of collagen. As collagen has a typical hierarchical structure, the introduction to its structure will also be divided into different structural levels, from primary structure to quaternary structure. Due to the diversity of collagen types, this chapter will mainly focus on type I collagen.

Key words Collagen, α-helix, Hierarchical structure

1 Introduction

Collagens are the most abundant fibrous proteins and widely existed in both soft and hard tissues, such as tendons, skin, cornea, bones, and teeth [3, 4, 6, 10]. For example, in mammals, collagens account for approximately 30% of the total protein [1–3]. These collagens not only can act as the structural unit of tissues themselves but also can combine with minerals through biomineralization to form bionanocomposites, which are structural motifs of most hard tissues in mammals [7–9].

So far, 29 types of collagen have been identified by their polypeptide sequences. According to the function and domain homology, these collagens can be classified to fibril forming collagen, fibril-associated collagen with interrupted triple helices; network-forming collagen, transmembrane collagen, endostatin-producing collagen; anchoring fibrils, and beaded-filament collagen [29]. Among them, the fibril forming collagens widely exist in mammals. Fibril-associated collagens with interrupted triple helices can be found at the surfaces of collagen fibrils, transmembrane

Shengjie Ling (ed.), *Fibrous Proteins: Design, Synthesis, and Assembly*, Methods in Molecular Biology, vol. 2347,
https://doi.org/10.1007/978-1-0716-1574-4_2, © Springer Science+Business Media, LLC, part of Springer Nature 2021

collagen that have triple-helical ectodomains are being identified in vertebrates and invertebrates [3, 29]. In addition, according to the structural characteristics, collagens are usually divided into type I, II, III, and IV. In these collagen classifications, type I, II, and III are the most common types, accounting for 80–90% of the total body collagen [11]. Collagen type I is a heterotrimer with two α1 (I) chains and one α2 (II) chain, featuring a triple helix structure. Type I collagen is the common collagen component in tendons, skin, ligaments, cornea, and many other interstitial tissues, account- ing for 25% of the dry protein mass, constituting more than 90% of the organic matrix of bone [3, 14–17]. Collagen type II is a homotrimer that consisted of three α1 (II) chains [12]. Type II collagen and its associated minor collagens tend to be glycosylated to a greater extent than type I–rich fibrils. Type II collagen is found almost exclusively in cartilage, where the presence of additional minor collagens and noncollagenous glycoproteins is crucial for modulating fibril diameter, surface properties, and interfibrillar interactions [28]. Collagen type III is also a homotrimer with a structure of three α1(III) chains. Besides, collagen type III has disulfide bonds that connect three chains in the C-terminal of the triple helix domain. In extensible tissues, such as skin, vessel walls, and reticular fibers of lungs, liver, and spleen, collagen type III often coexists with collagen type I in collagenous fibrils [5, 13].

Same as other biological materials, collagen is usually assem- bled into more complex and delicate structures in biological tissues (Fig. 1) [7, 18]. Each collagen molecule is made of three peptide chains that forming the ≈300 nm long triple helical collagen mole- cule. Collections of collagen molecules aggregate both in lateral and longitudinal directions to form fibrils. Fibrils in cornea are normally thin (≈30 nm) and uniform in diameter, while those in tissues such as tendon contain a wide-ranging distribution of dia- meters (100–500 nm). Fibrils include tiny hydroxyapatite crystals in bone tissue, which provide stiffness and compressive load resis- tance. In tendons and ligaments, multiple fibrils make up collagen fibers [19–21].

Therefore, in this chapter, we will introduce the hierarchical structures of collagens, including primary structure, secondary structure, tertiary structure (triple helix), and quaternary structure (fibrils). We particularly focus on the most extensive type of colla- gen, that is, type I collagen.

2   Structure of Collagen

2.1   Primary Structure

Type I collagen contains three α-polypeptide chains with 1014 amino acids in each chain, adding up to 3042 amino acids for a well-folded molecule [5]. The amino acids have been repeated permutation along each chain followed by the tripeptide sequence

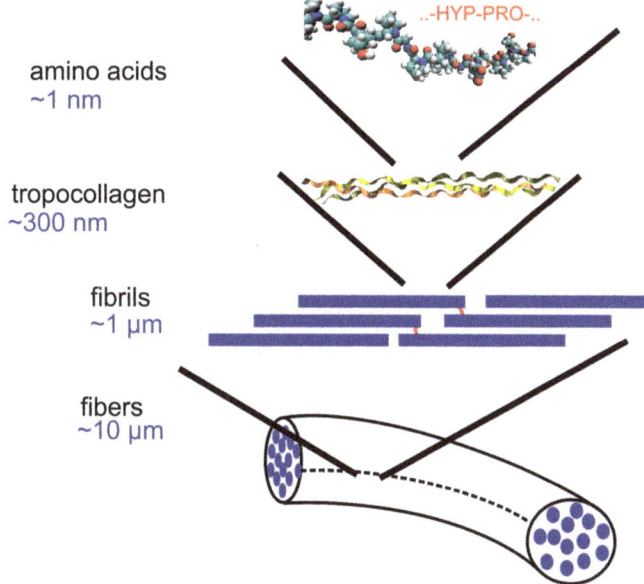

amino acids
~1 nm

..-HYP-PRO-..

tropocollagen
~300 nm

fibrils
~1 μm

fibers
~10 μm

Fig. 1 Schematic view of some of the hierarchical features of collagen. Adapted by permission from ref. [7], Copyright (2006) National Academy of Sciences, USA

of a N-terminus, a series of Gly-X-Y repeats, and a C-terminus. Gly represents glycine, which constitutes about 30% of the total amino acid content in collagen. X or Y can be any amino acid except Gly, but are usually proline and hydroxyproline, respectively. The hydroxyproline is commonly a 4-hydroxyproline or 5-hydroxyproline [1, 23]. In the tripeptide sequences, a high proportion of residues (about 20%) is frequently comprised of X and Y. Interestingly, hydroxyproline is only found in collagen protein, and its total content is over 50% compared with other amino acids, providing the binding sites for water molecules and playing a critical role in hydrogen bond formation and the stability of the triple-helical structure in the triple-helical domain [3, 20]. The proline and lysine can form hydroxyproline and hydroxylysine by hydroxylation of hydroxylase in the collagen α-polypeptide chain. These two amino acids participate in the formations of hydrogen bond, van der Waals force, and covalent bonds between polypeptide chains.

2.2 Secondary Structure

The secondary structure of all-natural collagens is left-handed α helix [23]. The formation of this structure results from the electrostatic repulsions between proline (X) and hydroxyproline (Y), highly depending on the posttranslation of proline hydroxylation [5, 19]. The structure stability in α-helix chains is due to the hydrogen bonds between amino acid residues. C-terminus and N-terminus are also nonhelical domains at the end of α chain.

C-terminus is the starting point of triple-helix formation, and N-terminus is involved in the regulation of primary fibril diameters [4, 25] Besides, the free amino acid residues in the side chains extend outward perpendicular to the axis of the spiral and assist in forming multiple hydrogen bonds in the spiral chain, which is essential for maintaining the stability of the helical structure [3].

2.3 Tertiary Structure

The tertiary structure of collagen is further coiled and folded by the secondary bond between the peptide chains in the molecule based on the secondary structure. The tertiary structure of the collagen refers to the triple-helical structure composed of three left-handed α helixes, two α1 (I) chains, and one α2 (II) chain, which assemble to form a right-handed triple-helical collagen called tropocollagen [5, 22]. It can be assembled in a sophisticated, hierarchical manner [3] (Fig. 2a). The average relative molecular weight of tropocollagen is about 300 kDa, and its size is approximately 300 nm in length and 1.5 nm in diameter (Fig. 2a) [3, 5]. The α chains form left-handed helices with 3.3 residues per turn and a pitch of 0.94 nm, compared with the common peptides which form right-handed α-helices with a pitch of 0.54 nm and 3.6 residues per turn [23], suggesting that α chain of collagen is stretched but narrowly twisted. In the triple helix, every third amino acid residue is located around the central axis, where the interspace along α chains is narrowest. Therefore, the glycine residues are suitable for the

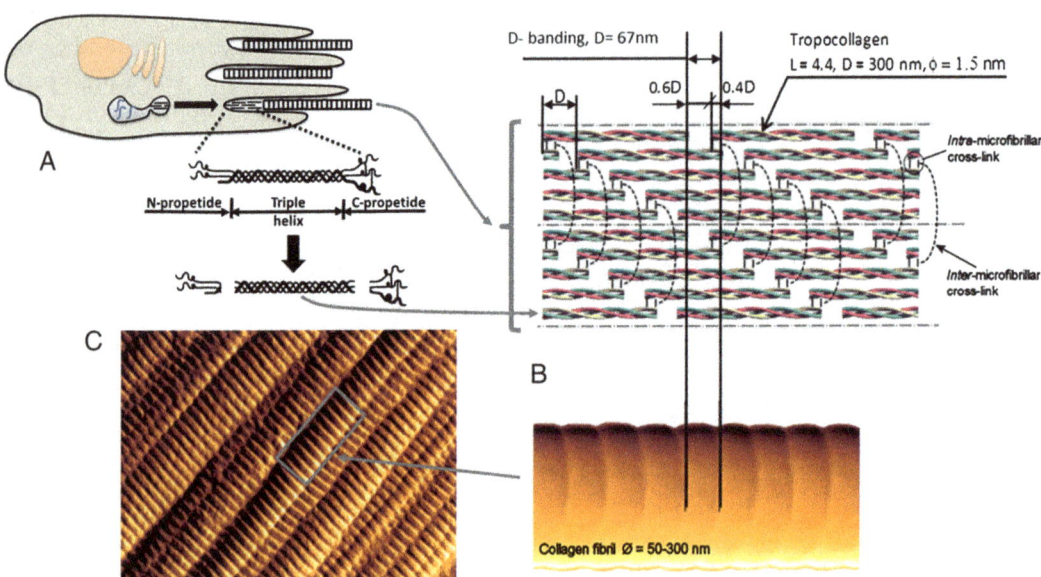

Fig. 2 Micro and ultra-structure of collagen. (**a**) Secretion of "procollagen" from osteoblast into extracellular matrix. (**b**) The structure of tropocollagen. (**c**) AFM image of tropocollagen. Adapted by permission from Elsevier [3], Copyright 2013

third amino acid residues because of its minimum volume. The stability of conformation and achievement of biomechanical and physiological functions mainly depend on the secondary links such as the hydrogen bonds between N-H group of Gly and adjacent O=C (X), as well as covalent bonds [23, 24].

2.4 Quaternary Structure

Tropocollagens are packed end-to-end in a line and parallel aligned to form stable microfibril, which cross-linked by covalent bonds that almost occurred between lysine and histidine of C-terminus or N-terminus. The insoluble fiber is formed by intermolecular or intermolecular cross-linking between collagen molecules [25]. The collagen molecules are staggered to adjacent molecular by approximately 64 nm or 67 nm (a unit that is referred to as "D" and changes depending on the hydration state of the aggregate), while the vertical distance between adjacent molecules is about 40 nm. Collagen microfibrils aligned laterally according to "1/4 staggered permutation rules" (Fig. 3) [3, 22]. The "overlap" or dense region consists of five tropocollagens in cross-section. The "gap" or sparse regions in each D-period repeat of the macrofibril are composed of four tropocollagens (Fig. 2b, c) [3], both of which are retained in collagen nanofibrils.

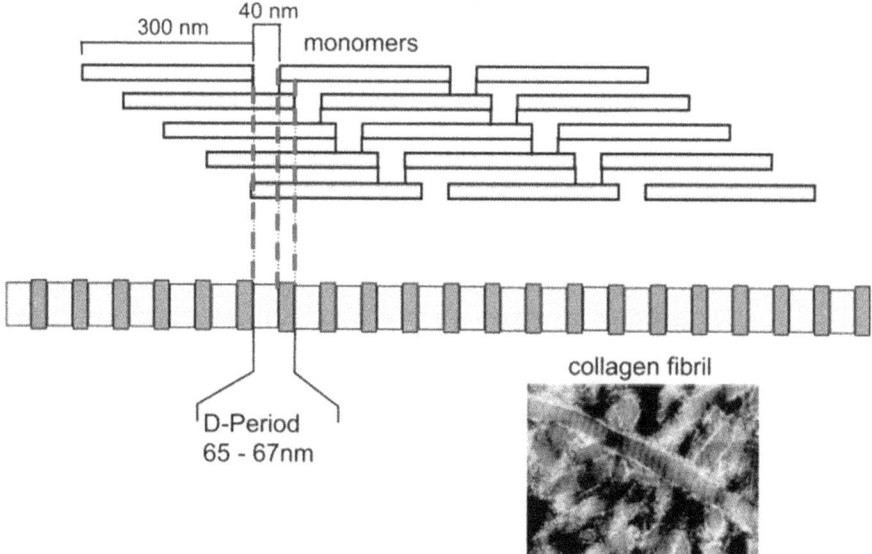

Fig. 3 Schematic representation of the supramolecular assembly of the collagen fibrils in the characteristic quarter-staggered form. Adapted by permission from Elsevier [21], Copyright 2019

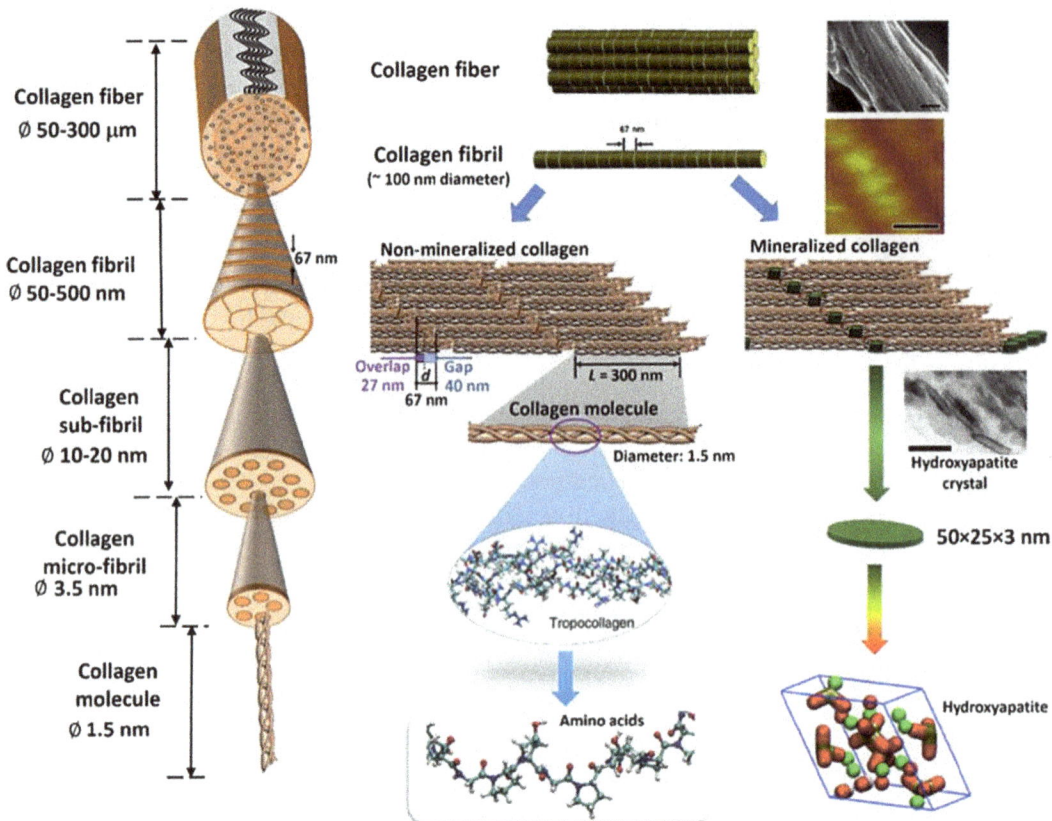

Collagen fiber
Ø 50-300 µm

Collagen fibril
Ø 50-500 nm

Collagen
sub-fibril
Ø 10-20 nm

Collagen
micro-fibril
Ø 3.5 nm

Collagen
molecule
Ø 1.5 nm

Collagen fiber

Collagen fibril
(~ 100 nm diameter)

Non-mineralized collagen

Mineralized collagen

Overlap 27 nm | d | Gap 40 nm
67 nm

L = 300 nm
Collagen molecule
Diameter: 1.5 nm

Tropocollagen

Amino acids

Hydroxyapatite
crystal

50×25×3 nm

Hydroxyapatite

Fig. 4 Hierarchical structure of a collagen fiber. Adapted by permission from Elsevier [25], Copyright 2019

Collagen microfibrils further orderly assemble into collagen fibrils with a rather crystallographic structure, and then aggregate and arrange into bundles to form collagen fibers [20, 21]. Finally, collagen fiber bundles interpenetrate and intertwine with one another to construct the basic organizational structure of the organism with a three-dimensional network. At the higher hierarchical level, for example, in the tendon and bone, the fibers are arranged into a helicoidal structure (Figs. 4, 5, and 6) [20, 24–27].

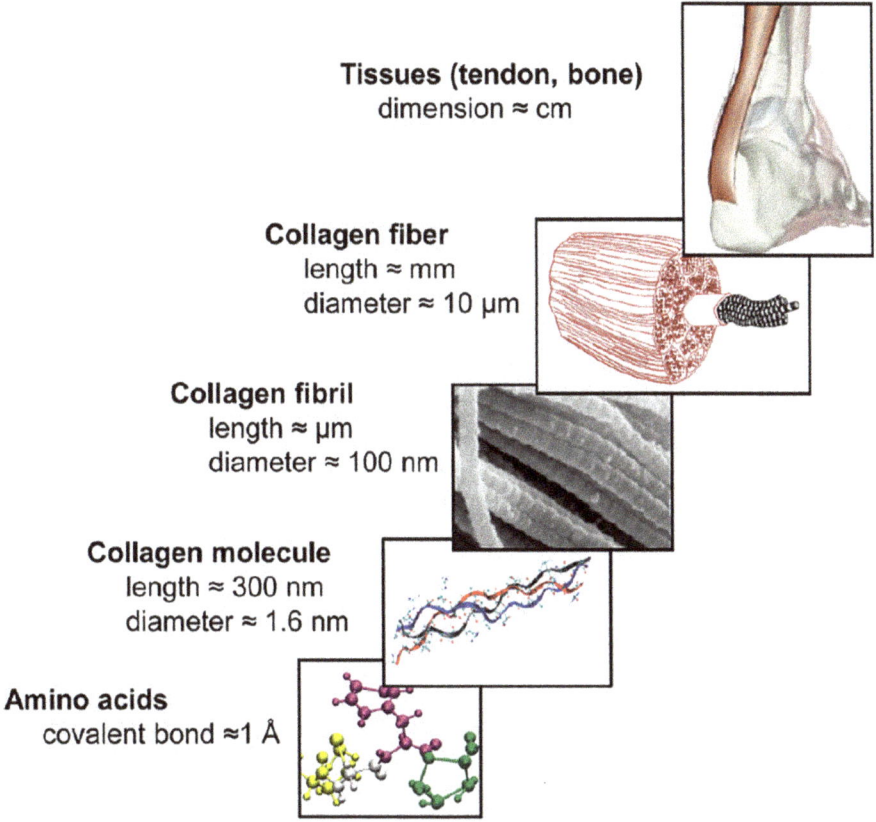

Tissues (tendon, bone)
dimension ≈ cm

Collagen fiber
length ≈ mm
diameter ≈ 10 μm

Collagen fibril
length ≈ μm
diameter ≈ 100 nm

Collagen molecule
length ≈ 300 nm
diameter ≈ 1.6 nm

Amino acids
covalent bond ≈1 Å

Fig. 5 Hierarchical structure of collagen protein materials. Adapted by permission from American Chemical Society [20], Copyright 2011

3 Summary

In summary, in this chapter, we take type I collagen as an example to systematically introduce its hierarchical structures, from primary structure to quaternary structure. These fundamental understanding of the structure construction of collagens, on the one hand, can provide insights into the structure–property–function relationship of the biological tissues, and on the other hand, can guide the utilization of these abundant fibrous proteins and even can inspire the new engineering material designs.

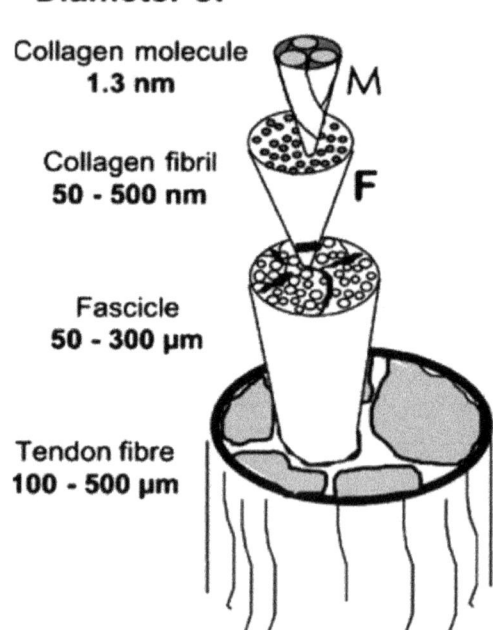

Diameter of

Collagen molecule
1.3 nm

M

Collagen fibril
50 - 500 nm

F

Fascicle
50 - 300 μm

Tendon fibre
100 - 500 μm

Fig. 6 Sketch of the hierarchical structure of tendon. Adapted by permission from ref. [18], Copyright CC-BY-NC-ND-4.0

Acknowledgments

This work was supported by the National Natural Science Foundation of China [grant numbers. 51973116, U1832109, 21935002], the Users with Excellence Program of Hefei Science Center CAS [grant number 2019HSC-UE003], the starting grant of ShanghaiTech University, and State Key Laboratory for Modification of Chemical Fibers and Polymer Materials.

References

1. Zhu S, Yuan Q, Yin T, You J, Gu Z, Xiong S, Hu Y (2018) Self-assembly of collagen-based biomaterials: preparation, characterizations and biomedical applications. J Mater Chem B 6(18):2650–2676

2. Liu DS, Nikoo M, Boran G, Zhou P, Regenstein JM (2015) Collagen and gelatin. Annu Rev Food Sci Technol 6:527–557

3. Abou Neel EA, Bozec L, Knowles JC, Syed O, Mudera V, Day R, Hyun JK (2013) Collagen-emerging collagen based therapies hit the patient. Adv Drug Deliv Rev 65(4):429–456

4. Gomes S, Leonor IB, Mano JF, Reis RL, Kaplan DL (2012) Natural and genetically engineered proteins for tissue engineering. Prog Polym Sci 37(1):1–17

5. Ling S, Chen W, Fan Y, Zheng K, Jin K, Yu H, Buehler MJ, Kaplan DL (2018) Biopolymer nanofibrils: structure, modeling, preparation, and applications. Prog Polym Sci 85:1–56

6. Barthelat F, Yin Z, Buehler MJ (2016) Structure and mechanics of interfaces in biological materials. Nat Rev Mater 1(4):16007

7. Markus JB (2006) Nature designs tough collagen: explaining the nanostructure of collagen fibrils. PNAS 103(33):12285–12290

8. Fratzl P (2008) Collagen: structure and mechanics. Springer, New York

9. Cisneros DA, Hung C, Franz CM, Muller DJ (2006) Observing growth steps of collagen self-assembly by time-lapse high-resolution atomic force microscopy. J Struct Biol 154 (3):232–245

10. Cen L, Liu W, Cui L, Zhang W, Cao Y (2008) Collagen tissue engineering: development of novel biomaterials and applications. Pediatr Res 63(5):492–496

11. Zeugolis DI, Raghunath M (2011) Collagen: materials analysis and implant uses, in comprehensive. Biomaterials 2:261–278

12. Ervin H, Epstein JR (1974) [α1(III)]3 human skin collagen. J Biol Chem 249 (10):3225–3231

13. Coppola D, Oliviero M, Vitale GA, Lauritano C, D'Ambra I, Iannace S, Pascale B (2020) Marine collagen from alternative and sustainable sources: extraction, processing and applications. Mar Drugs 18(4):214

14. Orgel JPRO, Irving TC, Miller A, Wess TJ (2006) Microfibrillar structure of type I collagen in situ. PNAS 103(24):9001–9005

15. Williams BR, Gelman RA, Poppke DC, Piez KA (1978) Collagen fibril formation. Optimal in vitro conditions and preliminary kinetic results. J Biol Chem 253(18):6578–6585

16. Di Lullo GA, Sweeney SM, Korkko J, Ala-kokko L, San Antonio JD (2002) Mapping the ligand-binding sites and disease-associated mutations on the most abundant protein in the human, type I collagen. J Biol Chem 277 (6):4223–4231

17. Hafner AE, Gyori NG, Bench CA, Davis LK, Šarić A (2020) Modeling fibrillogenesis of collagen-mimetic molecules. Biophys J 119 (9):1791–1799

18. Fratzl P, Weinkamer R (2007) Nature's hierarchical materials. Prog Mater Sci 52 (8):1263–1334

19. Ottani V, Martini D, Franchi M, Ruggeri A, Raspanti M (2002) Hierarchical structures in fibrillar collagens. Micron 33(7–8):587–596

20. Gautieri A, Vesentini S, Redaelli A, Buehler MJ (2011) Hierarchical structure and nanomechanics of collagen microfibrils from the atomistic scale up. Nano Lett 11(2):757–766

21. Liu X, Zheng C, Luo X, Wang X, Jiang H (2019) Recent advances of collagen-based biomaterials: multi-hierarchical structure, modification and biomedical applications. Mater Sci Eng C Mater Biol Appl 99:1509–1522

22. Tang K (2012) Collagen physics and chemistry. China Science Publishing & Media Ltd, Beijing

23. Shoulders MD, Raines RT (2009) Collagen structure and stability. Annu Rev Biochem 78:929–958

24. Sorushanova A, Delgado LM, Wu Z, Shologu N, Kshirsagar A, Raghunath R, Mullen AM, Bayon Y, Pandit A, Raghunath M, Zeugolis DI (2019) The collagen suprafamily: from biosynthesis to advanced biomaterial development. Adv Mater 31(1):1801651

25. Yang W, Meyers MA, Ritchie RO (2019) Structural architectures with toughening mechanisms in nature: a review of the materials science of type-I collagenous materials. Prog Mater Sci 103(6):425–483

26. Sanders JE, Goldstein BS (2001) Collagen fibril diameters increase and fibril densities decrease in skin subjected to repetitive compressive and shear stresses. J Biomech 34 (12):1581–1587

27. Sivakumar L, Agarwal G (2010) The influence of discoidin domain receptor 2 on the persistence length of collagen type I fibers. Biomaterials 31(18):4802–4808

28. Eyre D (2002) Collagen of articular cartilage. Arthritis Res 4(1):30–35

29. Kadler KE, Baldock C, Bella J, Boot-Handford RP (2007) Collagens at a glance. J Cell Sci 120 (12):1955–1958

Chapter 3

Structure of Elastin

Yuelong Xiao, Shengjie Ling, and Ying Pei

Abstract

As the extracellular matrix protein, elastin is a crucial component of connective tissue in life. It is responsible for the structural integrity and function of tissues undergoing reversible extensibility or deformability, even though it may make up only a small percentage of a tissue. The structure stability, elastic resilience, bioactivity, and ability of self-assembly make elastin a highly desirable candidate for the fabrication of biomaterials. Elastin's properties mainly depend on their special structure. As elastin can be obtained by the assembly and cross-linking of its soluble precursor, tropoelastin. This chapter centers on introducing the structure of those two materials.

Key words Elastin, Tropoelastin, Protein, Structure

1 Introduction

Elastin is a critical extracellular matrix protein that contributes to the properties of extensibility and elastic recoil of vertebrate tissues, including large arteries, lung, ligament, tendon, skin, and elastic cartilage [1–5]. The amount of elastin varies among different tissues, depending on the tissue structure and the required elastic properties. Elastin accounts for 30–57% of the aorta, 50% of elastin ligaments, 28–32% of major vascular vessels, 3–7% of lung, 4% of tendons, and 2–5% of the dry weight of skin [6–8]. It is an insoluble biopolymer and usually laid down only during development and subsequently does not turn over appreciably in healthy tissue [9, 10]. Elastin is produced by different types of cells, such as smooth muscle cells, fibroblasts, and endothelial cells [2, 3]. Mechanically, the polymeric elastin matrix is exceptionally durable, for example, elastin with a half-life of about 70 years can bear billions of loading and unloading cycles without failure in the large arteries [4, 11, 12]. In the lung, elastin is considered as a lattice to support the shape stability of the alveoli during tidal breathing [13]. In the skin, elastin fibers in the dermis endow skin flexibility and extensibility [14, 15].

Shengjie Ling (ed.), *Fibrous Proteins: Design, Synthesis, and Assembly*, Methods in Molecular Biology, vol. 2347, https://doi.org/10.1007/978-1-0716-1574-4_3, © Springer Science+Business Media, LLC, part of Springer Nature 2021

The fundamental building block of elastin is tropoelastin, which is a low complexity protein composed of partly respective primary sequence motifs within hydrophobic and hydrophilic domains and are arranged in an alternating manner [16, 17]. The formation of elastin has been explored through the native precursor tropoelastin, involving a remarkable process of hierarchical self-assembly at physiological temperatures through interactions principally between their hydrophobic sequences (Fig. 1). The tropoelastin molecule is "captured" by its molecular chaperone in the rough endoplasmic reticulum immediately after translation. The complex is localized on the cell surface in the extracellular space. When it comes into contact with a nascent elastin fiber, the chaperone and galactose of the microfibrillar component interact with each other leading to the local release of the tropoelastin molecule. Then the tropoelastin molecule is aligned and promptly modified by lysyl oxidase. Finally, it is incorporated in the elastin network by irreversible polymerization, thus growing within the microfibrillar scaffold [3, 18]. Tropoelastin is a highly elastic protein, capable of extending to approximately eight times its resting length with no evident hysteresis [19, 20]. Elastin and tropoelastin are difficult to isolate for investigation. They can be extracted from bovine, porcine, and humans, especially in equine elastin-rich tissues such as ligament and aorta [21, 22]. The structure of the elastin network is still not fully understood [23].

2   Structure of Elastin and Soluble Elastin (Tropoelastin)

Macroscopically, elastin appears to be an amorphous mass [23]. According to the results of ultrastructure electron microscopy, elastin has a fibrillar substructure comprised of parallel-aligned ~5 nm thick filaments with a twisted ropelike structure [24]. In terms of protein structure, elastin is rich in glycine, proline, alanine, leucine, and valine residues and is generally organized in short repeated sequences of three to nine amino acids with flexible and highly dynamic structures [25, 26]. With approximately 800 amino acid residues, elastin can be synthesized through the lysine-mediated cross-linking of tropoelastin as a soluble precursor [27, 28]. Tropoelastin is an approximately 60–70 kDa protein, is nonglycosylated and highly nonpolar, consisting essentially of 34 alternating hydrophobic and cross-linking domains that are the two major domain types of the tropoelastin amino acid sequence and determining the length of the tropoelastin [29, 30]. And tropoelastin is composed of a single polypeptide chain about 800 residues in size, approximately the size of a serum albumin molecule. The nonpolar amino acids glycine, proline, alanine, valine, phenylalanine, isoleucine, and leucine make up the majority of the residues [31]. Hydrophobic domains are rich in

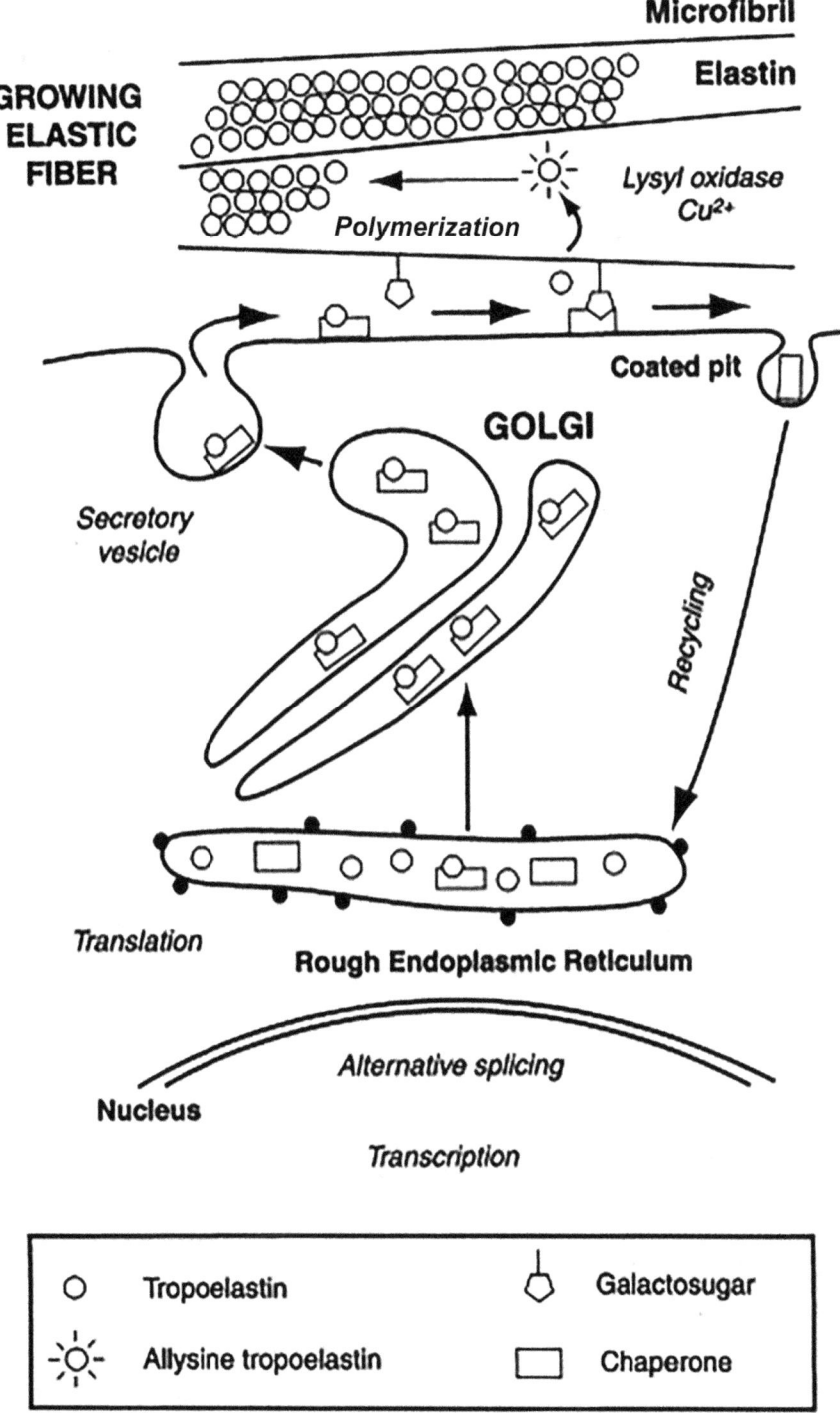

GROWING ELASTIC FIBER

Microfibril

Elastin

Lysyl oxidase Cu^{2+}

Polymerization

Coated pit

GOLGI

Secretory vesicle

Recycling

Translation

Rough Endoplasmic Reticulum

Alternative splicing

Nucleus

Transcription

○	Tropoelastin	⬠	Galactosugar
☀	Allysine tropoelastin	▭	Chaperone

Fig. 1 Schematic description of elastin deposition. Adapted by permission from Elsevier Ltd. [40], Copyright 1999

nonpolar amino acids, including glycine, valine, alanine, and pro-line, which are often arranged in repeats of tetra-, penta-, and hexapeptides such as GVGVP, GVPGV, and GVGVAP [32]. Cross-linking domains contain the lysine residues, which further form the covalent intermolecular cross-links to stabilize the polymer structure. In general, the hydrophobic domains are less well conserved than the cross-linking domains in size and composition. It is suggested that tropoelastin can tolerate the variations of the hydrophobic amino acids without destroying the molecular structure [33, 34]. In fact, these areas have diverged during evolution at twice the rate of the cross-link domains. Additionally, tropoelastin ends with a hydrophilic carboxy-terminal sequence containing two cysteine residues.

Furthermore, tropoelastin has a high content of nonpolar amino acids, including glycine, alanine, and proline, a particularly high value for valine, and a small amount of hydroxyproline [35]. In mature tissues, it also contains unique lysine-derived cross-linking amino acids, which are called desmosine and isodesmosine, as well as lysinonorleucine and merodesmosine, respectively [36]. Desmosine and isodesmosine are the two predominant cross-links of native elastin, and each of them involves four lysine residues cross-linked by lysyl oxidase (Fig. 2) [4]. Tropoelastin exists as a monomer in solution in two forms: an open globular molecular and a distended polypeptide [37]. The polar amino acids aspartate, glutamate, lysine, and arginine account for less than 5% of the amino acid composition of the insoluble fiber. An important difference between elastin and tropoelastin is the lysine content of the soluble precursor, resulting from the apparent direct conversion of lysine into the cross-links desmosine and isodesmosine during fiber formation [14, 16]. Thus, elastin appears a dual nature, an extensible portion containing hydrophobic amino acid residues, and a more compacted portion (rich in alanine)

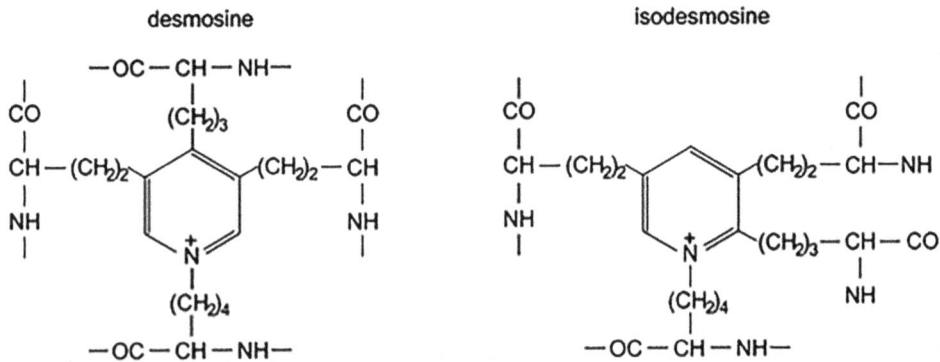

Fig. 2 Specific elastin intermolecular cross-links include the tetrafunctional desmosine and isodesmosine formed from four Lys residues from two different tropoelastin molecules. Adapted by permission from Elsevier Ltd. [4], Copyright 2007

containing the lysine residues that form the cross-link portions of the mature fiber. Besides, C-terminus is one of the most important and well-conserved regions in tropoelastin. In tropoelastin structure, only two cysteine residues are closely spaced in the final domain 36 and form a disulfide bond in the molecule, while the protein terminates with a positively charged RKRK sequence [19, 29].

However, the secondary structure of tropoelastin and its derivatives have been significantly investigated [38, 39]. It can have a high degree of flexibility and exist in either a globular or elongated form in solution. The analysis of individual domains of tropoelastin reveals secondary structure features that include polyproline II, compact β-turns and disordered structure. A structure feature that appears to be common to the repeating peptide sequence is the β-turn. The β-turn allows the elastin chain to fold back on itself. β spiral structure can be obtained with the formation of a repeat β-turn structure on the helical axis. The β spiral has a great influence on elastin function and fiber formation [40]. For example, human tropoelastin is encoded by a single gene containing 34 exons. The mRNA encodes a 72 kDa polypeptide and removal of a signal peptide that leaves a mature protein with a molecular weight of at least 60 kDa. As shown in Fig. 3, at least 11 human tropoelastin splice forms have been characterized, resulting from developmentally regulated alternative splicing of domains 22, 23, 24, 26A, 32, and 33. Hydrophilic cross-linking domains can be further divided into KP-and KA-rich regions. The tertiary structure of the expressed protein has not yet been definitively clarified due to the absence of aggregation structure [24].

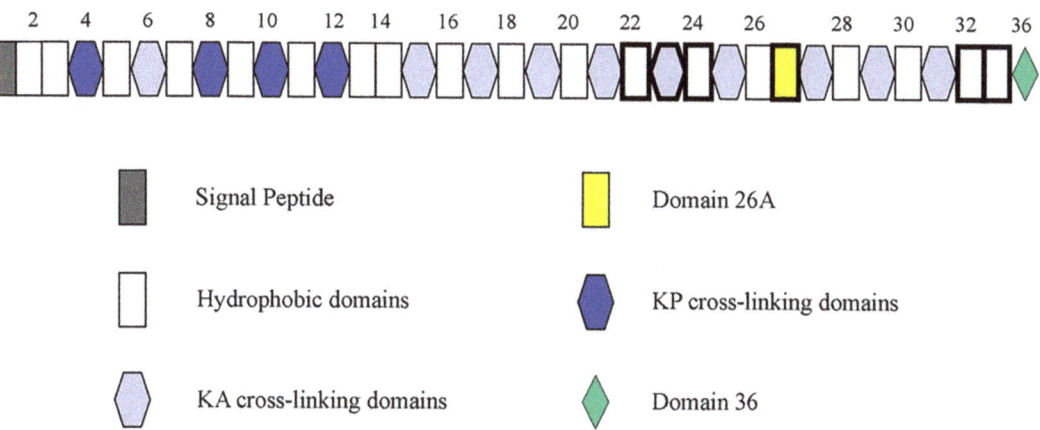

Fig. 3 Domain structure of human tropoelastin. Human tropoelastin consists of 34 domains and is dominated by alternating hydrophobic and cross-linking regions. Adapted by permission from ScienceDirect [16], Copyright 2009

3 Summary

In summary, in this chapter, we introduce the structure of elastin and tropoelastin. The fundamental understanding of the structure constructions of the proteins can provide insights into functions and performance-optimization mechanisms in biological materials.

Acknowledgments

This work was supported by the National Natural Science Foundation of China (No. 51603191).

References

1. Wen Q, Mithieux SM, Weiss AS (2020) Elastin biomaterials in dermal repair. Trends Biotechnol 38:280–291

2. Coenen AMJ, Bernaerts KV, Harings JAW, Jockenhoevel S, Ghazanfari S (2018) Elastic materials for tissue engineering applications: natural, synthetic, and hybrid polymers. Acta Biomater 79:60–82

3. Yeo GC, Mithieux SM, Weiss AS (2018) The elastin matrix in tissue engineering and regeneration. Curr Opin Biomed Eng 6:27–32

4. Daamen WF, Veerkamp JH, van Hest JC, van Kuppevelt TH (2007) Elastin as a biomaterial for tissue engineering. Biomaterials 28:4378–4398

5. John G, Margo L, Emily C, Paul G, Christine O, Ken S (2002) Elastic proteins: biological roles and mechanical propertie. Phil Trans R Soc Lond B 357:121–132

6. Lescan M, Perl RM, Golombek S, Pilz M, Hann L, Yasmin M, Behring A, Keller T, Nolte A, Gruhn F, Kochba E, Levin Y, Schlensak C, Wendel HP, Avci-Adali M (2018) De novo synthesis of elastin by exogenous delivery of synthetic modified mrna into skin and elastin-deficient cells. Mol Ther Nucleic Acids 11:475–484

7. Monfort DA, Koria P (2017) Recombinant elastin-based nanoparticles for targeted gene therapy. Gene Ther 24:610–620

8. Weihermann AC, Lorencini M, Brohem CA, de Carvalho CM (2017) Elastin structure and its involvement in skin photoageing. Int J Cosmet Sci 39:241–247

9. Mithieux SM, Anthony SW (2006) Elastin. Adv Protein Chem 70:437–461

10. Vindin H, Mithieux SM, Weiss AS (2019) Elastin architecture. Matrix Biol 84:4–16

11. Halabi CM, Mecham RP (2018) Elastin purification and solubilization. Methods Cell Biol 143:207–222

12. Fletcher EE, Yan D, Kosiba AA, Zhou Y, Shi H (2019) Biotechnological applications of elastin-like polypeptides and the inverse transition cycle in the pharmaceutical industry. Protein Expr Purif 153:114–120

13. Santos M, Serrano-Ducar S, Gonzalez-Valdivieso J, Vallejo R, Girotti A, Cuadrado P, Arias FJ (2019) Genetically engineered elastin-based biomaterials for biomedical applications. Curr Med Chem 26:7117–7146

14. Partridge SM (1963) Elastin. Adv Protein Chem 17:227–302

15. Kozma B, Candiotti K, Poka R, Takacs P (2018) The effects of heat exposure on vaginal smooth muscle cells: elastin and collagen production. Gynecol Obstet Investig 83:247–251

16. Wise SG, Weiss AS (2009) Tropoelastin. Int J Biochem Cell Biol 41:494–497

17. Wang R, Ozsvar J, Yeo GC, Weiss AS (2019) Hierarchical assembly of elastin materials. Curr Opin Chem Eng 24:54–60

18. Robert PM, Gary LJ, Barry S (1981) Elastin synthesis by ligamentum nuchae fibroblasts: effects of culture conditions and extracellular matrix on elastin production. J Cell Biol 90:332–338

19. Joel R (1982) Elastin: biosynthesis, structure, degradation and role in disease processes. Connect Tissue Res 10:73–91

20. Bataille L, Dieryck W, Hocquellet A, Cabanne C, Bathany K, Lecommandoux S, Garbay B, Garanger E (2016) Recombinant production and purification of short hydrophobic Elastin-like polypeptides with low

transition temperatures. Protein Expr Purif 121:81–87

21. Catherine MB, Margo AL, John MG, Glenda MW, BarryCS AJB, Kimberly AW, Fred WK (2003) Recombinant human elastin polypeptides self-assemble into biomaterials with elastin-like properties. Biopolymers 70:445–455

22. Tanzer ML (1989) Collagens and elastin: structure and interactions. Curr Opin Cell Biol 1:968–973

23. Hernandez B, Crowet JM, Thiery J, Kruglik SG, Belloy N, Baud S, Dauchez M, Debelle L (2020) Structural analysis of nonapeptides derived from elastin. Biophys J 118:2755–2768

24. Pang X, Wu JP, Allison GT, Xu J, Rubenson J, Zheng MH, Lloyd DG, Gardiner B, Wang A, Kirk TB (2017) Three dimensional microstructural network of elastin, collagen, and cells in Achilles tendons. J Orthop Res 35:1203–1214

25. Duca L, Floquet N, Alix AJ, Haye B, Debelle L (2004) Elastin as a matrikine. Crit Rev Oncol Hematol 49:235–244

26. William RG, Lawrence BS, Judith AF (1973) Molecular model for elastin structure and function. Nature 246:461–466

27. Numata K (2020) How to define and study structural proteins as biopolymer materials. Polym J 9:1043–1056

28. Urry DW, Hugel T, Seitz M, Gaub HE, Sheiba L, Dea J, Xu J, Parker T (2002) Elastin: a representative ideal protein elastomer. Philos Trans R Soc Lond Ser B Biol Sci 357:169–184

29. Parks WC, Pierce RA, Lee KA, Mecham RP (1993) Elastin. In: Bittar EE, Kleinman HK (eds) Extracellular matrix, advances in molecular and cell biology. JAI Press, Greenwich, pp 133–181

30. Jiménez Vázquez J, San Martín Martínez E (2019) Collagen and elastin scaffold by electrospinning for skin tissue engineering applications. J Mater Res 34:2819–2827

31. Dean YL, Benjamin B, Elaine CD, Robert PM, Lise KS, Beth BB, Ernst E, Mark TK (1998) Elastin is an essential determinant of arterial morphogenesis. Nature 393:276–280

32. Edward G, Cleary Mark AG (1983) Elastin-associated microfibrils and microfibrillar proteins. Tissue Res 10:97–207

33. Jeon WB, Park BH, Wei J, Park RW (2011) Stimulation of fibroblasts and neuroblasts on a biomimetic extracellular matrix consisting of tandem repeats of the elastic VGVPG domain and RGD motif. J Biomed Mater Res A 97:152–157

34. Mahesh B, Nanjundaswamy GS, Channe Gowda D, Siddaramaiah B (2017) Synthesis of elastin-based polymer and evaluation of its intermolecular interactions with hydroxypropyl methylcellulose. J Appl Polym Sci 134

35. Laurent D, Alain JP, Pierre L (1998) The secondary structure and architecture of human elastin. Eur J Biochem 258:533–539

36. Varanko AK, Su JC, Chilkoti A (2020) Elastin-like polypeptides for biomedical applications. Annu Rev Biomed Eng 22:343–369

37. Dan WU, Luan CH, Peng SQ (1995) Molecular biophysics of elastin structure, function and pathology. Ciba Found Symp 192:4–30

38. Godwin ARF, Singh M, Lockhart-Cairns MP, Alanazi YF, Cain SA, Baldock C (2019) The role of fibrillin and microfibril binding proteins in elastin and elastic fibre assembly. Matrix Biol 84:17–30

39. Lawrence BS, Norman TS, John GL (1981) Elastin structure, biosynthesis, and relation to disease states. Med Prog 304:566–579

40. Debelle L, Tamburro AM (1999) Elastin: molecular description and function. Int J Biochem Cell Biol 31:261–272

Chapter 4

Structure of Resilin

Yuelong Xiao, Shengjie Ling, and Ying Pei

Abstract

Resilin, an insect structural protein, exhibits rubberlike elasticity characterized by low stiffness, high extensibility, efficient energy storage, exceptional resilience, and fatigue lifetime. The outstanding mechanical properties of native resilin have motivated recent research about resilin-like biomaterials for a wide range of applications. The systematic understanding of the resilin structure provides theoretical guidance for its applications. In this chapter, we systematically introduce its special structure, providing useful information for the structure and elastic mechanism of native resilin protein.

Key words Resilin, Protein, Structure

1 Introduction

Resilin is a novel rubberlike protein found in specialized compartments of the cuticles of most insects, typically in exoskeletons of arthropods, that is, the jumping organs of arthropods such as froghopper and cat fleas, the wings of dragonflies, or the vibrating membrane of cicadas [1–5]. Resilin usually incorporate with other proteins and/or chitin fibers [6, 7]. Resilin also as pure masses of protein exists in continuous layers and can be extracted by separating thin chitinous lamellae. Resilin and proresilin usually refer to the fully cross-linked protein and the not yet cross-linked or not fully cross-linked protein, respectively [8, 9]. The proresilin is secreted in the subcuticular space where it undergoes rapid cross-linking involving tyrosine residues, through di- and trityrosine cross-link formations [10].

Moreover, resilin features exceptional rubberlike properties such as relatively low stiffness, rather pronounced long-range deformability, and nearly perfect elastic recovery due to its molecular structures. It even exhibits resilience of up to 92–97% and a fatigue limit of over 300 million cycles [11]. Resilin has outstanding material properties of high resilience and a very high fatigue lifetime [12, 13]. Today, resilin is known to by virtue of the

Shengjie Ling (ed.), *Fibrous Proteins: Design, Synthesis, and Assembly*, Methods in Molecular Biology, vol. 2347, https://doi.org/10.1007/978-1-0716-1574-4_4, © Springer Science+Business Media, LLC, part of Springer Nature 2021

generation of deformability and flexibility in the membrane and joint systems, the storage of elastic energy in jumping and catapulting systems, the enhancement of adaptability to uneven surfaces in attachment and prey catching systems, the reduction of fatigue and damage in reproductive, folding, and feeding systems, and the sealing of wounds in a traumatic reproductive system [8, 14]. Resilin exhibits a unique combination of different outstanding properties [15, 16]. This chapter focuses on the structure of resilin.

2 Structure of the Resilin

Resilin with high flexibility and mobility is a protein network of randomly orientated coiled polypeptide chains that are linked at regular intervals by stable covalent cross-links [3, 17]. The structure of swollen resilin is a random network of proteins in which almost all amino-acid residues are free to rotate relative to each other. This is attributed to the lack of stable hydrogen bonds within and between proteins [2, 12]. Three amino acids, glycine, dityrosine, and tyrosine, can be obtained by hydrolysis resilin. Among them, glycine has the highest proportion, while the other two form the cross-linking point between molecular chains. The primary sequence and molecular structure of resilin have been difficult to identify result from the limited amount of protein and the reduced stability during the process of purification [1, 3].

Therefore, we take an example of this protein to help to understand the relationship between structure and function of the unique mechanical properties of resilin [10, 18]. For example, a partial clone of resilin, from the first exon of *Drosophila melanogaster*, provided the major source of elasticity via unstructured amorphous features. And it was considered to be a resilin precursor due to its amino acid composition and the presence of an N-terminal signal peptide sequence for secretion. Full-length resilin in *Drosophila melanogaster* contains three significant domains (Fig. 1). The first exon with 323 amino acids (exon 1 reported as "proresilin") is composed of 18 pentadecapeptide repeats (GGRPSDSY-GAPGGGN) (Fig. 2), standing for a cuticular secretion signal peptide; the second exon with 62 amino acids (exon 2) determined to be a typical cuticular chitin - binding domain (ChBD) with high affinity to chitin, implying a role in the formation of the resilin–chitin composites in the cuticle, resilin binds to cuticle chitin via the ChBD and is further polymerized through oxidation of the tyrosine residues, resulting in the formation of dityrosine bridges, this sequence is similar to a region conserved in several matrix proteins from insect cuticle; and the third exon with 235 amino acids (exon 3) is composed of 11 tridecapeptide repeats (GYSGGRPGGQDLG) (Fig. 3) [9, 18]. Exon 1 and exon 3 are all rich in proline and glycine residues that providing high flexibility to resilin but lack

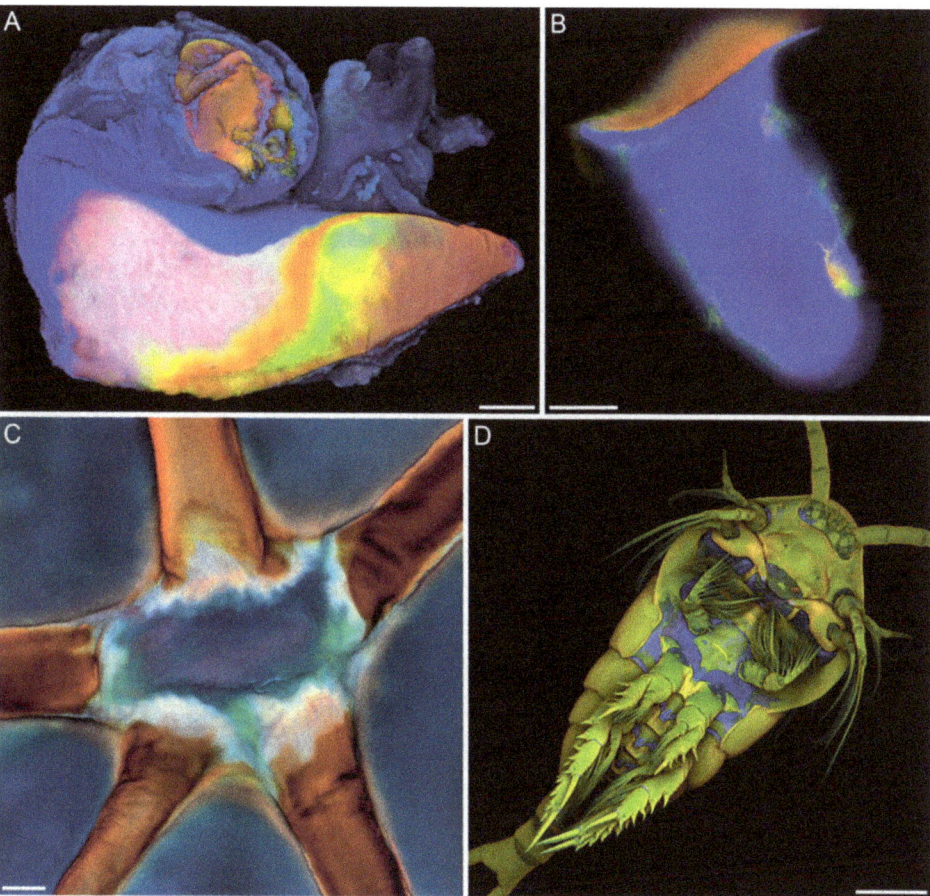

Fig. 1 Occurrence of resilin in insects, crustaceans, and monogeneans. Overlays of three different autofluorescences exhibited by the exoskeletons. Blue colors indicate large proportions of resilin, green structures consist mainly of non- or weakly sclerotized chitinous material, and red structures are composed of relatively strongly sclerotized chitinous material. (**a**) Wing hinge and (**b**) prealar arm of the migratory locust (*Locusta migratoria*). (**c**) Wing vein joint of the common darter (*Sympetrum striolatum*). (**d**) Ventral view of a female copepod of the species *Temora longicornis*. Blue = autofluorescence of resilin, red = Congo red fluorescence of stained chitinous exoskeleton parts, green = mixture of autofluorescence and Congo red fluorescence of stained chitinous exoskeleton parts. Scale bars (**a**, **b**), 20 μm (**c**), 200 μm (**d**), 25 μm. Adapted by permission from [9], Copyright 2016

sulfur-containing amino acid or tryptophan. The exon 1 repeats contain two proline residues and exon 3 repeats contain one. The strong conservation of the positions of proline and glycine residues encoded in exon 1 and exon 3 indicates that they are important for chain folding [12, 19]. Besides, the tyrosine residues they have can form intermolecular cross-links through di- and trityrosines connecting resilin polypeptides. The resilin sequence is determined by hydrophilic residues, indicating that hydrophobic interactions are minimal [6, 20]. Compared with exon 3, exon 1 is composed of more hydrophilic blocks and has a more flexible structure that

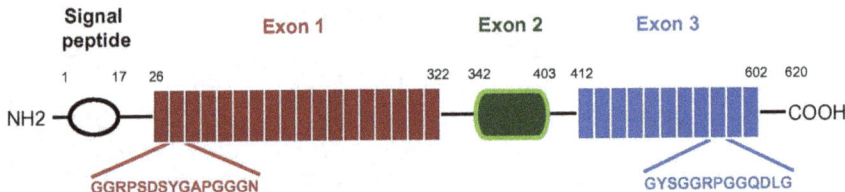

Fig. 2 The putative resilin sequence from the *Drosophila melanogaster* CG15290 gene product. The sequence consists of a signal peptide and three different exons (exons 1–3). The signal peptide is removed before secretion into the extracellular space. Exons 1 and 3 include 18 repeats of GGRPSDSYGAPGGGN and 11 copies of GYSGGRPGGQDLG, respectively. The sequence in exon 2 is involved in binding of chitin. Adapted by permission from Elsevier Ltd. [15], Copyright 2013

```
1      MVRPEPPVNS  YLPPSDSYGA  PGQSGPGGRP  SDSYGAPGGG  NGGRPSDSYG
51     APGQGQGQGQ  GQGGYAGKPS  DTYGAPGGGN  GNGGRPSSSY  GAPGGGNGGR
101    PSDTYGAPGG  GNGGRPSDTY  GAPGGGGNGN  GGRPSSSYGA  PGQGQGNGNG
151    GRSSSSYGAP  GGGNGGRPSD  TYGAPGGGNG  GRPSDTYGAP  GGGNNGGRPS
201    SSYGAPGGGN  GGRPSDTYGA  PGGGNGNGSG  GRPSSSYGAP  GQGQGGFGGR
251    PSDSYGAPGQ  NQKPSDSYGA  PGSGNGNGGR  PSSSYGAPGS  GPGGRPSDSY
301    GPPASGSGAG  GAGGSGPGGA  DYDNDEPAKY  EFNYQVEDAP  SGLSFGHSEM
351    RDGDFTTGQY  NVLLPDGRKQ  IVEYEADQQG  YRPQIRYEGD  ANDGSGPSGP
401    GGPGGQNLGA  DGYSSGRPGN  GNGNGNGGYS  GGRPGGQDLG  PSGYSGGRPG
451    GQDLGAGGYS  NGKPGGQDLG  PGGYSGGRPG  GQDLGRDGYS  GGRPGGQDLG
501    ASGYSNGRPG  GNGNGGSDGG  RVIIGGRVIG  GQDGGDQGYS  GGRPGGQDLG
551    RDGYSSGRPG  GRPGGNGQDS  QDGQGYSSGR  PGQGGRNGFG  PGGQNGDNDG
601    SGYRY
```

Fig. 3 Conceptual amino acid sequence of *Drosophila* gene product CG15920 after cleavage of the signal peptide. The regions corresponding to exon 2 conserved in a number of matrix proteins from insect solid cuticle are indicated by bold type, and the remaining, exon 1 parts of the native resilin are indicated by italics; exon 3 parts of native resilin are indicated by single underlining. Adapted by permission from Elsevier Ltd. [18], Copyright 2011

promotes self-aggregation to fibrillar structures in water. Exon 3 consists of hydrophobic and hydrophilic regions that tend to form micelles in water. Different from exon 1 and exon 3, exon 2 forms micelles of different sizes in water because of its properties of relatively hydrophobic [21, 22].

3 Summary

In this chapter, we take an example of this protein to introduce its special structure, the systematic understanding of the resilin structure, on the one hand, can help to understand and optimize the structure–property–function relationship of the resilin, on the other hand, can provide a blueprint for the de novo design of features and functions that match or even exceed these native materials.

Acknowledgments

This work was supported by the National Natural Science Foundation of China (No. 51603191).

References

1. Chae SK, Kang E, Khademhosseini A, Lee SH (2013) Micro/nanometer-scale fiber with highly ordered structures by mimicking the spinning process of silkworm. Adv Mater 25:3071–3078

2. Niazov-Elkan A, Sui X, Kaplan-Ashiri I, Shimon LJW, Leitus G, Cohen E, Weissman H, Wagner HD, Rybtchinski B (2019) Modular molecular nanoplastics. ACS Nano 13:11097–11106

3. McKee JR, Huokuna J, Martikainen L, Karesoja M, Nykanen A, Kontturi E, Tenhu H, Ruokolainen J, Ikkala O (2014) Molecular engineering of fracture energy dissipating sacrificial bonds into cellulose nanocrystal nanocomposites. Angew Chem Int Ed Engl 53:5049–5053

4. Hassanzadeh P, Sun W, de Silva JP, Jin J, Makhnejia K, Cross GLW, Rolandi M (2014) Mechanical properties of self-assembled chitin nanofiber networks. J Mater Chem B 2:2461–2466

5. Theberge AB, Courtois F, Schaerli Y, Fischlechner M, Abell C, Hollfelder F, Huck WT (2010) Microdroplets in microfluidics: an evolving platform for discoveries in chemistry and biology. Angew Chem Int Ed Engl 49:5846–5868

6. Zhou L, Li N, Shu J, Liu Y, Wang K, Cui X, Yuan Y, Ding B, Geng Y, Wang Z, Duan Y, Zhang J (2018) One-pot preparation of carboxylated cellulose nanocrystals and their liquid crystalline behaviors. ACS Sustain Chem Eng 6:12403–12410

7. Elvin CM, Carr AG, Huson MG, Maxwell JM, Pearson RD, Vuocolo T, Liyou NE, Wong DC, Merritt DJ, Dixon NE (2005) Synthesis and properties of crosslinked recombinant pro-resilin. Nature 437:999–1002

8. Li L, Kiick KL (2012) Resilin in the engineering of elastomeric biomaterials. Polymer Sci 9:105–116

9. Jan M, Esther A, Stanislav NG (2016) Functional diversity of resilin in Arthropoda. Beilstein J Nanotechnol 7:1241–1259

10. Tamburro AM, Panariello S, Santopietro V, Bracalello A, Bochicchio B, Pepe A (2010) Molecular and supramolecular structural studies on significant repetitive sequences of resilin. Chembiochem 11:83–93

11. Li L, Kiick KL (2013) Resilin-based materials for biomedical applications. ACS Macro Lett 2:635–640

12. Michels J, Appel E, Gorb SN (2016) Functional diversity of resilin in Arthropoda. Beilstein J Nanotechnol 7:1241–1259

13. Su RS, Kim Y, Liu JC (2014) Resilin: protein-based elastomeric biomaterials. Acta Biomater 10:1601–1611

14. Kim M, Elvin C, Brownlee A, Lyons R (2007) High yield expression of recombinant pro-resilin: lactose-induced fermentation in E. coli and facile purification. Protein Expr Purif 52:230–236

15. Renay S, Su C, Kim YJ, Liu JC (2013) Resilin: protein-based elastomeric biomaterials. Acta Biomater 4:1601–1611

16. Tatham AS, Shewry PR (2002) Comparative structures and properties of elastic proteins. Philos Trans R Soc Lond Ser B Biol Sci 357:229–234

17. Andersen SO, Weis-Fogh T (1964) Resilin. A rubberlike protein in arthropod cuticle. Adv Insect Physiol 2:1–65

18. Qin G, Rivkin A, Lapidot S, Hu X, Preis I, Arinus SB, Dgany O, Shoseyov O, Kaplan DL (2011) Recombinant exon-encoded resilins for elastomeric biomaterials. Biomaterials 32:9231–9243

19. Balu R, Whittaker J, Dutta NK, Elvin CM, Choudhury NR (2014) Multi-responsive biomaterials and nanobioconjugates from resilin-like protein polymers. J Mater Chem B 2:5936–5947

20. Andersson M, Jia Q, Abella A, Lee XY, Landreh M, Purhonen P, Hebert H, Tenje M, Robinson CV, Meng Q, Plaza GR, Johansson J, Rising A (2017) Biomimetic spinning of artificial spider silk from a chimeric minispidroin. Nat Chem Biol 13:262–264

21. Andersen SO (2003) Structure and function of resilin. In: Shewry PR, Tatham AS, Bailey AJ (eds) Elastomeric proteins, structures, biomechanical properties, and biological roles. Cambridge University Press, Cambridge, pp 259–278

22. Costa F, Silva R, Boccaccini AR (2018) Fibrous protein-based biomaterials (silk, keratin, elastin, and resilin proteins) for tissue regeneration and repair. In: Barbosa MA, Martins CL (eds) Peptides and proteins as biomaterials for tissue regeneration and repair. Woodhead Publishing, Cambridge, pp 175–204

Chapter 5

Structure of Keratin

Wenwen Zhang and Yimin Fan

Abstract

Keratins, as a group of insoluble and filament-forming proteins, mainly exist in certain epithelial cells of vertebrates. Keratinous materials are made up of cells filled with keratins, while they are the toughest biological materials such as the human hair, wool and horns of mammals and feathers, claws, and beaks of birds and reptiles which usually used for protection, defense, hunting and as armor. They generally exhibit a sophisticated hierarchical structure ranging from nanoscale to centimeter-scale: polypeptide chain structures, intermediated filaments/matrix structures, and lamellar structures. Therefore, more and more attention has been paid to the investigation of the relationship between structure and properties of keratins, and a series of biomimetic materials based on keratin came into being. In this chapter, we mainly introduce the hierarchical structure, the secondary structure, and the molecular structure of keratins, including α- and β-keratin, to promote the development of novel keratin-based biomimetic materials designs.

Key words Keratins and keratinous materials, Hierarchical structure, Secondary structure, Primary structure

1 Introduction

Keratin, the structural protein of epithelial cells, is a ubiquitous biological material. It is formed by the keratinization process and considered as "dead tissues" that epidermal cells die and build up at the outermost layer [1]. The amount of keratins varies significantly among different body tissues, individuals, and species.

Keratin consists of two classes: type I (acidic) keratin and type II (basic) keratin [2]. According to the different mechanisms of biosynthesis, it is typically divided into soft and hard keratins. The content of sulfur in soft and hard keratins is 1% and 5%, respectively. Soft keratins found in cellular tissues, such as stratum corneum in the skin, are classified as Ib (acidic-soft) and IIb (basic-soft) [3, 4], while the hard keratins are the main component of hair, feathers, hooves, shells, claws, horns, and so on, and are divided into Type Ia (acidic-hard) and IIa (basic-hard) [5, 6]. Compared with soft keratins, hard keratins possess high cysteine content and low glycine content and have better toughness and strength [7].

Shengjie Ling (ed.), *Fibrous Proteins: Design, Synthesis, and Assembly*, Methods in Molecular Biology, vol. 2347,
https://doi.org/10.1007/978-1-0716-1574-4_5, © Springer Science+Business Media, LLC, part of Springer Nature 2021

Concerning the molecular structure, the keratin can be further classified into α- and β-keratin [8]. Both α- and β-keratins exhibit the hierarchical structure. α-Keratin, commonly known as mammalian keratin, is the primary constituent of skin, wool, hair, nails, hooves, horns, and the stratum corneum (outermost layer of skin). And the β-form is the major component of hard avian and reptilian tissues, such as claws, feathers, and beaks of birds, and scales, and claws of reptiles [9]. In this chapter, the hair and feathers were taken as representatives of α- and β-keratins to introduce the structure of keratin. Their delicate hierarchical structure is particularly highlighted.

2 Hierarchical Structure of Keratins

2.1 Animal Hair

Figure 1 shows the hierarchical structure of hair fibers [10]. At the macroscopic fiber scale, hair fibers are mainly composed of three layers, including the outermost cuticle layer (~10% weight of the total fiber), middle cortex layer (~90% weight of the total fiber) as well as a certain amount of central porous structure. They have an

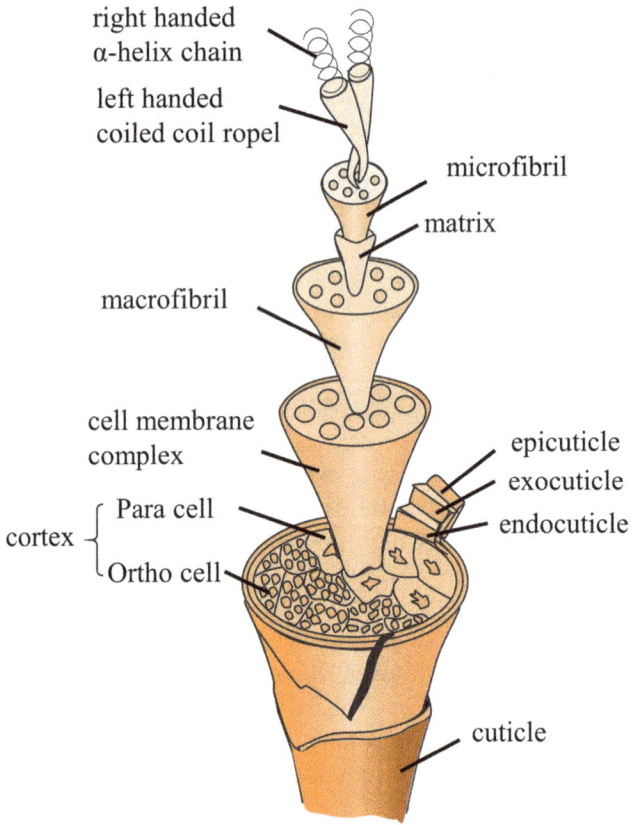

Fig. 1 Schematic of the hierarchical structure of animal hair

average density of 1.3 g/cm^3 and have cross-sectional areas around tens to hundreds of microns. The cross section of a single hair fiber is not necessarily circular; it may be elliptic.

The outermost cuticle layer is a chemically resistant region surrounding the cortex in hair fibers, and it consists of flat over-lapping scales [11]. The cuticle in human hair is five to ten scales thick, while one to two scales thick in wool fibers [12, 13]. One such scale cell generally has three layers, including epicuticle, exo-cuticle, and endocuticle. The shape and orientation of the cuticle cells are responsible for the differential friction effect in hair. In general, the fibers with larger diameters tend to have multiple layers of cuticle cells, such as the fibers from kangaroo, platypus, and seal guard have a single layer of cuticle cells whereas the layers in human hair and pigs bristle are about 6 and 15, respectively [14]. The thickness of individual cuticle cells is also variable: 0.3µm in Merino wool, 0.5–0.66µm in Lincoln wool, and 0.33µm in human hair [15, 16]. Besides, the cells from human hair and wool fibers are approximately rectangular, while those from kangaroo, seal, and platypus are much more elongated [14].

The middle cortex layer is the major part of the hair fibers and consists of cells and intercellular binding materials. The cortical cells are generally 1–6µm thick and ~100µm long [17], consist of orthocortex, paracortex, and mesocortex, which have different assemblies of structural components and decide the curly nature of the hair fibers [18]. The sulfur content of orthocortical cells and paracortical cells is ~3% and ~5%, respectively. Differently, the cells in orthocortex are more densely stained than those in paracortex from optical observation, and the orthocortical cells and paracor-tical cells have different intermediate filaments (IFs)/matrix pack-ing arrangements [19–21].

As illustrated in Fig. 2a–c, the IFs in the paracortex are more uniform and show clear hexagonal packing; whereas in the cross section of orthocortex, the IFs are organized into discrete bundles with a characteristic appearance resembling the "whorl" of a finger-print. The amount of matrix in the paracortex appears to be greater than that in the orthocortex [22]. And the diameter of paracortical cells is smaller than that of orthocortical cells. In addition, the cortical cells of hair fibers contain spindle-shaped macrofibrils with a diameter of 0.1–0.4µm [11, 23]. Each macrofibrils consist of high organized microfibrils (also called intermediate filaments (IFs)) in spiral formation with a diameter of about 7–10 nm and a high-sulfide amorphous matrix [24]. At the nanoscale, four right-handed α-helix chains form two dimer and then compose a proto-filament, and then two such protofilaments associate into a proto-fibril. Four protofibrils further combine into one IF [25, 26] (Fig. 2c). The composition of the microfibril is protofibrils with a diameter of ~20 Å.

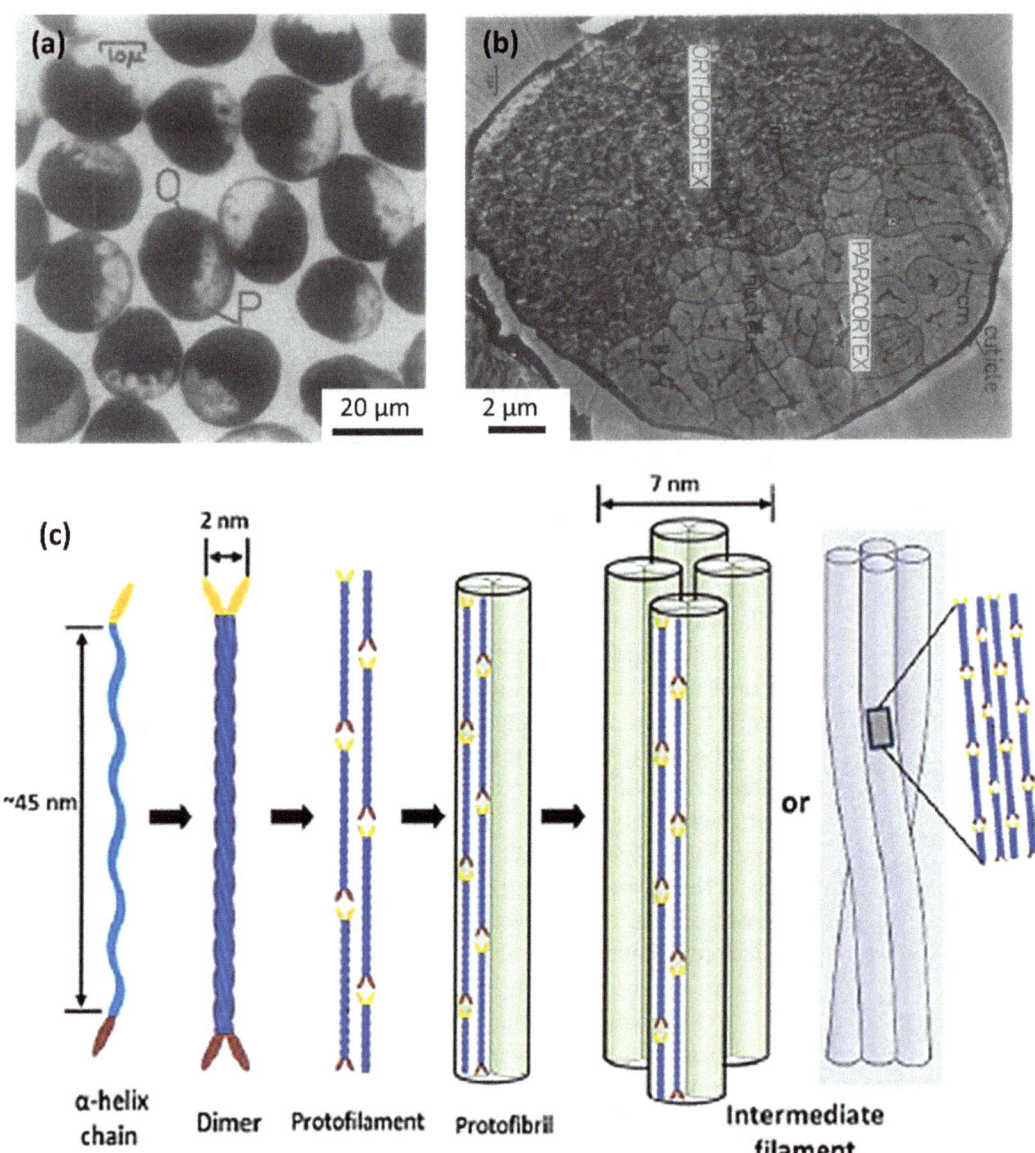

Fig. 2 The hierarchical structure of animal hairs. Light micrograph (**a**) and TEM image (**b**) of cross sections of Merino wool. Adapted by permission from Elsevier [20], Copyright 1959. (**c**) Schematic representation of the formation of intermediate filament. Adapted by permission from Elsevier [1], Copyright 2016

The central porous region of the hair fiber is the medulla, and its content varies greatly among different types of hair fibers. In fine hair fibers, such as fibers from merino wool and polar bear, the medulla (hollow core) is not present, which results in excellent thermal insulation [27]. While in thick animal hairs such as fibers from *Oryctolagus cuniculus* (rabbit), *Cervus elaphus* (elk), horse, and human, the mass of medulla comprises a large percentage of the whole fibers [28]. Besides, the medulla can be either completely

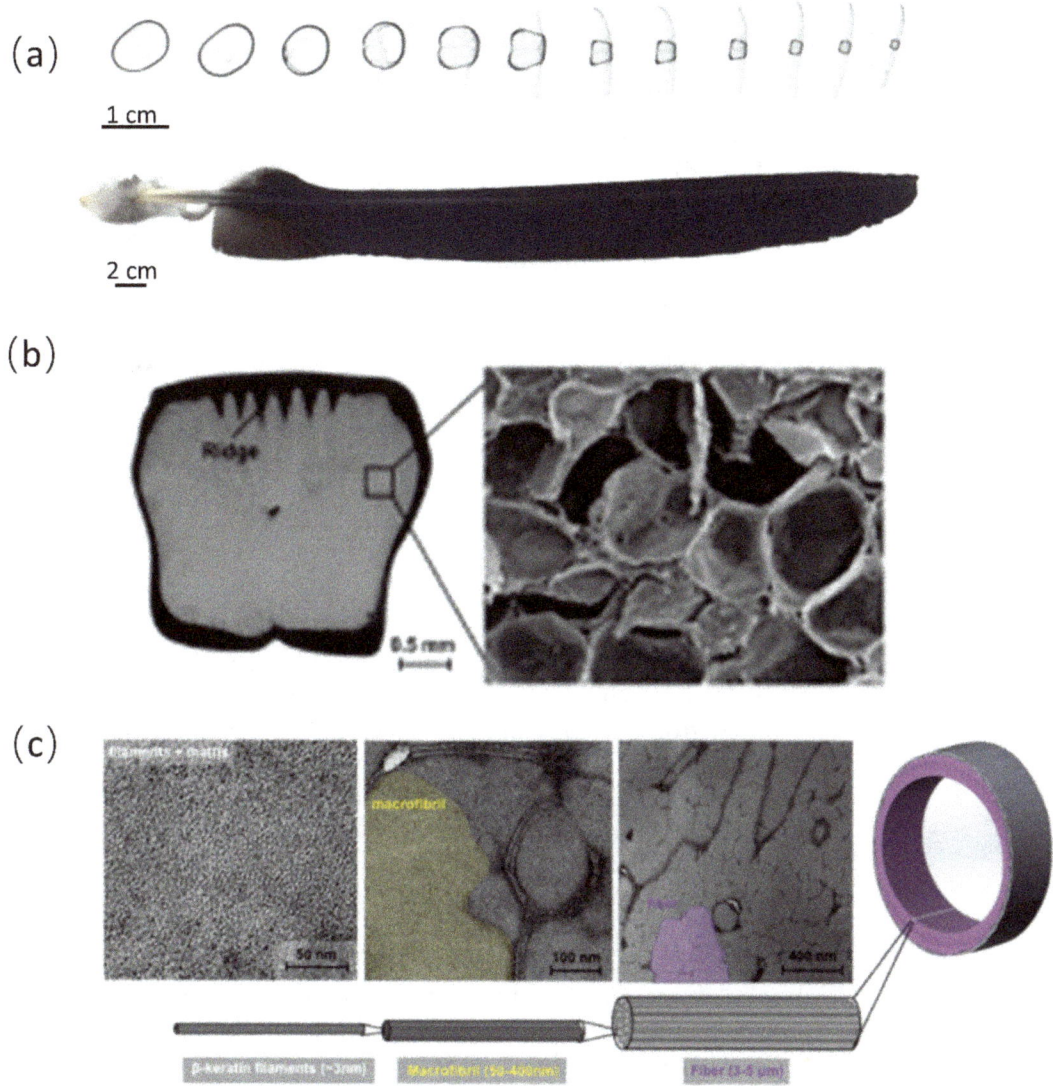

Fig. 3 The hierarchical structure of the flight feather. Adapted by permission from Elsevier [30], Copyright 2017. (**a**) The flight feather is divided into rachis, calamus, and asymmetrical vanes. And the cross-sectional shape from the calamus to feather rachis changes from circular to rectangular. (**b**) the cross-sectional image and the medullary foam of rachis. (**c**) Schematic of the hierarchical structure of the flight feather

absent, continuous along the fiber axis, or discontinuous. The cross-sectional images of the anter velent hair from extreme tip to the root showed that the distribution of medulla in a single fiber is also vary greatly [29].

2.2 Flight Feather

Flight feathers are mainly composed of rachis, calamus, and asymmetrical vanes [30] (Fig. 3a). The cross-sectional shape of rachis from the calamus to feather tip changes from circular to rectangular (Fig. 3b). The rachis supports barbs, which are secondary keratinous features that form the herringbone pattern of the vane [31]. It is composed of uniformly sized cells of ~20μm in diameter and mainly consists of three components: cuticle, cortex, and medulla, which is similar to hair fibers (Fig. 3c). The most superficial layer, cuticle, consists of circumferentially oriented fibers [32, 33]. The feather can be described as a paradigm of a sandwich-structured composite [34], and the cortex itself is a hierarchical, bilaminate, fiber-reinforced composite. The cortex is constructed of fibers with a diameter of 6μm and aligned predominantly along the length of the shaft. These fibers are comprised of macrofibrils (~200 nm in diameter), which are surrounded by amorphous intermacrofibrillar material. The crystalline β-keratin filaments (~3–4 nm in diameter) are embedded within amorphous matrix proteins to form macrofibrils [1, 28, 35, 36]. The feather core composed of closed-cell foam-like medullae and the cell (about 20–30μm in diameter) walls exhibit a porous and fibrous structure with curved fibrils piling up with spaces, further down in the structural hierarchy [1].

2.3 Horns

Horns belong to the Bovidae family, which includes sheep, antelope, cattle, waterbuck, and goats are hard keratinous materials. They are tough enough to protect themselves from predators and other stronger wild animals. Unlike antlers, horns are permanent structures and will not grow back when broken [37]. Horns are made up of a sheath of α-keratin and a hollow interior or a core of cancellous bone. The hierarchical structure of rhinoceros horn, shown in Fig. 4 indicates that the horn sheath is a three-dimensional laminated composite that consists of longitudinally aligned lamellae with a thickness of 2–5μm and lamellae elliptical dark tubules with a diameter of 40–100μm [38]. The long hollow tubules are spread between the lamellae and extend along the length of the horn in the growth direction. The circular lamellae stack in the radial direction with tubules and are composed of the IFs embedded in the amorphous matrix.

3   Secondary Structure of Keratins

At the nanoscale, both the α- and β-keratin show a characteristic filament-matrix structure. A major difference between α- and β-keratins is IF. The secondary structure of the IF in α- and β-keratin is α-helix and β-sheet, respectively.

3.1 α-Keratin

α-Keratin proteins are organized as coiled-coil structures. The α-helices in keratins are right-handed, and the helical structure is stabilized by the interchain hydrogen bonds, which cause the chain to twist and form a helical shape. The wide-angle X-ray diffraction

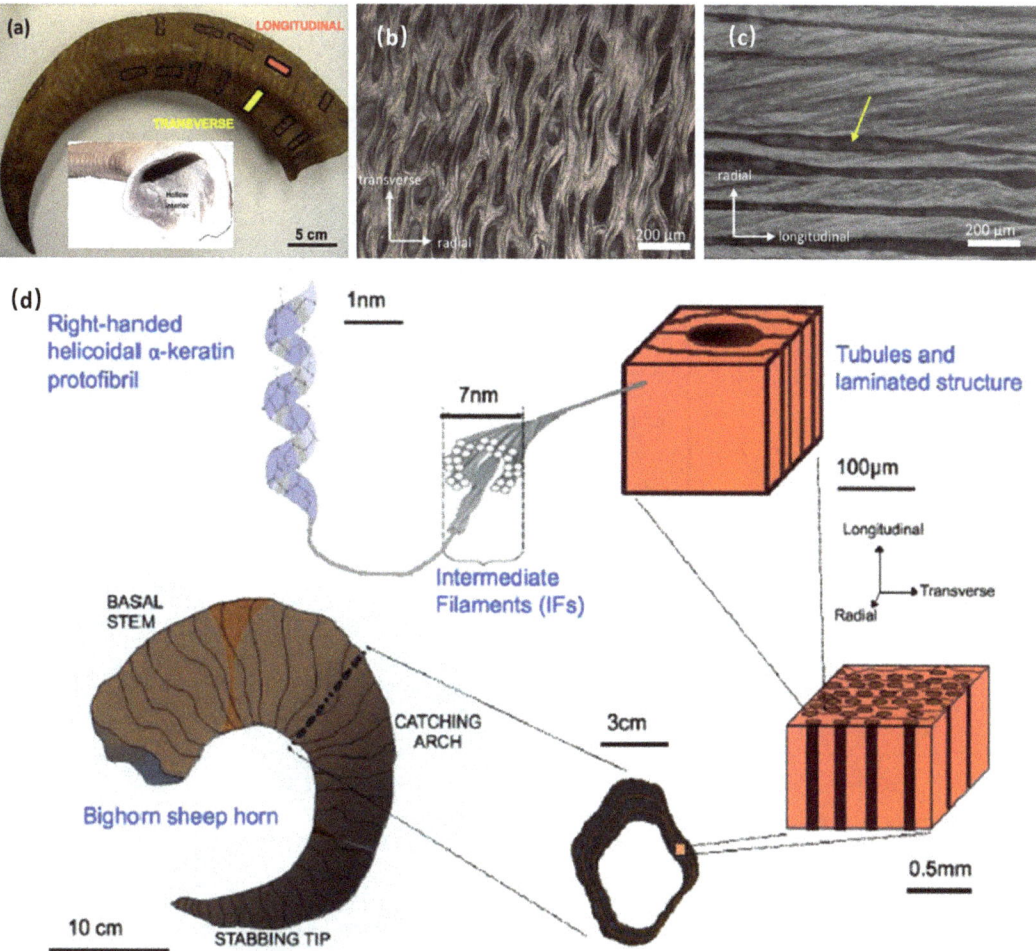

Fig. 4 Hierarchical structure of the bighorn sheep horn. Adapted by permission from Elsevier [37], Copyright 2010. (**a**) Photograph of the bighorn sheep horn that shows a hollow interior. (**b**) Cross-sectional optical micrograph of the bighorn sheep horn shows dark, elliptical tubules. (**c**) Longitudinal section of the bighorn sheep horn shows the outline of the parallel tubules in which the yellow arrow points to a tubule. (**d**) Schematic of the hierarchical structure of the bighorn sheep horn

pattern of α-keratin indicated that the distance between α-helical axes in α-keratin is 0.98 nm and the α-helix pitch projection is 0.51 nm [39, 40] (Fig. 5a).

3.2 β-Keratin

β-Keratin exhibits pleated-sheet structure, which is stabilized by the hydrogen bonds between β-strands and the planarity of the peptide bond [41]. The distorted β-sheet is organized by four parallel and antiparallel β-strands, which formed by folding of one polypeptide chain. These chains are held together by intermolecular hydrogen bonds. The inter-sheet distance in β-keratin is 0.97 nm, and the distance between residues along the chain and the distance between the chains in a β-sheet is 0.31 nm and 0.47 nm, respectively [42] (Fig. 5b).

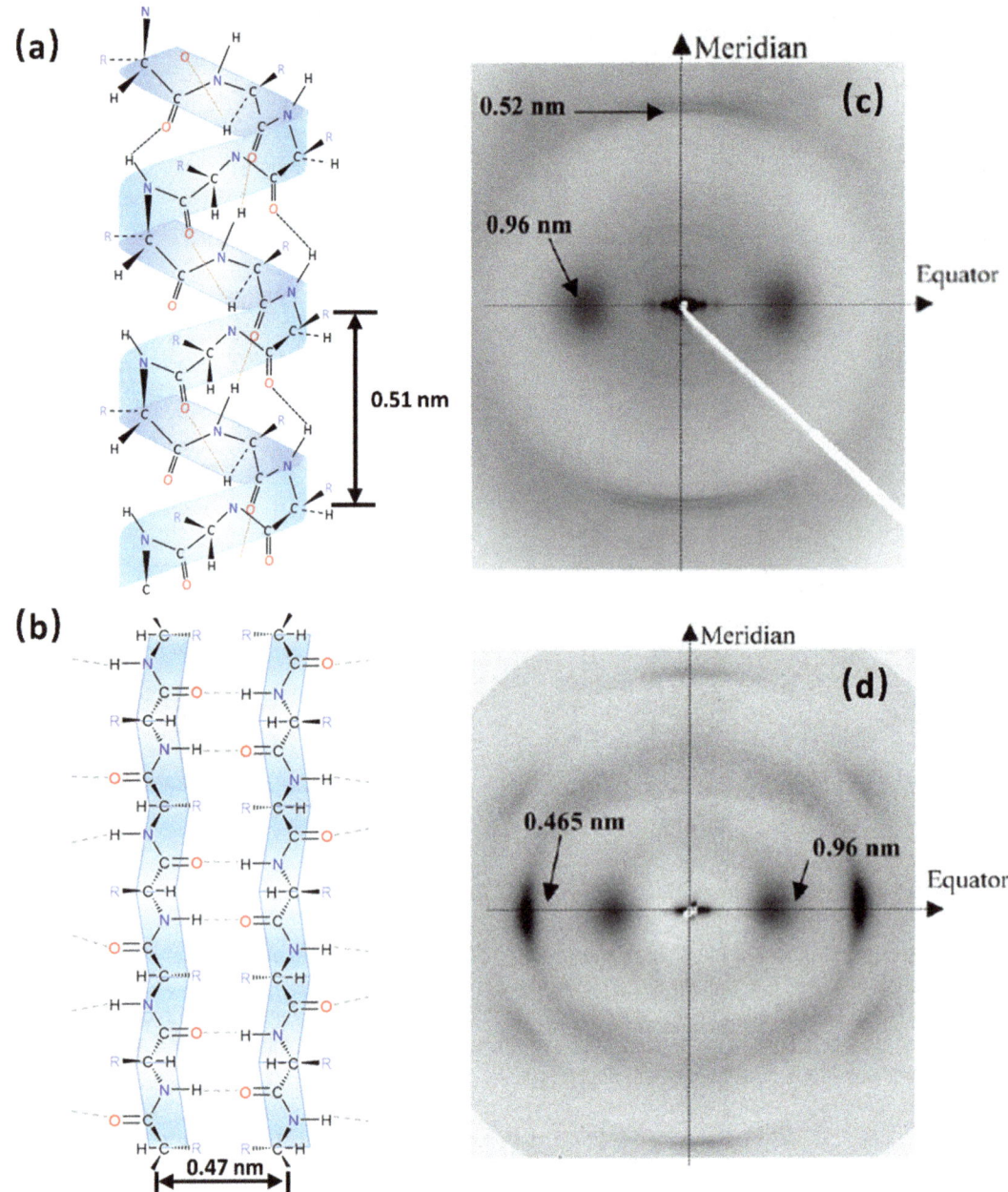

Fig. 5 The secondary structure of keratins. Structure of the polypeptide chain of the α-helix (**a**) and pleated β-sheet (**b**). Typical WAXS pattern of hard a-keratin (**c**) and β-keratin (**d**) obtained from horsehair before and after stretch. Adapted by permission from Elsevier [45], Copyright 2004

3.3 α–β Transition   X-ray diffraction experiments reveal that hair fibers can undergo an α–β phase transformation under tensile load: the 3.6 residues of one turn in α-helix elongated from 0.52 nm to 1.2 nm in β-sheet structures, which is reversible up to approximately 30% strain and led to the increase of elongation [43] (Fig. 5c, d). The stretching process includes the stretching of α-helices, progressive unraveling of the α-helical coiled-coil and refolding to β-sheet and the spatial expansion of the β-structured zones [44].

In particular, the amount of water has a significant influence on both α- and β-keratins. The structure transition in wet horsehair fiber under stretching indicates that the water may act as a polymer plasticizer in the structural transformation. At low humidity conditions, the chains remain unraveled while the β-sheet formed when stretching in water [45]. Also, the water may break down and reform the intermolecular hydrogen bonds related to the shape memory process of keratins [46, 47]. Overall, the water has effects on both the IFs and matrix in keratins [48].

4   Primary Structure of Keratins

4.1   α-Keratin

Another important feature to distinguish α-and β-keratin is the difference of primary structure, which is also known as molecular structures or the amide acid sequences [28, 41, 49, 50]. Three distinct regions can be identified in the molecular structure of α-keratin: the crystalline fibrils, the matrix, and the terminal domains of the filaments (Fig. 6a) [51]. The hydrogen bond inside the α-helix chain causes the chain to twist and exhibit a helical

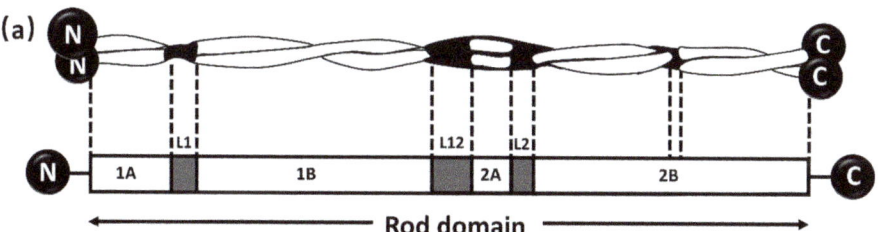

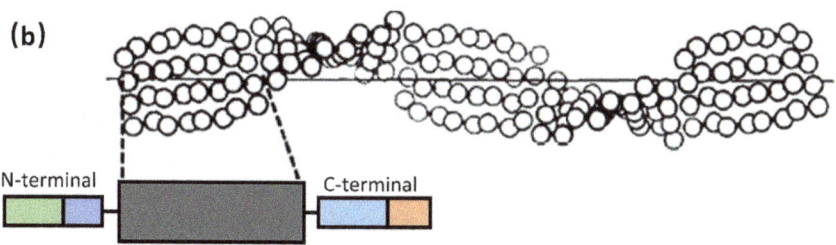

Fig. 6 The secondary structure of keratins. (**a**) The molecular structure of α-keratin filament. Adapted by permission from Elsevier [57], Copyright 2002. (**b**) The molecular structure of β-keratin filament. Adapted by permission from Elsevier [61], Copyright 2011

shape. With the sulfur cross-links, the dimer formed by α-helix chains assembles to form protofilaments. These protofilaments contain central α-helical rod with nonhelical N- and C-termini that are rich in cysteine residues [52, 53]. The central rod region contains nonhelical links at L1, L12, L2, and a stutter. The C- and N-terminal domains are involved in bonding with other IF molecules and matrix. The protofilaments then polymerize to form IFs with acidic (type I) or basic (type II) types [54, 55]. In α-keratins, the IFs of α-keratin is composed of several kinds of low-sulfur proteins [56], while the matrix consists of high-sulfur and high-glycine-tyrosine proteins [57]. The central helical region of the α-keratin molecular unit contains about 33–35 residues, and the nonhelical N- and C-terminal domains include about 136 residues [58]. The length of the central region is about 45 nm [59], and the diameter about 2 nm [60].

4.2 β-Keratin

For β-keratin, the molecule structure of the filaments also consists of three domains: the central domain with residues forming β-sheet and the N- and C-terminal domains [61, 62] (Fig. 6b). The central domain as the focus for the molecular structure of β-keratin filament is the central part to form the pleated sheet structure: two or more β- strand link through hydrogen bonding and position side by side, then the linked β- strands form small rigid planar surfaces that are slightly bent with respect to each other, forming a pleated sheet arrangement. For β-keratin, the length of the central region is about 2.3 nm and the diameter about 2 nm, and there are no different types of proteins; the filament and matrix are incorporated into one single protein [61]. Therefore, the molecular mass of α-keratin ranges from 40 to 68 kDa, which is much larger than that of β-keratin, 10–22 kDa [63].

5 Summary

In summary, this chapter introduces the hierarchical structure of keratin. The detailed secondary structure and molecular structure of α- and β-keratin are highlighted. The structure of keratin is a perfect example of understanding the structure–property relationship of biological materials. For instance, although both animal hairs and horns are composed of keratins, their physical properties, mechanical properties, for example, are completely different. Animal hairs usually are soft and supple, and have beauty and warm functions. They are also the tactile organ of some animals. Horns, instead, are very rigid and are often used by animals as an attack weapon or a shield to defend against attacks.

References

1. Wang B, Yang W, McKittrick J, Meyers MA (2016) Keratin: structure, mechanical properties, occurrence in biological organisms, and efforts at bioinspiration. Prog Mater Sci 76:229–318

2. Lee H, Noh K, Lee SC, Kwon I-K, Han D-W, Lee I-S, Hwang Y-S (2014) Human hair keratin and its-based biomaterials for biomedical applications. Tissue Eng Regen Med 11:255–265

3. Yu J, Yu DW, Checkla DM, Freedberg IM, Bertolino AP (1993) Human hair keratins. J Invest Dermatol 101:56–59

4. Karthikeyan R, Balaji S, Sehgal PK (2007) Industrial applications of keratins—a review. J Sci Ind Res 66:710–715

5. Dhouailly D, Cong X, Manabe M, Schermer A, Sun TT (1989) Expression of hair-related keratins in a soft epithelium: subpopulations of human and mouse dorsal tongue keratinocytes express keratin markers for hair-, skin- and esophageal-types of differentiation. Exp Cell Res 181:0–158

6. Heid HW, Moll I, Franke WW (1988) Patterns of expression of trichocytic and epithelial cytokeratins in mammalian tissues. Differentiation 37:137–157

7. Strnad P, Usachov V, Debes C, Grater F, Parry DAD, Omary MB (2011) Unique amino acid signatures that are evolutionarily conserved distinguish simple-type, epidermal and hair keratins. J Cell Sci 124:4221–4232

8. Spearman RIC (2010) The nature of the horny scales of the pangolin. Lancet 310:267–273

9. Chen PY, Mckittrick J, Meyers MA (2012) Biological materials: functional adaptations and bioinspired designs. Prog Mater Sci 57:1492–1704

10. Mercer EH (1957) The fine structure of keratin. Text Res J 27:860–866

11. Harland DP, Walls RJ, Vernona JA, Dyer JM, Woods JL, Bell F (2014) Three-dimensional architecture of macrofibrils in the human scalp hair cortex. J Struct Biol 185:397–404

12. Robbins CR (1988) The physical properties and cosmetic behavior of hair. Chemical and physical behavior of human hair. Springer, New York

13. Buckley JH, Boyle P, Burdett A, Gordo JB, Zweerink J (1997) Multiwavelength observations of Markarian 421. PhD dissertation

14. Bradbury J, Leeder J (1970) Keratin fibres IV. Structure of cuticle. Aust J Biol Sci 23:843–854

15. Bradbury JH, King N (1967) The chemical composition of wool. IV. The quantity of each histological component. Aust J Chem 20:2803–2807

16. Appleyard HM, Greville CM (1950) The cuticle of mammalian hair. Nature 166:1031–1031

17. Deutsch (1988) Chemical and physical behavior of human hair. Springer, New York

18. Fan J, Yu WD (2011) Fractal analysis of the ortho-cortex and para-cortex of wool fiber. Adv Mater Res 197-198:86–89

19. Rogers GE (2010) Electron microscope studies of hair and wool. Ann N Y Acad Sci 83:378–399

20. Rogers GE (1959) Electron microscopy of wool. J Ultrastruct Res 2:309–330

21. Rogers GE, Filshie M (1963) Some aspects of the ultrastructure of α-keratin, bacterial flagella, and feather keratin. Ultrastruct Protein Fibers:123–138

22. Dobb MG (1970) Electron-diffraction studies of keratin cells. J Text Inst 61:232–234

23. Fortier P, Suei S, Kreplak L (2012) Nanoscale strain-hardening of keratin fibres. PLoS One 7: e41814

24. Wegst UGK, Ashby MF (2004) The mechanical efficiency of natural materials. Philos Mag 84:2167–2186

25. Pauling L, Corey RB, Branson HR (1951) The structure of proteins; two hydrogen-bonded helical configurations of the polypeptide chain. Proc Natl Acad Sci U S A 37:205–211

26. Fraser RDB, MacRae TP (1973) The structure of α-keratin. Polymer 14:61–67

27. Cui Y, Gong H, Wang Y, Li D, Bai H (2018) A thermally insulating textile inspired by polar bear hair. Adv Mater 30:e1706807

28. McKittrick J, Chen PY, Bodde SG, Yang W, Novitskaya EE, Meyers MA (2012) The structure, functions, and mechanical properties of keratin. JOM 64:449–468

29. Woods JL, Harland DP, Vernon JA, Krsinic GL, Walls RJ (2011) Morphology and ultrastructure of antler velvet hair and body hair from red deer (Cervus elaphus). J Morphol 272:34–49

30. Sullivan TN, Wang B, Espinosa HD, Meyers MA (2017) Extreme lightweight structures:

avian feathers and bones. Mater Today 20:377–391

31. Bodde SG, Meyers MA, Mckittrick J (2011) Correlation of the mechanical and structural properties of cortical rachis keratin of rectrices of the Toco Toucan (Ramphastos toco). J Mech Behav Biomed Mater 4:723–732

32. Earland C, Blakey PR, STELL JGP (1962) Molecular orientation of some keratins. Nature 196:1287–1291

33. Earland C, Blakey PR, Stell JGP (1962) Studies on the structure of keratin IV. The molecular structure of some morphological components of keratins. Bba. Biochim Biophys Acta 56:268–274

34. Crenshaw DG (1980) Design and materials of feather shafts: very light, rigid structures. J Biomech 13:199

35. Lingham-Soliar T, Bonser RH, Wesley-Smith J (2010) Selective biodegradation of keratin matrix in feather rachis reveals classic bioengineering. Proc Biol Sci 277:1161–1168

36. Wang B, Meyers MA (2017) Seagull feather shaft: correlation between structure and mechanical response. Acta Biomater 48:270–288

37. Tombolato L, Novitskaya EE, Chen PY, Sheppard FA, McKittrick J (2010) Microstructure, elastic properties and deformation mechanisms of horn keratin. Acta Biomater 6:319–330

38. Hieronymus TL, Witmer LM, Ridgely RC (2006) Structure of white rhinoceros (Ceratotherium simum) horn investigated by X-ray computed tomography and histology with implications for structure and external form. J Morphol 267:1172–1176

39. Fraser RDB, Macrae TP, Rogers GEA (1972) Keratins: their composition, structure, and biosynthesis. Springfield, Ill.: Charles C. Thomas

40. Yang F-C, Zhang Y, Rheinstädter MC (2014) The structure of people's hair. PeerJ 2:e619

41. Lodish H, Berk A, Kaiser CA, Krieger M (2004) Molecular cell biology. W.H. Freeman and Company, New York

42. Bear RS, Rugo HJ (1951) THE results of X-ray diffraction studies on keratin fibers. Ann N Y Acad Sci 53:627–648

43. Fraser RDB, Macrae TP, Stewart FHC, Suzuki E (1965) Poly-L-alanylglycine. J Mol Biol 11:706–712

44. Kreplak L, JD BF (2001) Unraveling double stranded-helical coiled coils: an X-ray diffraction study on hard-keratin fibers. Biopolymers 58:526–533

45. Kreplak L, Doucet J, Dumas P, Briki F (2004) New aspects of the α-helix to β-sheet transition in stretched hard a-keratin fibers. Biophys J 87:640–647

46. Xiao XL, Hu JL, Gui XT, Qian K (2017) Shape memory investigation of alpha-keratin fibers as multi-coupled stimuli of responsive smart materials. Polymers 9:87

47. Liu ZQ, Jiao D, Zhang ZF (2015) Remarkable shape memory effect of a natural biopolymer in aqueous environment. Biomaterials 65:13–21

48. Fudge DS, Gosline JM (2004) Molecular design of the alpha-keratin composite: insights from a matrix-free model, hagfish slime threads. Proc Biol Sci 271:291–299

49. Jones LN, Simon M, Watts NR, Booy FP, Parry DAD (1999) Intermediate filament structure: hard α-keratin. Exp Dermatol 8:83–93

50. Nelson DL, Cox MM (2005) Lehninger principles of biochemistry. W.H. Freeman, New York

51. Parry DAD, North ACT (1998) Hard α-keratin intermediate filament chains: substructure of the N- and C-terminal domains and the predicted structure and function of the C-terminal domains of Type I and Type II chains. J Struct Biol 122:0–75

52. Dowling LM, Crewther WG, Parry DAD (1986) Secondary structure of component 8c-1 of alpha keratin; an analysis of the amino acid sequence. Biochem J 236:705

53. Fraser RDB, MacRae TP, Parry DAD, Suzuki E (1986) Intermediate filaments in α-keratins. Proc Natl Acad Sci U S A 83:1179–1183

54. Stewart M (1990) Intermediate filaments: structure, assembly and molecular interactions. Curr Opin Cell Biol 2:91–100

55. Moll R, Divo M, Langbein L (2008) The human keratins: biology and pathology. Histochem Cell Biol 129:705–733

56. Gillespie JM (1990) The proteins of hair and other hard α-keratins. In: Goldman RD, Steinert PM (eds) Cellular and molecular biology of intermediate filaments. Springer, Boston, MA

57. Coulombe PA, Omary MB (2002) 'Hard' and 'soft' principles defining the structure, function and regulation of keratin intermediate filaments. Curr Opin Cell Biol 14:110–122

58. Welham AC (2013) The coloration of wool and other keratin fibres. Wiley, New York

59. Steinert PM, Rice RH, Roop DR, Trus BL, Steven AC (1983) Complete amino acid sequence of a mouse epidermal keratin subunit and implications for the structure of intermediate filaments. Nature 302:794–800

60. Squire J, Vibert PJ (1987) Fibrous protein structure. Academic, New York

61. Fraser RDB, Parry DAD (2011) The structural basis of the filament-matrix texture in the avian/reptilian group of hard β-keratins. J Struct Biol 173:0–405

62. Fraser RDB, Parry DAD (2008) Molecular packing in the feather keratin filament. J Struct Biol 162:0–13

63. Alibardi L, Valle LD, Toffolo V, Toni M (2010) Scale keratin in lizard epidermis reveals amino acid regions homologous with avian and mammalian epidermal proteins. Ana Rec Part A 288:734–752

Part II

Methods in Synthesis of Fibrous Proteins

Chapter 6

General Methods to Produce and Assemble Recombinant Spider Silk Proteins

Na Kong

Abstract

Orb-weaving spiders are known to spin up to seven types of silks/glues from different silk glands. The inherent mechanical variety of these silks makes them attractive models for a variety of biomaterial design, from superglues to extremely strong and/or extendible fibers. Spider silk spinning is a process in which spinning dope stored in specific glands assembles into fibrils upon chemical and mechanical stimuli. The exploration of silk protein assembly into controllable filaments is vital for both uncovering biological functions and molecular structure relationship, as well as fabricating new biomaterials. This chapter describes the methods for biosynthesis and assembly of recombinant spider silk proteins, which will provide insights into the mechanism exploration of fiber formation and spider silk-based material manufacture.

Key words Spider silks, Recombinant proteins, Self-assembly, Kinetics, Characterization

1 Introduction

Spider silks, featuring with distinguishing hierarchical structure, high mechanical performance as well as functional diversity, have been explored as one of the most promising biomaterial candidates [1–9]. Protein engineering and rational design with multifunctional properties further expanded its applications [10–12]. For this reason, the underlying mechanism of silk protein assembly and its accompanying kinetic models play critical roles in the investigation of molecular structure–function relationship and fabrication of novel biomaterials.

Biochemically, spider silks are mainly composed of proteins (spidroins) that are secreted in abdominal glands. As shown in Fig. 1a, these spidroins are primarily composed of a distinctive repetitive structure flanked by highly conserved nonrepetitive amino-terminal (NT) and carboxyl-terminal (CT) domains [1, 13, 14]. The repetitive sequences usually composed of more than 90% of the whole silk protein, of which short polypeptide stretches composed of distinct domains with functional features

Shengjie Ling (ed.), *Fibrous Proteins: Design, Synthesis, and Assembly*, Methods in Molecular Biology, vol. 2347,
https://doi.org/10.1007/978-1-0716-1574-4_6, © Springer Science+Business Media, LLC, part of Springer Nature 2021

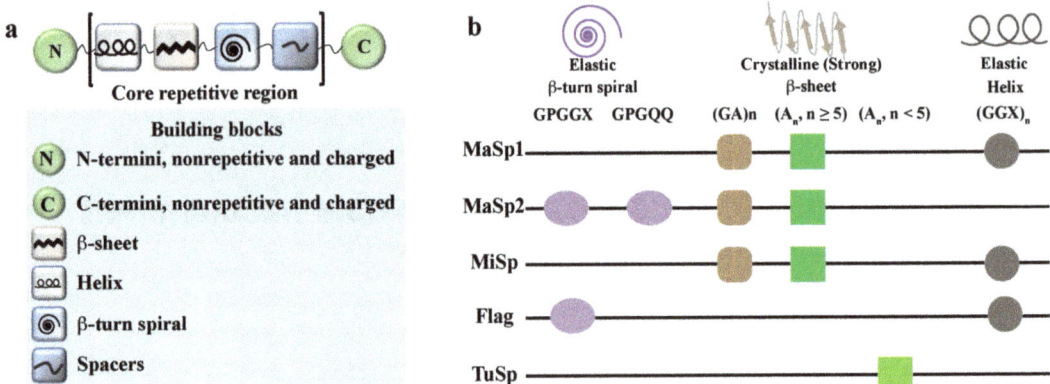

Fig. 1 Spidroin composition and their structural motifs and secondary structures in spider silks

(Fig. 1a, b) [3, 15]. It is proposed that these domains play vital roles in polypeptide folding and assembly during the spinning process [16]. Such as polyalanine (A_n) and glycine-rich (GGX) repeats in the dragline silks give rise to the high tensile strength by forming β-sheets as the backbone and representative amorphous domains contributing to extensibility to silk fibers [3, 15, 17–22], respectively.

Spider silk spinning is a process that spinning dope stored in the silk glands shuttled through the ducts and spinnerets to assemble into silk fibers upon chemical and mechanical stimuli. The spidroin assembly is triggered by water extraction, simultaneous ion exchange from sodium, chloride to potassium, phosphate, and pH decrease from 6.9 to 6.3, during the passage of silk dope through the spinning duct (Fig. 2) [23, 24]. The assembly relies on both the properties of precursor peptide sequences and the nature of microenvironment-derived solute–solvent interactions. Successful mimics of the spiders' spinning processing and precise control over protein folding and assembly allow both understanding of the molecular structure–function relationship and fabricating of artificial engineered spider silk-based materials.

Several recombinant spider silk polypeptides have been produced to emulate diverse natural silk fiber and study their assembly mechanism through biotechnology [25–28]. Especially the recombinant proteins derived from dragline silk, due to its excellent toughness higher than steel and Kevlar 49 fiber [1, 29]. For example, engineered polyalanine and glycine-rich segments derived from *Nephila clavipes* [28, 30–32], *Araneus diadematus* (ADF3/4) [23, 33–38], and *Euprosthenops australis* (4RepCT) [39, 40] are all explored as building blocks for specific-task recombinant proteins. The following protocols introduce the biosynthesis and assembly of recombinant spider silk proteins and aim to provide a general method with the design and implementation of a variety of

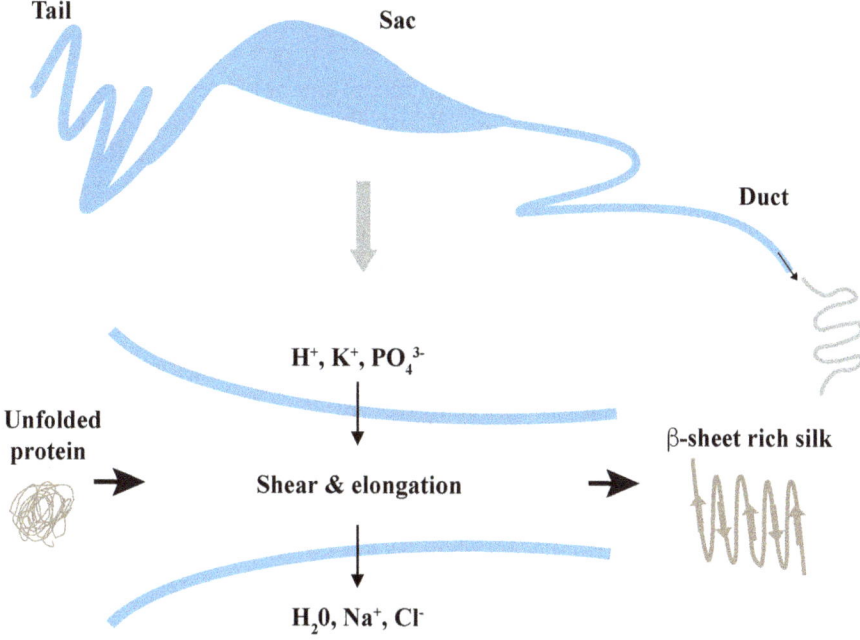

Fig. 2 Schematic diagram of the proposed natural spinning process of spider silk

spider silk–based protein assembly for potential biomedical and materials engineering applications.

2   Materials

Prepare all solutions using ultrapure water (resistivity of 18 MΩ·cm at 25 °C) from a Milli-Q ultrapure water purification system, and analytical grade reagents. Filter all solutions with a 0.22μm filter before use.

1. Buffer A (10 mM Tris-HCl, pH = 7.5, 500 mM NaCl): Take 5 mL of commercial 1 M of Tris-HCl (pH 8.8), and dissolve in 0.5 L of ultrapure water. Add 14.63 g NaCl to the solution and titrate the pH to 7.5 using 0.1 M HCl solution.

2. Buffer B (10 mM Tris-HCl, pH = 7.5, 500 mM NaCl, 500 mM imidazole): weigh 17.02 g of imidazole, and dissolve in 0.5 L of buffer A.

3. 1 M of potassium phosphate (K_3PO_4): weigh 21.2 g of K_3PO_4, and dissolve in 0.1 L of ultrapure water.

4. 6 M of guanidine thiocyanate (GdmSCN): weigh 354.5 g of GdmSCN, and dissolve in 0.5 L of ultrapure water.

3 Methods

3.1 Synthesis of Recombinant Spider Silk Proteins

3.1.1 Recombinant Spider Silk Protein Expression

1. Insert the designed recombinant protein-coding sequence into a pET28a (Fig. 3) and transform the resulting construct into *Escherichia coli* (*E. coli*) strain BL21(DE3) for protein expression.

2. Inoculate a single colony into 10-mL Luria broth (LB) medium containing 50μg/mL kanamycin. Incubate at 37 °C with constant shaking (160 rpm) for 2–4 h until the culture becomes apparently turbid ($OD_{600} > 0.6$).

3. Transfer the seed culture into 1-L LB medium containing 50μg/mL kanamycin (in 2-L baffled flask, prewarmed to 37 °C). Incubate at 37 °C with constant shaking (160 rpm) until the OD_{600} reaches 0.5–0.6.

4. Add isopropylthiogalactoside (IPTG) to a final concentration at 0.25 mM. Incubate at 37 °C with constant shaking overnight.

5. Pellet the bacteria by centrifugation at $5000 \times g$ for 10 min.

6. Resuspend the pellet with 30-mL buffer A containing 15 mM of imidazole and 1 mg/mL of lysozyme. Disrupt the cells using a French press at 900–1000 bar for 10 min.

7. Centrifuge the lysate at $16,000 \times g$ for 30 min at 4 °C. Filter the lysate with a 0.22-μm membrane to obtain the protein solution.

3.1.2 Purification of Recombinant Spider Silk Protein

8. Equilibrate a 5-mL HisTrap column (GE Healthcare) with 30-mL buffer A containing 15-mM imidazole.

9. Load all the lysate onto the equilibrated column.

10. Wash the HisTrap column with 30-mL buffer A containing 50-mM imidazole to wash unbound proteins.

11. Elute the target protein with 20-mL buffer B to obtain target protein. Elution of the target protein could be fractioned with 2-mL tubes to get pure protein.

12. Merge fractions with the highest protein concentration. Dialyze the protein against buffers containing 6-M, 4-M, 2-M, 1-M, and 0-M guanidine hydrochloride.

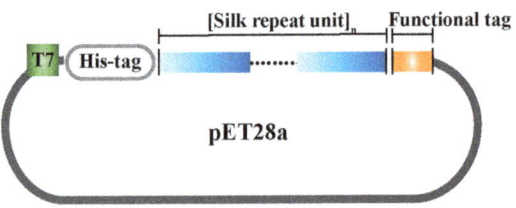

Fig. 3 Strategy to clone the engineered synthetic silk-initiated sequences in the pET28a expression vector

13. Add 1-mL His-tagged HRV3c proteinase (1 mg/mL) to the resulting protein solution, and incubate at 4 °C overnight to remove His-tag of the target protein (*see* **Note 1**).

14. Equilibrate a 5-mL HisTrap column with 30-mL buffer A without imidazole (*see* **Note 1**).

15. Load all of the reaction mixtures onto the equilibrated column (*see* **Note 1**).

16. Wash the column with 30-mL buffer A containing 100 mM of imidazole to elute the tag-free target protein (*see* **Note 1**).

17. Concentrate the eluted protein by ultrafiltration (*see* **Note 1**).

18. Purify the concentrated protein (volume < 100 μL) using a Superdex 200 10/300 GL column (GE Healthcare) with an isocratic flow of 20-mM phosphate buffer (pH = 7) containing 150 mM of NaCl.

19. Merge three to four tubes with the highest protein concentration and add into a 50-mL Ultrafiltration tubes. Centrifuge at $5800 \times g$ for 40 min under 4 °C. Concentrate the protein until the volume reaches about 2 mL, and lyophilize the solution.

20. Determine the purity of protein by sodium dodecyl sulfate–polyacrylamide gel electrophoresis (SDS-PAGE, 4–20%) and MALDI-TOF mass spectrometry.

3.2 Assembly of Recombinant Spider Silk Protein

1. Add the recombinant silk protein (1–10 mg) into 6 M GdmSCN, and dialyze extensively against 10 mM Tris/HCl, pH = 8 to obtain protein solution (*see* **Notes 2–5**).

2. Prepare fresh protein solutions and filter with 0.22 μm membrane to exclude the presence of oligomers before assembly.

3. Test the protein concentration with the BCA assay kit.

4. Transfer 1 mL of silk solution into a glass vial (*see* **Note 6**). This glass vial for protein assembly is for a smaller volume of silk solution. It is not necessary to use glass vials.

5. Initiate the protein self-assembly by addition of 100–300 mM potassium phosphate buffer (pH = 8), while gently stirring on a stir plate. Hours and days are needed to complete the assembly, and size-controllable fibrils could be obtained through time control.

3.3 Determination of Assembly Kinetics

3.3.1 Turbidity Measurement

1. Prepare samples in triplicate at distinct assembly protein concentrations (*see* **Note 4**).

2. Add 100–300 mM potassium phosphate buffer, pH 8 to initiate the assembly.

3. Record absorption at 340 nm every 10–30 min during the aggregation course for turbidity measurement until a final plateau was obtained.

4. Centrifuge samples at the specified time point at $185,000 \times g$ at 4 °C for 30 min, and determine the concentration of monomeric species in the supernatant by UV spectroscopy at 280 nm.

5. Normalize the results through $A = (A_i - A_0)/(A_{max} - A_0)$. A_0, A_i, and A_{max} is the absorbance intensity of the initial monomer solution, sample solution at a different time point, and sample solution at maximum, respectively (*see* **Note 7**).

6. Plot the kinetic curve through normalized absorbance vs time points (*see* **Note 5**).

3.3.2 Thioflavin T Binding Experiments

1. Prepare 1 mM of Thioflavin T (ThT) stock solutions by dissolving 1 mg of ThT dry powder in 3.14 mL ultrapure water. During the preparation, ultrasonication is needed for dissolution.

2. Filter the solution through 0.22μm syringe filters.

3. Add 10μL of stock solution to 0.99 mL of ethanol and measure the absorbance at 416 nm.

4. Calculate the concentration through absorption measurement at 416 nm by the Beer-Lambert equation, using a molar extinction coefficient 26,620 M^{-1} cm^{-1}. The concentration should be determined routinely to compensate for change from the loss or evaporation during the filtration.

5. The stock solution is covered with foil at 4 °C. The stock solution can be stored for up to a month.

6. Dilute the stock solution to 20μM.

7. Add 200μL of assembly solution to 96-well plate, and subsequently add 20 mM ThT each.

8. Place the 96-well plate to SpectraDrop™ Micro-Volume Microplate reader.

9. Shake for 5 s every 10–30 min, and record the spectrum between 465 nm and 565 nm with excitation at 438 nm.

10. Normalize the results through $I = (F_i - F_0)/(F_{max} - F_0)$. F_0, F_i, and F_{max} is the fluorescence intensity of background solution, sample solution at a different time point, and sample solution at maximum, respectively (*see* **Note 7**).

11. Plot the kinetic curve through normalized fluorescence intensity vs. time points (*see* **Notes 5, 8**).

3.4   Characterization

3.4.1   Circular Dichroism (CD) Spectroscopy

1. Dilute the assembly solution to a concentration of 10–40μM (0.5–1 mg/mL) of recombinant protein at a temperature of 20 °C (*see* **Note 4**).

2. Record a background spectrum of instruments from 190 to 240 nm.

3. Place the background solvent in a quartz cuvette with 0.05 or 0.1 cm path length, and record the solvent spectrum from 190 to 240 nm.

4. Collecting data points at 1 nm step, 1 nm bandwidth, with four acquisitions, and 500 nm/min scan rate.

5. Place the diluted assembly solution in the same quartz cuvette, and record the spectrum from 190 to 240 nm. The position of the cuvette to place should be the same as the solvent.

6. Record the CD spectra at the assembly time point of 5–30 min for structure transformation measurement.

7. Subtract the background and average over five to ten consecutive scans. Smooth the spectrum to 4 cm^{-1} and save the result (*see* **Note 8**).

3.4.2 Attenuated Total Reflection Fourier Transform Infrared Spectroscopy (ATR-FTIR)

1. Cast about 200 μL protein solution at distinct assembly time point on freshly cleaved muscovite mica stuck to iron disc, and allow the sample to dry in a fume hood for 30 min (*see* **Note 9**).

2. Record a background spectrum with the parameters at 4 cm^{-1} resolution from 4000 cm^{-1} to 400 cm^{-1} and 256 coadded scans.

3. Record the spectra in absorption mode at 4 cm^{-1} resolution from 4000 cm^{-1} to 400 cm^{-1} with 256 coadded scans.

4. Smooth the spectrum to 9 cm^{-1} by convolution.

5. Deconvolute the amide I (1700 cm^{-1} to 1600 cm^{-1}) from the second derivative spectra using Peakfit 4.12.

6. Assign peak absorption bands in the amide I band (1600 cm^{-1} to 1700 cm^{-1}) as follows: the peaks at 1620 cm^{-1} to 1630 cm^{-1} to β-sheet, at 1640 cm^{-1} to 1650 cm^{-1} to random-coil, 1650 cm^{-1} to 1660 cm^{-1} to α-helix, and 1690 cm^{-1} to 1700 cm^{-1} to β-turn.

7. Fix the peak position.

8. Fit the peak using numerical fitting to iteration 7 or 12.

9. Average secondary structure percent composition of the assembled fibril by integrating the area of each deconvoluted curve.

3.4.3 Atomic Force Microscopy (AFM)

1. Dilute the assembled solution to about 0.5–1 mg/mL.

2. Spot or spin-coat 20 μL aliquot of the assembled solution onto freshly cleaved mica or clean silicon wafers.

3. Incubate the samples for 5 min, and subsequently wash three times using 50 μL of ultrapure water.

4. Allow the samples dry overnight at ambient temperature.

5. Load the nanofibril sample to a Bruker Dimension FastScan atomic force microscope equipped with an Icon scanning module.

6. Select peak force feedback control under the mode of ScanAsyst in the air using a ScanAsyst-Air probe. AFM tips: silicon tip on nitride lever (Bruker) with resonant frequency f_0: 70 kHz, 2 nm radius of curvature, force constant k: 0.4 N/m.

7. Scan the sample and collect a height AFM image.

3.4.4 Transmission Electron Microscopy (TEM)

1. Prepare 2% (w/v) stain solution: Weigh 0.02 g uranyl acetate, and dissolve in 1 mL water in 1.5 mL tube.

2. Vortex the stain solution for 10 min, and filter the solution using a 0.22μm membrane. Protect the stain solution using an aluminum foil from light exposure, and the solution can be stored up to 3 months at room temperature.

3. Put the carbon-coated grids (carbon-coated 200-mesh copper grids) in a petri dish with carbon side facing up, and irradiate under glow-discharge plasma cleaner for 30 s under 25 mA current to produce a hydrophilic surface.

4. Drop 30μL of stain solution on top of a clean parafilm.

5. Spot 5μL of the assembled solution (0.5–1 mg/mL) on the treated grids, and incubate for about 10 min. Subsequently, rinse the grid gently three times using 10μL of ultrapure water to remove excess ions.

6. Incubate the grid for about 1 min. Attention should be paid to keep the grid from drying up.

7. Flip the grid with a tweezer into the top of each stain droplet for about 10 s sequentially.

8. Incubate for about 2 min, and blot residual sample solution from the grid edge using a filter paper. A little stain solution could be left to obtain a deeper stain.

9. Allow the samples dry overnight at ambient temperature and store in a grid storage box before imaging.

10. Perform TEM analysis using bright-field TEM micrographs operating at 120 kV.

4   Notes

1. If the removal of his-tag is not necessary, **steps 13–17** could be omitted.

2. A freshly prepared protein is preferred for self-assembly to avoid the solubility problem of protein in aqueous solution.

3. If the protein is lyophilized as a solid formulation and can be directly dissolved in water. Then the protein assembly could be directly triggered by K_3PO_4 buffer with pH 8 without the involvement of GdmSCN.

4. Dilution of protein samples is generally carried out in either gradually ascending or descending order of protein concentration.

5. For the protein self-assembly, a range of a few milligram recombinant proteins (1–10 mg) all could be used for assembly, which relies on the protein solubility, and will induce different assembly kinetics.

6. Care must be taken to maintain cuvettes/glass vials clean for the protein measurement and self-assembly.

7. For protein assembly kinetics and details of kinetic models, you could see references [38, 41].

8. All sample measurements should be carried out in triplicate at a minimum.

9. It should be noted that the amide I band is extremely sensitive to atmospheric water vapor, and must try to keep the samples as dry as possible.

Acknowledgments

The author acknowledges insightful discussion and suggestions from Prof. Shengjie Ling and thanks Dr. Zhaowei Wu from ShanghaiTech University for the helpful suggestions with the production of recombinant spider silk proteins. Financial support was provided by the National Natural Science Foundation of China (51703128).

References

1. Omenetto FG, Kaplan DL (2010) New opportunities for an ancient material. Science 329:528

2. Vollrath F (2000) Strength and structure of spiders' silks. Rev Mol Biotechnol 74:67–83

3. Lewis RV (2006) Spider silk: ancient ideas for new biomaterials. Chem Rev 106:3762–3774

4. Kluge JA, Rabotyagova O, Leisk GG, Kaplan DL (2008) Spider silks and their applications. Trends Biotechnol 26:244–251

5. Leal-Egaña A, Scheibel T (2010) Silk-based materials for biomedical applications. Biotechnol Appl Biochem 55:155–167

6. Schacht K, Jüngst T, Schweinlin M, Ewald A, Groll J, Scheibel T (2015) Biofabrication of cell-loaded 3d spider silk constructs. Angew Chem Int Ed 54:2816–2820

7. Sponner A (2007) Spider silk as a resource for future biotechnologies. Entomol Res 37:238–250

8. Vepari C, Kaplan DL (2007) Silk as a biomaterial. Prog Polym Sci 32:991–1007

9. Vollrath F, Porter D (2006) Spider silk as a model biomaterial. Appl Phys A Mater Sci Process 82:205–212

10. Norn CH, André I (2016) Computational design of protein self-assembly. Curr Opin Struct Biol 39:39–45

11. Saric M, Scheibel T (2019) Engineering of silk proteins for materials applications. Curr Opin Biotechnol 60:213–220

12. Huang P-S, Boyken SE, Baker D (2016) The coming of age of de novo protein design. Nature 537:320–327

13. Rising A, Hjälm G, Engström W, Johansson J (2006) N-terminal nonrepetitive domain common to dragline, flagelliform, and cylindriform spider silk proteins. Biomacromolecules 7:3120–3124

14. Gatesy J, Hayashi C, Motriuk D, Woods J, Lewis R (2001) Extreme diversity, conservation, and convergence of spider silk fibroin sequences. Science 291:2603

15. Tokareva O, Jacobsen M, Buehler M, Wong J, Kaplan DL (2014) Structure–function–property–design interplay in biopolymers: spider silk. Acta Biomater 10:1612–1626

16. Lin Z, Huang W, Zhang J, Fan J-S, Yang D (2009) Solution structure of eggcase silk protein and its implications for silk fiber formation. Proc Natl Acad Sci U S A 106:8906

17. Römer L, Scheibel T (2008) The elaborate structure of spider silk: structure and function of a natural high performance fiber. Prion 2:154–161

18. Garb JE, Haney RA, Schwager EE, Gregorič M, Kuntner M, Agnarsson I, Blackledge TA (2019) The transcriptome of Darwin's bark spider silk glands predicts proteins contributing to dragline silk toughness. Commun Biol 2:275

19. Hayashi CY, Shipley NH, Lewis RV (1999) Hypotheses that correlate the sequence, structure, and mechanical properties of spider silk proteins. Int J Biol Macromol 24:271–275

20. Rising A, Nimmervoll H, Grip S, Fernandez-Arias A, Storckenfeldt E, Knight DP, Vollrath F, Engström W (2005) Spider silk proteins–mechanical property and gene sequence. Zool Sci 22:273–281

21. Lin A, Chuang T, Pham T, Ho C, Hsia Y, Blasingame E, Vierra C (2015) 2—advances in understanding the properties of spider silk. In: Basu A (ed) Advances in silk science and technology. Woodhead Publishing, Cambridge, pp 17–40. https://doi.org/10.1016/B978-1-78242-311-9.00002-1

22. Jenkins JE, Creager MS, Lewis RV, Holland GP, Yarger JL (2010) Quantitative correlation between the protein primary sequences and secondary structures in spider dragline silks. Biomacromolecules 11:192–200

23. Rammensee S, Slotta U, Scheibel T, Bausch AR (2008) Assembly mechanism of recombinant spider silk proteins. Proc Natl Acad Sci U S A 105:6590

24. Yarger JL, Cherry BR, van der Vaart A (2018) Uncovering the structure–function relationship in spider silk. Nat Rev Mater 3:18008

25. Rising A, Johansson J (2015) Toward spinning artificial spider silk. Nat Chem Biol 11:309–315

26. Chung H, Kim TY, Lee SY (2012) Recent advances in production of recombinant spider silk proteins. Curr Opin Biotechnol 23:957–964

27. Teulé F, Cooper AR, Furin WA, Bittencourt D, Rech EL, Brooks A, Lewis RV (2009) A protocol for the production of recombinant spider silk-like proteins for artificial fiber spinning. Nat Protoc 4:341–355

28. Dai B, Sargent CJ, Gui X, Liu C, Zhang F (2019) Fibril self-assembly of amyloid–spider silk block polypeptides. Biomacromolecules 20:2015–2023

29. Ling S, Kaplan DL, Buehler MJ (2018) Nanofibrils in nature and materials engineering. Nat Rev Mater 3:18016

30. Winkler S, Szela S, Avtges P, Valluzzi R, Kirschner DA, Kaplan D (1999) Designing recombinant spider silk proteins to control assembly. Int J Biol Macromol 24:265–270

31. Aich P, An J, Yang B, Ko YH, Kim J, Murray J, Cha HJ, Roh JH, Park KM, Kim K (2018) Self-assembled adhesive biomaterials formed by a genetically designed fusion protein. Chem Commun 54:12642–12645

32. Prince JT, McGrath KP, DiGirolamo CM, Kaplan DL (1995) Construction, cloning, and expression of synthetic genes encoding spider dragline silk. Biochemistry 34:10879–10885

33. Eisoldt L, Hardy JG, Heim M, Scheibel TR (2010) The role of salt and shear on the storage and assembly of spider silk proteins. J Struct Biol 170:413–419

34. Borkner CB, Lentz S, Müller M, Fery A, Scheibel T (2019) Ultrathin spider silk films: Insights into spider silk assembly on surfaces. ACS Appl Polym Mater 1:3366–3374

35. Molina A, Scheibel T, Humenik M (2019) Nanoscale patterning of surfaces via DNA directed spider silk assembly. Biomacromolecules 20:347–352

36. Humenik M, Mohrand M, Scheibel T (2018) Self-assembly of spider silk-fusion proteins comprising enzymatic and fluorescence activity. Bioconjug Chem 29:898–904

37. Zha RH, Delparastan P, Fink TD, Bauer J, Scheibel T, Messersmith PB (2019) Universal

nanothin silk coatings via controlled spidroin self-assembly. Biomater Sci 7:683–695

38. Humenik M, Magdeburg M, Scheibel T (2014) Influence of repeat numbers on self-assembly rates of repetitive recombinant spider silk proteins. J Struct Biol 186:431–437

39. Nilebäck L, Arola S, Kvick M, Paananen A, Linder MB, Hedhammar M (2018) Interfacial behavior of recombinant spider silk protein parts reveals cues on the silk assembly mechanism. Langmuir 34:11795–11805

40. Nilebäck L, Hedin J, Widhe M, Floderus LS, Krona A, Bysell H, Hedhammar M (2017) Self-assembly of recombinant silk as a strategy for the chemical-free formation of bioactive coatings: a real-time study. Biomacromolecules 18:846–854

41. Morris AM, Watzky MA, Agar JN, Finke RG (2008) Fitting neurological protein aggregation kinetic data via a 2-step, minimal/"ockham's razor" model: the finke–watzky mechanism of nucleation followed by autocatalytic surface growth. Biochemistry 47:2413–2427

Chapter 7

Self-Assembly of *Bombyx mori* Silk Fibroin

Na Kong

Abstract

Silk fibroin from *Bombyx mori* (silkworm) distinguishes for its unique mechanical performance, controllable degradation rates, and easily large-scale production, making it attractive models for a variety of biomaterial design. These outstanding properties of silk fibroin originate from its unique modular composition of silk proteins. To exploit the structure–function relationship and fabricate silk fibroin–based biomaterials, comprehensive strategies to uncover assembly behaviors of fibrous proteins are necessary. This chapter describes methods to produce regenerated silk fibroin protein from *Bombyx mori* silk and their self-assembly strategies. This could provide insight into the fabrication of various silk fibroin–based biomaterials, such as hydrogels, tubes, sponges, fibers, microspheres, and diverse thin film patterns, which can be used for textiles, electronics and optics, environmental engineering, and biomedical applications.

Key words Silk fibroin, Self-assembly, Heat, Annealing, Characterization

1 Introduction

Bombyx mori (*B. mori*) silk distinguishes with high strength and elasticity, controllable degradation rates, and easily large-scale production from synthetic materials, and have been widely applied in textile, electronic, optical, and biomedical fields [1–8]. The natural spinning of *B. mori* silk is an externally triggered self-assembly process (Fig. 1), of which spinning dope in the silk gland at a very high concentration (up to 30% wt/vol) extrudes along ducts and spinnerets under solvent environmental and mechanical stimuli [9]. This is a process that fibrous polypeptides interact with each other to organize into ordered supramolecular structures, and optimize with unique structural, functional, and mechanical properties [2, 6, 9, 10].

B. mori silk is composed of silk fibroin and coated sericin proteins (Fig. 2a). Sericin, as adhesive hydrophilic proteins, is usually removed by a degumming process of boiling silk cocoons in an alkaline solution for robust silk fibroin. The resulting silk fibroin proteins, referred to as *B. mori* silk, biochemically consist of a heavy

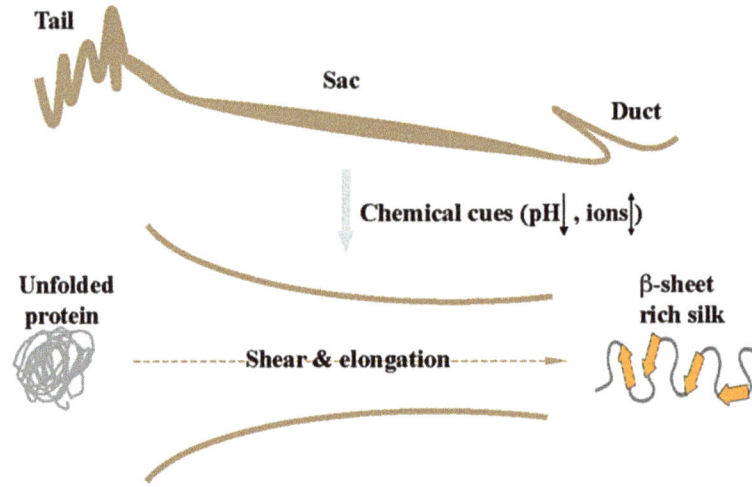

Fig. 1 Schematic diagram of the silk gland and silk production process of *B. mori* larvae

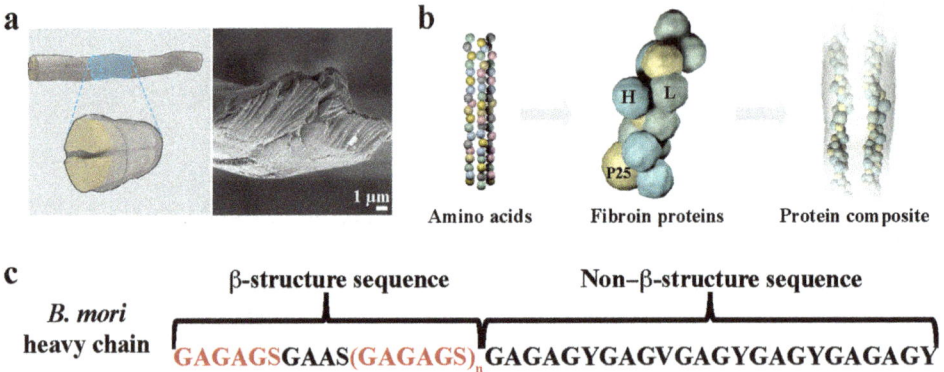

Fig. 2 Hierarchical structure of *B. mori* silk fiber. (**a**) Morphology and SEM image of silk fiber. (**b**) Fibroin protein composition. (**c**) The primary structure of heavy chain protein

(H) chain, a light (L) chain, and a glycoprotein, P25 (Fig. 2b). The H and L chains are linked by the formation of disulfide bonds between cysteines, and P25 associates with disulfide-linked H-L complex through noncovalent interactions to maintain the integrity of complex [11]. The hydrophobic domains of H chain contribute to the outstanding mechanical properties of silk fibers, which can form antiparallel β-sheet crystalline regions to give rise to high tensile strength [12]. As shown in Fig. 2c, the crystalline regions mainly comprise of repeating peptide GAGAGX (G is glycine, A is alanine, X is A, serine, and tyrosine), which are interspersed with nonrepetitive amorphous domains in the fibers responsible for the extensibility of silk fibers [12, 13].

Because of this unique primary structure of *B. mori* silk fibroin, it endows its intrinsic propensity to form β-sheet structures under

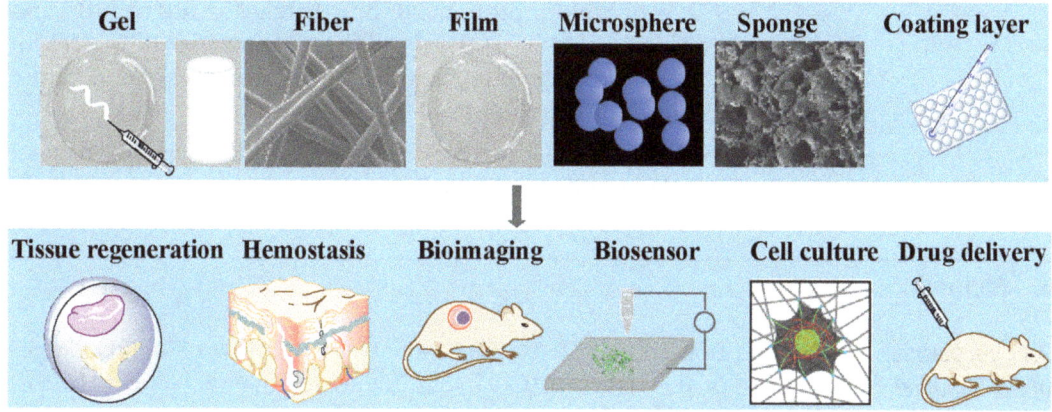

Fig. 3 Multiple silk fibroin-based material forms and their potential applications

favorable conditions [14–22]. Regenerative silk fibroin (RSF) as building blocks could be obtained from either aqueous solution or organic solvent treatment of silk fibroin fibers to mimic the natural silk fiber assembly [7, 23]. The most commonly used method for RSF assembly is the introduction of low dielectric constant alcohol solvents (methanol, ethanol) [15, 16, 24], mechanical treatment [25, 26], changes in temperature [18], or temperature-controlled water vapor annealing [22]. As shown in Fig. 3, multiple silk fibroin-based materials with nano/microstructure have been fabricated through the assembly of RSF in the forms of hydrogels, films, fibers, microspheres, and tubes for biomaterials [14, 19, 23, 27–31].

Assembly of RSF relies on both the properties of precursor modular motifs, protein concentration, and external stimuli conditions [4, 6, 9, 32]. Successful control over protein assembly allows both understandings of the molecular structure–function relationship and fabricating of artificially engineered biomaterials. The following methods introduce assembly of RSF and try to provide a general method with the implementation of a variety of RSF-based protein assembly for the potential fabrication of various materials forms and applications.

2 Materials

Prepare all solutions using ultrapure water (resistivity of 18 MΩ·cm at 25 °C), and analytical grade reagents. Filter all solutions with a 0.22 μm filter before use.

1. *B. mori* silkworm cocoons.

2. 0.5 wt% of sodium bicarbonate ($NaHCO_3$) solution: weigh 10 g of $NaHCO_3$, and dissolve in 2 L of water.

3. 9.3 M of lithium bromide (LiBr): weigh 161.5 g of LiBr, and slowly dissolve in 200 mL of water under stirring. Use of protective equipment is necessary due to high temperature during LiBr dissolution.

4. 1 M of sodium hydroxide (NaOH) solution: weigh 40 g of NaOH, and dissolve in 1 L of water.

3   Methods

3.1 Preparation of RSF Solution

1. Boil 10 g of the *B. mori* cocoons in 2 L of two 30 min changes of 0.5 wt% $NaHCO_3$ solution to remove the sericin. In this process, it is necessary to constantly stir the cocoons with a heat insulation rods, thereby destroying the structure of the cocoons, and allowing the silk to be fully dispersed.

2. Rinse the degummed cocoon fibers using boiling distilled water for three times to remove attached $NaHCO_3$ and sericin (*see* **Note 1**), and allow the silk fibers to dry at a 60 °C oven for more than 12 h. This drying process can also be carried out at room temperature, but the time required is usually longer, generally more than 24 h.

3. Dissolve 10 g of degummed silk fibroin (SF) in 100 mL of 9.3 M LiBr aqueous solution (10% w/v) at 60 °C for 1–4 h until the complete dissolution of the fibers (*see* **Note 2**). Here the solution needs to be sealed, and a freshly prepared solution is preferred.

4. Cool the solution to room temperature and dialyze the resulting solution for three days against ultrapure water using a cellulose dialysis membrane (M_w: 8000–14,000 D) at room temperature. During the dialysis process, the ultrapure water needs to be replaced at least three times a day.

5. Centrifuge the dialyzed solution at $6000 \times g$ for about 5 min to remove insoluble residues.

6. Collect the supernatant and carefully stored at 4 °C (Fig. 4) for no more than 3 days (*see* **Note 3**). When collecting the supernatant, gauze filtration can be further used to remove insoluble residues.

7. Determine the concentration by lyophilizing a volume of solution and measure the final dried weight.

3.2 Assembly of RSF

3.2.1 Alcohol-Initiated Assembly

1. Slowly dilute the freshly prepared RSF solution to a concentration of 0.05–0.4 wt% (*see* **Note 4**). Here must try to avoid stirring because the shearing effect may cause protein flocculation.

2. Adjust the solution to pH 9.5 using 1 M NaOH solution. In this process, in order to mix the solution quickly and evenly, the flask can be gently shaken, but it is still necessary to avoid vigorous stirring.

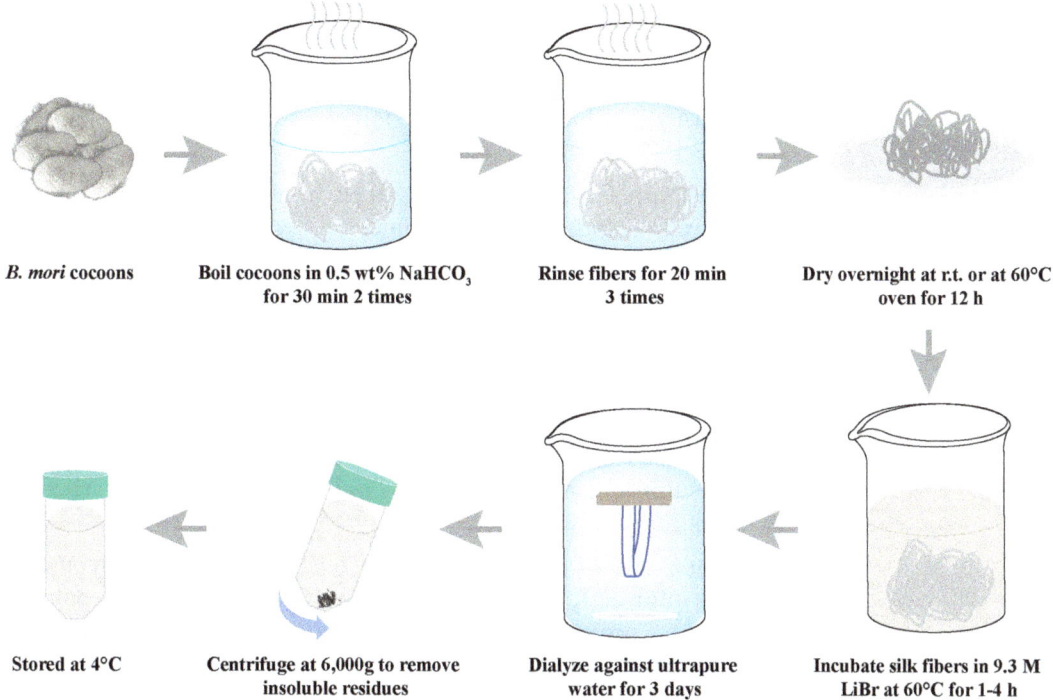

B. mori cocoons   **Boil cocoons in 0.5 wt% NaHCO₃**   **Rinse fibers for 20 min**   **Dry overnight at r.t. or at 60°C**
                             for 30 min 2 times   **3 times**   **oven for 12 h**

Stored at 4°C   **Centrifuge at 6,000g to remove**   **Dialyze against ultrapure**   **Incubate silk fibers in 9.3 M**
                           insoluble residues   **water for 3 days**   **LiBr at 60°C for 1-4 h**

Fig. 4 Schematic diagram of RSF solution preparation procedure

3. Initiate the protein self-assembly by the gradual addition of ethanol to form 7 vol% (v/v) ethanol solution (*see* **Note 5**), while gently shaking the flask (*see* **Note 6**).

4. Incubate the resulting solution for 1–3 days at room temperature to allow the protein assembly. Sometimes, it takes longer for silk protein to assemble, so one way to judge whether silk protein is assembled is to see if the clear solution has become opalescent. Another method uses a laser pointer laser to irradiate the solution to see that the Tyndall effect has already occurred.

3.2.2 Thermal-Initiated Assembly

1. Slowly dilute the freshly prepared RSF solution to a concentration of 0.05–0.4 wt% (*see* **Note 4**). Same as the alcohol-initiated assembly, all the operations must try to avoid vigorously stirring and shaking.

2. Incubate without any perturbation at 60 °C over 3–24 h, or a few days to obtain nanofibrils. During this process, the solution needs to be kept still, without stirring and shaking. Again, it may take longer for silk protein to assemble, so observing the color change and the Tyndall effect of the solution are also effective methods for monitoring whether the assembly has occurred.

3.2.3 Water Vapor
Annealing–Initiated
Assembly

1. Dilute the freshly prepared RSF solution to 1–2 wt% (*see* **Note 4**). Same as the methods mentioned above, all the operations must try to avoid vigorously stirring and shaking.

2. Cast the solution onto polystyrene dishes to form 5–15μm of thick films.

3. Keep the films in a dry vacuum environment at room temperature to avoid structural change.

4. Set the constant temperature humidity chamber at the desired annealing temperature (e.g., 60 °C) until homogenous water vapor (RH ≥ 90%) was obtained after several hours.

5. Transfer the sample into the chamber on an aluminum plate.

6. Anneal the silk samples at temperature values between 4 °C and 150 °C for 12 h. Sometimes, it takes longer for silk protein to assemble, and FTIR analysis could be used to judge whether silk protein is assembled.

7. Take out of the sample from the chamber and dry under vacuum at room temperature oven for 1–2 days to remove surface moisture. It should be noted that the constant temperature humidity chamber could be replaced using a setup device from reference [22].

3.2.4 Concentration
Dilution–Initiated Assembly

1. Prepare a fresh silk fibroin solution of 6 wt%. Attention should be paid to avoid vigorously stirring and shaking.

2. Slowly concentrate the solution to about 20 wt% over 24 h at 60 °C through the evaporation of water.

3. Dilute the resulting solution to 0.5–2 wt% with ultrapure water.

4. Incubate the dilute silk solution for about 24 h at 60 °C to induce the nanofibril formation. Again, it may take longer for silk protein to assemble. Color change and the Tyndall effect of the solution could be used to monitor the assembly process.

3.3 Characterization

3.3.1 Circular Dichroism
(CD) Spectroscopy

1. Dilute the RSF solution to a concentration of 0.5–1 mg/mL.

2. Record a background spectrum of instruments from 190 to 240 nm.

3. Place the solvent for assembly in a quartz cuvette with 0.05 or 0.1 cm path length (*see* **Note 6**), and record the solvent spectrum from 190 to 240 nm.

4. Place the sample in the same quartz cuvette, and record the solvent spectrum from 190 to 240 nm. The position of the cuvette to place should be totally the same with the solvent.

5. Collecting data points at 1 nm step, 1 nm bandwidth, with four acquisitions, and 500 nm/min scan rate.

6. Record the CD spectra every 5–30 min during the assembly course for structure transformation measurement. The interval time can be changed through the fibroin assembly time.

7. Subtract the background and average over five to ten consecutive scans. Smooth the spectrum to 4 cm^{-1} and save the result.

3.3.2 Attenuated Total Reflection Fourier Transform Infrared Spectroscopy (ATR-FTIR)

1. Cast about 100–200µL RSF solution at different assembly time point on freshly cleaved muscovite mica stuck to iron disc, and allow the sample to dry in a fume hood for 30 min (*see* **Note** 7).

2. Record a background spectrum with the parameters at 4 cm^{-1} resolution from 4000 cm^{-1} to 400 cm^{-1} and 256 coadded scans.

3. Scan the spectra in absorption mode at 4 cm^{-1} resolution from 4000 cm^{-1} to 400 cm^{-1} with 256 coadded scans.

4. Store the spectrum files in a format compatible with your analysis program.

5. Smooth the spectrum to 9 cm^{-1} by convolution.

6. Deconvolution the amide I (1700 cm^{-1} to 1600 cm^{-1}) or amide III band (1300 cm^{-1} to 1200 cm^{-1}) from the second derivative spectra using Peakfit 4.12.

7. Assign peak absorption bands in amide I band (1600 cm^{-1} to 1700 cm^{-1}) as follows: the peaks at 1620 cm^{-1} to 1630 cm^{-1} to β-sheet, at 1640 cm^{-1} to 1650 cm^{-1} to random-coil structure, 1650 cm^{-1} to 1660 cm^{-1} to α-helix, and 1690 cm^{-1} to 1700 cm^{-1} to β-turn; amide III band as follows: 1222 cm^{-1} to β-sheet, 1242 cm^{-1} to random coil, and 1265 cm^{-1} to α-helix, respectively.

8. Fix the peak position.

9. Fit the peak using numerical fitting to iteration 7 or 12.

10. Average secondary structure percent composition of the assembled fibril by integrating the area of each deconvoluted curve (details *see* ref. 33, 34).

3.3.3 Atomic Force Microscopy (AFM) Imaging

1. Dilute the assembled solution to about 0.5–1 mg/mL (*see* **Note 5**).

2. Spot or spin-coat 20µL aliquot of the assembled solution onto freshly cleaved mica or clean silicon wafers.

3. Incubate the samples for 5 min, and allow the samples dry overnight at ambient temperature before imaging.

4. Load the nanofibril sample.

5. Perform AFM analysis using a Bruker Dimension FastScan atomic force microscope equipped with an Icon scanning module utilizing peak force feedback control at ambient conditions under the mode of ScanAsyst in the air using a ScanAsyst-Air

probe (*see* **Note 8**). AFM tips: silicon tip on nitride lever (Bruker) with resonant frequency f_0: 70 kHz. 2 nm radius of curvature, force constant k: 0.4 N/m.

3.3.4 Peakforce QNM for Mechanical Measurement

1. Load a sapphire sample (SAPPHIRE-12 M).

2. Open "Nanoscope" software.

3. Load experiment through the sequential selection of "Mechanical Properties," "Quantitative Nanomechanical Mapping," followed by "Peak force QNM in the air" from experimental category.

4. Calibrate "Deflection Sensitivity."

    (a) Set scan parameters and make sure the "ScanAsyst Auto Control" is on.

    (b) Engage the probe, and scan a relatively flat region to obtain the force curve.

    (c) Open the "Ramp" control interface, and set one of the channels to "Deflection error and Z."

    (d) Collect and save three force curves for the reference standard.

    (e) Open the three force curve using "NanoScope Analysis," and perform baseline correction to each curve.

    (f) Set the cursor on the contact curve, and update the sensitivity value.

    (g) Repeat the calibration and calculation of the other two curves.

    (h) Calculate the average results of the three measurements, and type the new "deflection sensitivity" value into the "Detector" button from the "Calibrate" menu.

    (i) Return the scan interface.

5. Calibrate "Thermal Tune."

    (a) Withdraw the probe from the sapphire sample, and select "Thermal Tune" from the toolbar.

    (b) Fit the curve and calculate spring constant, and update it.

6. Calibrate "Tip Radius."

    (a) Replace the sample to Ti roughness sample (RS-12 M).

    (b) Engage probe and collect a clean sample image.

    (c) Level the image using 1st flatten order using "NanoScope Analysis."

    (d) Select "Tip Qualification" from the toolbar, and click "Estimate tip."

    (e) Withdraw the probe from the Ti sample, and place soft "PDMS Gel (PDMS-SOFT-1-12 M).

(f) Engage the probe and optimize parameters to obtain proper tracking. Make sure the "ScanAsyst Auto Control" off.

(g) Adjust peakforce setpoint manually to produce noticeable deformation data on the sample scan.

(h) Click "Capture line" from "Peakforce monitor window" to open high-speed data capture window to "Upload data."

(i) Use "HSDS Force Curve Analysis" to display the curve, and export the selected force curve.

(j) Open the exported curve, and select "peak deflection vs. separation" in display mode.

(k) Place the cursor on the approach curve and locate the initial deflection point, and note the separation value.

(l) Select the previously collected Ti roughness image, and type the separation value into "height form Apex" window.

(m) Click the "Qualify Tip" button to carry out the calculation.

(n) Note the tip radius result.

(o) Input the new calculated tip radius to "cantilever parameter."

7. Capture sample force curve.

(a) Select "Peakforce Capture" from the toolbar.

(b) Collect and save the scan.

8. Process images offline.

(a) Open "NanoScope Analysis."

(b) View force measurements over specific points or over an area.

(c) Average measurements using the "Roughness" button.

3.3.5 Transmission Electron Microscopy (TEM)

1. Weigh 0.02 g uranyl acetate, and dissolve into 1 mL water in 1.5 mL tube to prepare 2% (w/v) stain solution.

2. Vortex the stain solution for 10 min, and filter the solution using a 0.22μm membrane. An aluminum foil is necessary to protect the stain solution from light exposure, and the solution can be stored up to three months at room temperature.

3. Put the carbon-coated grids (carbon-coated 200-mesh copper grids) in a petri dish with carbon side facing up, and irradiate under glow-discharge plasma cleaner for 30 s under 25 mA current to produce a hydrophilic surface.

4. Drop 30μL of stain solution on top of a clean Parafilm.

5. Spot 5μL of the assembled solution (0.5–1 mg/mL) on the treated grids, and incubate for about 1 min.

6. Blot residual sample solution from the grid edge. Keep the grid from drying up.

7. Flip the grid with a tweezers into the top of each stain droplet for about 10 s sequentially.

8. Incubate for about 2 min, and remove the excess stain by blotting using filter paper. A little stain solution could be left to obtain a deeper staining.

9. Allow the samples dry overnight at ambient temperature and store in a grid storage box before imaging.

10. Perform TEM analysis using bright-field TEM micrographs operating at 120 kV.

3.3.6 Scanning Electron Microscopy (SEM)

1. Spot or spin-coat 20μL aliquots of the assembled solution onto freshly cleaned silicon wafers. Incubate the samples for 5 min, and allow the samples dry overnight at ambient temperature before imaging.

2. Alternatively, freeze the assembly solution in liquid nitrogen for about 20 min, and subsequently lyophilize the solution for 48 h to obtain solid nanofibril powder. Stick the sample power to carbon adhesive. If too much ions were involved in the assembly, dialysis of assembly solution is necessary to remove excess ions before freezing.

3. Coat the samples with a 5-nm-thick Au/Pt layer to prevent electrical charging.

4. Observe the sample at an acceleration voltage of 1–5 kV for SEM imaging.

3.3.7 Nano IR

1. Dilute the assembled solution to about 0.5–1 mg/mL.

2. Spin-coat 20μL aliquots of the assembled solution onto cleaned ultra-flat-gold-films or clean silicon wafers (*see* **Note 9**). A sample thickness of 20–500 nm is ideal.

3. Incubate the samples for 5 min, and allow the samples dry overnight at room temperature. Dry the sample with nitrogen before the measurements (*see* **Note 7**).

4. Initialize the Analysis Studio software and select the Tapping AFM-IR mode.

5. Set up the AFM.
   (a) Install a gold-coated tapping AFM-IR probe (model PR-EX-TnIR-A) with a spring constant of 1–7 N/m into the magnetic probe mount on the AFM scan head.
   (b) Place the sample substrate onto the AFM scanning stage.
   (c) Load the probe close to the sample (approach only).

(d) Move AFM Laser onto the end of the probe, and align the AFM laser with the end of the cantilever to obtain a maximum laser sum value.

(e) Adjust the cantilever deflection value to 0 V.

(f) Auto-Tune Cantilever. Set Auto-Tune parameters for the probe: Search Center = 65 kHz, Search Range = 50 kHz, Mode 2 Center: Auto = checked (on), Mode 2 Range = 100 kHz. In an ideal cantilever tune, each mode is dominated by a single peak (*see* **Note 10**).

(g) Update the mode frequencies and drive strength to the new values in the AFM Scan/Tapping AFM-IR panel. Pulse Rate (Pulse Rate = $f_{2nd\ mode} - f_{1st\ mode}$) in the panel is usually set automatically to the difference in frequency between the modes, and sometimes manually update is necessary.

6. Engage the probe to the sample surface. A slow approaching speed between 2 and 10μm/min is recommended until the cantilever edges become visible, but blurred with the probe tip (*see* **Note 11**).

7. Collect a height AFM image. IR Imaging Enabled should be unchecked (off) in the NanoIR panel.

8. IR alignment.

(a) In the AFM scan panel, stop the scan and target icon to the desired sample location.

(b) Focus and align the visible IR laser. Before the IR laser alignment, the probe must be engaged in the sample (*see* **Note 12**).

(c) Set initial nanoIR/General parameters: wavenumber 1640 cm^{-1}, Co-averages 128× (disabled), Power ~30%, and Pulse Rate was set automatically when Cantilever Tune was completed.

(d) Start the IR laser on.

(e) Increase the Power gradually until some probe oscillation is observed, up to a maximum power of 20%.

(f) Optimize the IR beam position at the current wavenumber. Use the slider bar to start with the size of 400μm, and decrease to 200μm to choose the final spot position (*see* **Note 12**).

(g) Set Tune Range of 50 kHz to tune the Pulse Rate.

9. Calibrate IR background (five times of average).

10. Run the Optimization function (at 200μm zoom) for multiple wavenumbers corresponding to the major absorption peaks of the sample (*see* **Note 12**).

11. Run a final Pulse Tune. Adjust the Pulse Rate using the green vertical cursor if needed.

12. Acquire a spectrum. IR Imaging Enabled should be unchecked (off) in the NanoIR panel.

13. Add Tap-IR Amplitude to the AFM Meter, and adjust the power gradually to make Tap-IR Amplitude between 0.5–10 mV.

14. Enable IR Imaging with channels of height, phase, and Tap-IR Amplitude. Check the box next to IR Imaging Enabled in the NanoIR/General panel.

15. Collect the imaging results at the desired wavenumber.

16. Data analysis.

    (a) Average all the collected spectra. Click on "Analysis-Process-Average in the software to average all the collected spectra.

    (b) Smooth the spectra by using the "Savitzky-Golay" function using a polynomial function of 2–5 cm^{-1}.

    (c) Label the individual peaks in the spectra. Click on "Analysis-Analyze-Peak find or show all peaks" in the software to automatically or manually identify peaks.

4 Notes

1. During the preparation of RSF solution, make sure the complete rinsing of ion and sericin on the silk fibroins.

2. For the dissolution of silk in 9.3 M LiBr, a freshly prepared aqueous solution and instant adding once the solution temperature decreased to 60 °C are necessary.

3. A freshly prepared RSF solution is preferred to avoid the solubility problem of protein in aqueous solution and if storage at 4 °C, no more than 3 days.

4. Dilution of protein samples is generally carried out in either gradually ascending or descending order of protein concentration.

5. For alcohol-initiated assembly, methanol also could be used. As a hazardous organic solvent, proper safety equipment, and operation in the fume hood is necessary.

6. Care must be taken to maintain cuvettes/glass vials clean for the protein measurement and self-assembly.

7. It should be noted that the amide I bands are extremely sensitive to atmospheric water vapor, and must try to keep the samples as dry as possible.

8. For AFM analysis, the scan mode and probe are alternative.

9. For Nano-FTIR, mica and silicon wafers are also available, if the sample thickness is more than 20 nm.

10. To try to improve cantilever tune, adjust the position of the probe in the mount slightly, or take the probe out and place it back.

11. It should be noted that the approach speed of the scanning head for engaging the tip is important. A fast and uncontrolled contact of the tip with the surface may cause damage to the tip.

12. Proper laser alignment and optimization before AFM-IR spectra acquisition are critical for obtaining accurate spectra. When the IR signal is collected without focusing the IR laser on the cantilever, a number of peaks may be attenuated or not present, which produces inaccurate data.

Acknowledgments

The author acknowledges insightful discussion and suggestions from Prof. Shengjie Ling and thanks Dr. Xinyang Wang from ShanghaiTech University, Dr. Linlin Wei from Bruker, and Ph.D. student Yawen Liu for the helpful suggestion with NanoIR characterization method. Financial support was provided by the National Natural Science Foundation of China (51703128).

References

1. Kundu B, Rajkhowa R, Kundu SC, Wang X (2013) Silk fibroin biomaterials for tissue regenerations. Adv Drug Deliv Rev 65:457–470

2. Omenetto FG, Kaplan DL (2010) New opportunities for an ancient material. Science 329:528

3. Ren J, Wang Y, Yao Y, Wang Y, Fei X, Qi P, Lin S, Kaplan DL, Buehler MJ, Ling S (2019) Biological material interfaces as inspiration for mechanical and optical material designs. Chem Rev 119:12279–12336

4. Wang Y, Guo J, Zhou L, Ye C, Omenetto FG, Kaplan DL, Ling S (2018) Design, fabrication, and function of silk-based nanomaterials. Adv Funct Mater 28:1805305

5. Mehrotra S, Chouhan D, Konwarh R, Kumar M, Jadi PK, Mandal BB (2019) Comprehensive review on silk at nanoscale for regenerative medicine and allied applications. ACS Biomater Sci Eng 5:2054–2078

6. Mason TO, Shimanovich U (2018) Fibrous protein self-assembly in biomimetic materials. Adv Mater 30:1706462

7. Vepari C, Kaplan DL (2007) Silk as a biomaterial. Prog Polym Sci 32:991–1007

8. Tao H, Kaplan DL, Omenetto FG (2012) Silk materials—a road to sustainable high technology. Adv Mater 24:2824–2837

9. Jin H-J, Kaplan DL (2003) Mechanism of silk processing in insects and spiders. Nature 424:1057–1061

10. Rammensee S, Slotta U, Scheibel T, Bausch AR (2008) Assembly mechanism of recombinant spider silk proteins. Proc Natl Acad Sci U S A 105:6590

11. Inoue S, Tanaka K, Arisaka F, Kimura S, Ohtomo K, Mizuno S (2000) Silk fibroin of bombyx mori is secreted, assembling a high molecular mass elementary unit consisting of h-chain, l-chain, and p25, with a 6:6:1 molar ratio. J Biol Chem 275:40517–40528

12. Zhou CZ, Confalonieri F, Jacquet M, Perasso R, Li ZG, Janin J (2001) Silk fibroin: structural implications of a remarkable amino acid sequence. Protein Struct, Funct. Bioinf 44:119–122

13. Zhou C-Z, Confalonieri F, Jacquet M, Perasso R, Li Z-G, Janin J (2001) Silk fibroin: structural implications of a remarkable amino acid sequence. Proteins 44:119–122

14. Martel A, Burghammer M, Davies RJ, Di Cola E, Vendrely C, Riekel C (2008) Silk fiber assembly studied by synchrotron radiation saxs/waxs and raman spectroscopy. J Am Chem Soc 130:17070–17074

15. Chen X, Shao Z, Marinkovic NS, Miller LM, Zhou P, Chance MR (2001) Conformation transition kinetics of regenerated bombyx mori silk fibroin membrane monitored by time-resolved ftir spectroscopy. Biophys Chem 89:25–34

16. Canetti M, Seves A, Secundo F, Vecchio G (1989) Cd and small-angle x-ray scattering of silk fibroin in solution. Biopolymers 28:1613–1624

17. Bai S, Liu S, Zhang C, Xu W, Lu Q, Han H, Kaplan DL, Zhu H (2013) Controllable transition of silk fibroin nanostructures: an insight into in vitro silk self-assembly process. Acta Biomater 9:7806–7813

18. Ling S, Qin Z, Huang W, Cao S, Kaplan DL, Buehler MJ (2017) Design and function of biomimetic multilayer water purification membranes. Sci Adv 3:e1601939

19. Bai S, Zhang X, Lu Q, Sheng W, Liu L, Dong B, Kaplan DL, Zhu H (2014) Reversible hydrogel-solution system of silk with high beta-sheet content. Biomacromolecules 15:3044–3051

20. Zhong J, Ma M, Li W, Zhou J, Yan Z, He D (2014) Self-assembly of regenerated silk fibroin from random coil nanostructures to antiparallel β-sheet nanostructures. Biopolymers 101:1181–1192

21. Greving I, Cai M, Vollrath F, Schniepp HC (2012) Shear-induced self-assembly of native silk proteins into fibrils studied by atomic force microscopy. Biomacromolecules 13:676–682

22. Hu X, Shmelev K, Sun L, Gil E-S, Park S-H, Cebe P, Kaplan DL (2011) Regulation of silk material structure by temperature-controlled water vapor annealing. Biomacromolecules 12:1686–1696

23. Rockwood DN, Preda RC, Yücel T, Wang X, Lovett ML, Kaplan DL (2011) Materials fabrication from bombyx mori silk fibroin. Nat Protoc 6:1612–1631

24. Ling S, Li C, Adamcik J, Shao Z, Chen X, Mezzenga R (2014) Modulating materials by orthogonally oriented β-strands: composites of amyloid and silk fibroin fibrils. Adv Mater 26:4569–4574

25. Ishida M, Asakura T, Yokoi M, Saito H (1990) Solvent- and mechanical-treatment-induced conformational transition of silk fibroins studies by high-resolution solid-state carbon-13 nmr spectroscopy. Macromolecules 23:88–94

26. Dubey P, Murab S, Karmakar S, Chowdhury PK, Ghosh S (2015) Modulation of self-assembly process of fibroin: an insight for regulating the conformation of silk biomaterials. Biomacromolecules 16:3936–3944

27. Xiao L, Liu S, Yao D, Ding Z, Fan Z, Lu Q, Kaplan DL (2017) Fabrication of silk scaffolds with nanomicroscaled structures and tunable stiffness. Biomacromolecules 18:2073–2079

28. Gong Z, Huang L, Yang Y, Chen X, Shao Z (2009) Two distinct β-sheet fibrils from silk protein. Chem Commun. https://doi.org/10.1039/B914218E:7506-7508

29. Callone E, Dirè S, Hu X, Motta A (2016) Processing influence on molecular assembling and structural conformations in silk fibroin: elucidation by solid-state NMR. ACS Biomater Sci Eng 2:758–767

30. Ma M, Zhong J, Li W, Zhou J, Yan Z, Ding J, He D (2013) Comparison of four synthetic model peptides to understand the role of modular motifs in the self-assembly of silk fibroin. Soft Matter 9:11325–11333

31. Nguyen AT, Huang Q-L, Yang Z, Lin N, Xu G, Liu XY (2015) Crystal networks in silk fibrous materials: From hierarchical structure to ultra performance. Small 11:1039–1054

32. Zhang W, Yu X, Li Y, Su Z, Jandt KD, Wei G (2018) Protein-mimetic peptide nanofibers: motif design, self-assembly synthesis, and sequence-specific biomedical applications. Prog Polym Sci 80:94–124

33. Ling S, Dinjaski N, Ebrahimi D, Wong JY, Kaplan DL, Buehler MJ (2016) Conformation transitions of recombinant spidroins via integration of time-resolved ftir spectroscopy and molecular dynamic simulation. ACS Biomater Sci Eng 2:1298–1308

34. Ling S, Qi Z, Knight DP, Shao Z, Chen X (2011) Synchrotron ftir microspectroscopy of single natural silk fibers. Biomacromolecules 12:3344–3349

Chapter 8

Synthesis and Assembly of Recombinant Collagen

Chenxi Zhao, Yuelong Xiao, Shengjie Ling, Ying Pei, and Jing Ren

Abstract

Collagen represents the major structural protein of the extracellular matrix. The desired mechanical and biological performances of collagen that have led to its broad applications as a building block in a great deal of fields, such as tissue engineering, drug delivery, and nanodevices. The most direct way to obtain collagen is to separate and extract it from biological tissues, but these top-down methods are usually cumbersome, and the structure of collagen is usually destroyed during the preparation process. Moreover, there is currently no effective method to separate some scarce collagens (such as collagen from human beings). Alternatively, bottom-up assembly methods have been developed to obtain collagen assembly or their analogs. The collagen used in this type of method is usually obtained by genetic recombination. A distinct advantage of gene recombination is that the sequence structure of collagen can be directly customized, so its assembly mode can be regulated at the primary structure level, and then a collagen assembly with a predesigned configuration can be achieved. Additionally, insights into the assembly behavior of these specific structures provide a rational approach to understand the pathogenic mechanisms of collagen-associated diseases, such as diabetes. In this chapter, Type I collagen is used as an example to introduce the key methods and procedures of collagen recombination, and on this basis, we will introduce in detail the experimental protocols for further assembly of these recombinant proteins to specific structures, such as fibril.

Key words Recombinant collagen, Self-assembly, In vitro

1 Introduction

Collagen, as a structural protein, widely exists in animal skin, bone, and connective tissue, playing a dominant role in maintaining the biological and structural integrity of tissues due to its molecular architectures and mechanical features. Twenty-nine kinds of collagen have been identified, which account for approximately 25–35% of all body proteins. Among them, Type I, II, III, IX, and V are the most common types [1–3], and Type I collagen is the most common one. Type I collagen is one of the primary composites in bones, tendons, corneas, and ligaments, reaching up to 90% of all collagen in the body.

Shengjie Ling (ed.), *Fibrous Proteins: Design, Synthesis, and Assembly*, Methods in Molecular Biology, vol. 2347,
https://doi.org/10.1007/978-1-0716-1574-4_8, © Springer Science+Business Media, LLC, part of Springer Nature 2021

Type I collagen is a heterotrimer, including two α1(I) chains and one α2(II) chain. Three polypeptide chains assemble into a supramolecular with triple-helical structure. Each chain contains three α-polypeptide chains, each containing 1014 amino acids and added up to 3042 amino acids for a well-folded molecule. Each α-chain is composed of N-terminus, Gly-X-Y repeats, and C-terminus, where X and Y mostly refer to proline and hydroxyproline, respectively [4]. Proline residues (4-hydroxyproline or 5-hydroxyproline) eliminate the free rotation of the angle, increasing the rigidity of the chain segment, while the hydroxyproline residues provide the binding sites for water molecules and also play an essential role in the formation of hydrogen bonds [5–7].

The terminal peptide is composed of 16 N-terminus and 26 C-terminus and has periodic transverse striations. Each polypeptide chain of Type I collagen has a molecular weight of 100,000. The central region of the α1(I) chains and one α2(II) chain has approximately 338 Gly-X-Y repeats. There is a short amino acid sequence at each end of the α1(I) chains, which is arranged in a random order, different from the main body. Each α1(I) chain is a left-handed helix, where the pitch is 0.87 nm, with 3.3 amino acids in each circle. The formation of this helical structure is mainly caused by the electrostatic repulsion between the proline at the X position and the 4-hydroxyproline at the Y station.

The three polypeptide chains are intertwined to form a triple-helical strand, which pitch is 9.6 nm, and with 36 amino acids in each circle. The amino acid residues in the side chains are all outward of the helix axis, which can help to form hydrogen bonds in the helical chain. The helical structure is highly stable due to these specific arrangements of the amino acids, forming the unique three-helix structure of collagen molecules [8–10], and is also termed as tropocollagen. The average relative molecular mass of this collagen is about 300 kDa, the length is about 300 nm, and the diameter is 1.5 nm. The three strands of tropocollagen present an extended conformation, which is challenging to stretch. The stability of the helix depends on the hydrogen bonds between polypeptide chains. Since the direction of rotation of the triple helix is opposite to the direction of rotation of the polypeptide chains, the structure is difficult to unwinding, thus giving high strength to match the physiological functions of collagen [3, 11–14].

Collagen can assemble into nanofibrils in vitro. The fibrils formation process of Type I collagen is dependent on the source and extraction method of collagen, as well as medium conditions such as temperature, pH, ionic strength, and collagen concentration [15–20]. The process of the self-assembly can occur spontaneously at neutral pH and physiologic temperature with the thermodynamic aggregation of collagen molecules. The molecules initially pack into small fibrillar segments with interactions that are subsequently enhanced by covalent cross-linking. These short segments grow by the end to end fusion to form a highly interconnected random network of long continuous fibrils. The resulting microscopic fibrils, fibril bundles, and macroscopic fibers can exhibit D periodicity banding virtually indistinguishable from native collagen fibers [21–24].

In these years, collagen and their assemblies have become an attractive biopolymer for biomedical applications due to its unique fibrillar structure as well as superior biological performances (such as low immunogenicity, outstanding biocompatibility and biodegradability) [11, 12, 15, 25–33]. Collagen with a supramolecular structure presents diverse morphologies in different tissues, leading to different biological functions. The investigation of collagen assembly and corresponding kinetic models can reveal performance-optimization mechanisms and acquire inspiration for designing new engineering material [15, 16, 26, 34, 35].

For biomedical applications, high-purity and sequence-specific collagen is preferred. Therefore, in addition to purifying naturally sourced collagen, another solution is to synthesize recombinant collagen. Indeed, gene recombination is superior to natural collagen extraction technology in terms of sequence control. Using gene recombination, other functional peptide sequences can be fused with collagen motifs to generate collagen analogs. These fusion proteins, combined advantages of both collagen and functional peptides, have shown promising applications in orthopedic implants, controlled drug delivery systems, scaffolds design, and so on. Accordingly, this chapter introduces in detail how to synthesize, separate, and purify recombinant collagen and its analogs, and also summarize how to assemble these recombinant proteins into highly ordered supramolecular aggregations.

2   Materials

1. Ultra-pure water is produced from a Milli-Q ultrapure water purification system.

2. Filter membranes(pore size, 0.22μm).

3. Buffered glycerol complex medium (BMGY, pH 6.0): Yeast extract 1%, peptone 2%, 100 mM potassium phosphate, pH 6.0, YNB 1.34%, biotin 4×10^{-5}%, glycerol 1%.

4. Buffered methanol complex medium (BMMY, pH 6.0): Yeast extract 1%, peptone 2%, 100 mM potassium phosphate, pH 6.0, YNB 1.34%, biotin 4×10^{-5}%, methanol 1%.

5. 50 mM sodium phosphate buffer, pH 7.4.

6. Protein standard solution sample buffer solution (containing bromophenol blue indicator).

3 Methods

3.1 Recombination of Collagen

3.1.1 Construction of Expression Vector and Production of Recombinant Pichia Pastoris Strain

All the recombinant strains in this experiment are produced by electroporation (Table 1), and the specific operation method can be found in the *Manual of Methods for Expression of Recombinant Proteins in Pichia pastoris* (Invitrogen) [24].

The expression of 4-PH by the strain α/PDI-αMF has been described in 1977 by Vuorela et al. [36]. The general preparation method of recombinant plasmid is to use double enzyme digestion method to excise the original fragment from the plasmid, and then use DNA ligase to join two fragments with the same restriction site on one end. These fragments are then coligated into the empty vector plasmid. The restriction sites used and the processing of plasmids using PCR have been described in detail in the article by Vuorela and Myllyharju [24, 36–38].

3.1.2 Expression

1. The primary cultures of the clones are grown in 25 mL BMGY media at 30 °C in a 250 mL shaker flask for 12–16 h until OD_{600nm} reaches 3 to 4. Cells are harvested by centrifugation for 10 min at $8000 \times g$ at 22 °C.

2. Resuspend cells with BMMY culture medium (pH 6.0) until OD_{600} reaches 1. The cell pellet is grown at 30 °C in a 1 L shaker flask agitated at 250 rpm. For the next 60 h, the culture is induced with 0.5% methanol and an amino acid of 100μg/L at an interval of 12 h.

3. Cells are collected by centrifugation (4 °C) at $9000 \times g$ and the cell pellet is resuspended in a mixed solution of 5% glycerol, 1 mM Pefabloc, and 50 mM sodium phosphate buffer (pH 7.4) at 4 °C.

3.1.3 Purification

1. The cells are lysed by glass beads vortexing in 0.1 M HCl and centrifuged at $10,000 \times g$ for 30 min. Protein concentrations are determined using the Bio-Rad Protein Assay (Bio-Rad).

2. Aliquots of the soluble fractions of the cell lysates are analyzed by SDS-PAGE under reducing conditions, followed by Western blotting with polyclonal antibodies 1675 and 1669 (Fibro-Gen), recognizing the N propeptide of human proα1(I) and the C propeptide of human proα2(I) chains, respectively, or a monoclonal antibody 95D1A recognizing the collagenous regions of various collagen chains.

3. The supernatant is digested with pepsin for 2 h at 22 °C and analyzed by SDS-PAGE followed by silver staining. The amounts of the collagen chains are estimated by densitometry of the silver-stained bands using a GS-710 Calibrated Imaging Densitometer (Bio-Rad). 4-PH activity is assayed by a method

Table 1
Recombinant Pichia strains

Strain	Expression vectors	Selection	Polypeptides expressed
Proα1(I)	pPICZBproα1(I), pARG815α, pPIC9PDI	Zeo⁺, Arg⁺, His⁺	proα1(I) chain, 4-PH α subunit, PDI
Proα2(I)	pBLADEIXproα2(I), pBLARGIXα, pPIC9PDI	Ade⁺, Arg⁺, His⁺	Proα2(I) chain, 4-PH α subunit, PDI
Proα1(I)+ Proα2 (I)	pPICZBproα1(I), pBLADEIXproα2(I), pBLARGIXα, pPIC9PDI	Zeo⁺, Ade⁺, Arg⁺, His⁺	proα1(I) and proα2(I) chains, 4-PH α subunit, PDI
Proα2(I)/Proα1 (I)	PAO815proα2(I)/proα1(I), pBLADEIXα, p BLARGSXPDI	His⁺, Ade⁺, Arg⁺	proα1(I) and proα2(I) chains, 4-PH α subunit, PDI
PCα1(I)	pPICZBpCα1(I), pARG815α, pPIC9PDI	Zeo⁺, Arg⁺, His⁺	pCα1(I) chain, 4-PH α subunit, PDI
PCα2(I)	pBLADEIXpCα2(I), pBLARGIXα, pPIC9PDI	Ade⁺, Arg⁺, His⁺	pCα2(I) chain, 4-PH α subunit, PDI
PCα1(I)+ PCα2 (I)	pPICZBpCα1(I), pBLADEIXpCα2(I), pBLARGIXα, PPIC9PDI	Zeo⁺, Ade⁺, Arg⁺, His⁺	pCα1(I) and pCα2(I) chains, 4-PH α subunit, PDI

based on the hydroxylation-coupled decarboxylation of 2-oxo [1-^{14}C]glutarate.

4. The collagen is precipitated by adding NaCl to a final concentration of 3.0 M and acetic acid to 0.5 M in the presence of 0.05 M PB buffer (pH 7.4).

5. Resulting sediment is collected by centrifugating at $10,000 \times g$ for 30 min. The purified collagen is redissolved in minimal amount of 0.1 M HCl, and dialyzed against 0.1 M acetic acid in the presence of 0.05 M PB buffer (pH 7.4), stored at 4 °C.

6. Chromatographed on a Sephacryl S-500HR column in the AKTA explorer system. The purest fractions identified by SDS-PAGE is centrifugate and concentrate using ultrafiltration tubes.

3.2 Assembly of Recombinant Collagen

1. The purified collagen is dialyzed in 5 mM acetic acid with three diffusate changes over 48 hr

2. The dialyzed collagen clarified by centrifugation (2×10^4 g, 1 h) is diluted with 5 mM acetic acid to a final concentration of 0.2 mg/mL, then stored at 4 °C.

3. Transfer 1 mL of collagen solution into a glass vial after filtered with membranes (pore size, 0.22μm) to remove oligomers .

4. The self-assembly of collagens with various concentrations is performed according to a modified method [14] employing 30 mM K_2HPO_4 and 135 mM NaCl, pH 7.4. Fibril formation is induced by an increased temperature to 34 °C.

3.3 Characterization

3.3.1 Electrophoretic Profiling of the Recombinant Collagen

The cells are harvested after 60 hr methanol induction at 30 °C, washed once, and suspended in cold (4 °C) 5% glycerol, 1 mM Pefabloc SC, and 50 mM sodium phosphate buffer, pH 7.4. Then the cells are harvested and broken, protein concentrations are determined using the Bio-Rad Protein Assay (Bio-Rad), and the soluble part of the cell extract is analyzed by SDS-PAGE and Western blotting. Western blotting is used to recognize the N propeptide of human proα1(I), the C propeptide of human proα2 (I) chains, the collagenous regions of a variety of collagen chains.

The aliquot of the soluble fraction is further analyzed by radioimmunoassay to detect the trimeric C propeptide (PICP-RIA) or N propeptide (PINP-RIA) of human type I procollagen. The expression level of at least 10 transformants of each strain is determined by radioimmunoassay. The other part is digested with pepsin at 22 °C for 2 h, then analyzed by SDS-PAGE, and followed by silver staining. The number of collagen chains is estimated by using a GS-710 calibrated imaging densitometer (BioRad) to measure the density of the silver stained band. Determination of 4-PH activity is based on 2-oxo[1-^{14}C]glutarate hydroxylation-coupled decarboxylation method.

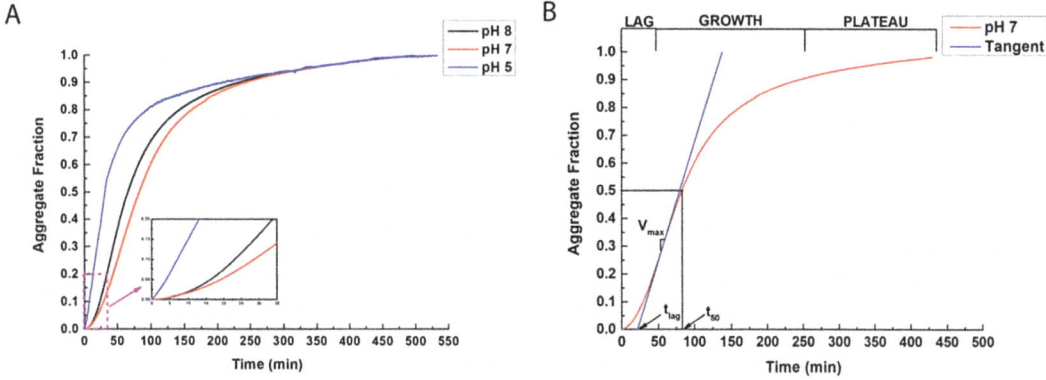

Fig. 1 Turbidity curves [26]. (Adapted by permission from Elsevier, Copyright 2019)

3.3.2 UV Measurements: Kinetics of Collagen Self-Assembly

Measurements

In the process of self-assembly, the collagen monomers progressively assembled into collagen fibers, which caused the turbidity and absorbance of the collagen solution to increase, thereby forming the gel. Consequently, it is good to record the kinetics of collagen self-assembly from observing the turbidity and the absorbance of recombinant collagen solution in the concentration of 0.1 mg mL^{-1} at 34 °C at 313 nm by dynamic tracking changes. The absorbance is recorded as a function of time at 310 nm. The measurement is carried out at 300 K, using the UV-visible spectrophotometer Jasco 7850, equipped with a thermostat, collecting data every 10 s.

Kinetics of Collagen Self-Assembly

Figure 1a shows the turbidity curve of the same recombinant protein in the measurement of the UV absorbance of the solution under changing pH. Figure 1b shows, under the condition of pH = 7, the UV absorbance with three distinct phases typical for filamentous protein aggregation. The curves are normalized to the maximum absorbance to show the relationship between aggregate fraction (AF) and time. The first step is the lag phase which the absorbance does not change: the fibers begin to nucleate. The second step is the growth phase which the absorbance increases: the cores of fibrils grow. The last step is the plateau in which the absorbance tends to be constant: the three-dimensional fiber network is formed at this time.

3.3.3 Circular Dichroism (CD) Spectroscopy

Measurements

CD spectroscopy can be used to analyze the triple helix structure of collagen. Here, rScl protein is used as an example to mainly describe the processing method of the sample and the key information of the analysis data. Protein samples are dialyzed with 1% phosphate buffered of pH 7.4. The path length of the cuvette where the sample is placed is 0.5 cm. The data is integrated for 1 s with a bandwidth of 1 nm at 0.2 nm intervals. The instrument for recording the far-ultraviolet spectrum is Jasco J720 spectropolarimeter, and it is thermostatically controlled.

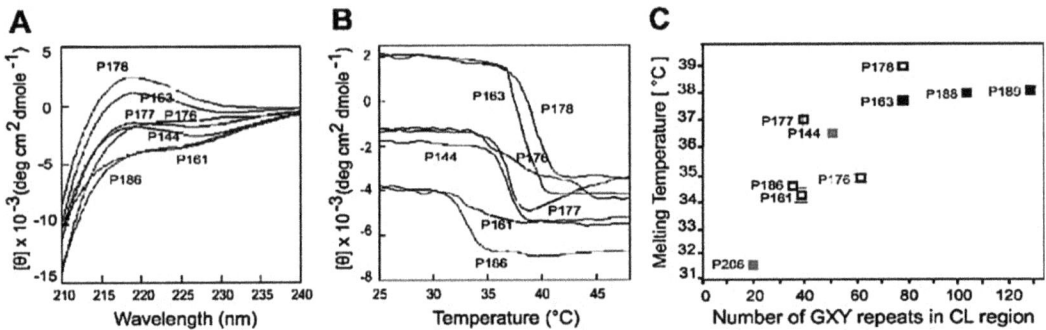

Fig. 2 Circular dichroism spectra and thermal stability [23]. (Adapted by permission from Springer Nature, Copyright 2006)

Each rScl protein underwent a wavelength scan at each endpoint temperature before unfolding (4 °C, 25 °C) and after unfolding (50 °C) or subsequent renaturation (25 °C). At a fixed wavelength of 220 nm, monitoring the change $[\theta]_{220}$ (the mean molar residue ellipticity), recorded thermal transition profiles, as a function of temperature. The sample is heated at a heating rate of 10 °C/h until 50 °C. The melting temperatures (T_m) are provided as the average value of T_m from several measurements.

Triple Helix Formation by rScl Proteins

Figure 2 shows the triple helix formation in each recombinant rScl construct studied by CD. The design of the primary sequence of the expressed recombinant protein and the length of the CL region (34-79 GXY repeats) are different. The formation of the triple helix structure of these proteins with different sequences can be evaluated by 220 nm, $[\theta]_{220}$. At 25 °C, the rScl protein exhibits a CD spectrum similar to the collagen triple helix, that is, the ellipticity increases at ~200 nm (Fig. 2a). At 50 °C, the CD spectrum does not increase, indicating that the triple helix structure disappears. When the sample recovered from 50 °C to 4 °C, the ellipticity gradually increases and the triple helix is restored.

Thermal Stability of rScl Proteins

The thermal stability of the triple helix of recombinant collagen is observed by detecting the change of $[\theta]_{220}$ when the temperature is raised (Fig. 2b). Most rScl variants exhibit a transition from a triple helix to a random coil within a limited temperature range, and this thermal unfolding is reversible, because the CD spectrum increases when the sample is cooled to 4 °C. Generally, rScl proteins with longer CL regions have higher T_m values, and even noncollagen domains seem to promote the stability of these proteins.

3.3.4 AFM

The morphology of nanofibrils is observed using the Dimension ICON AFM fast scanning system (Bruker, Germany) in the tapping mode. After diluting the resulting suspension to ~0.2‰ with ultrapure water, 10μL of the diluted suspension is instilled on the mica substrate and dry overnight (avoid dust deposition).

Jiang et al. [22] established the condition (pH and electrolyte) under which collagen molecules are adsorbed and arranged on mica under the action of externally applied hydrodynamic force. Collagen molecules aggregated into fibers to form a two-dimensional ultra-thin ($\approx$3 nm) matrix. Characteristic nanostructures of the matrix, such as the spacing and orientation of collagen fibers, can be produced with high precision and repeatability (>95%). In the presence of K^+ ions, the collagen fibers in the matrix only form the unique D-band periodicity ($\approx$67 nm).

When the pH is close to the pI value of collagen (9.3), the individual collagen fibers forming the collagen matrix can be clearly broken down. In order to study the self-assembly of individual collagen molecules into a fiber network, we conducted experiments in an aqueous solution with a pH of 9.2 and the presence of K^+ ions. Collagen molecules are injected into the buffer solution to start the growth of collagen fibers and monitored by AFM in time-lapse tap mode (TM) (Fig. 3). Collagen molecules at a concentration of $12\mu g$ mL^{-1} are dissolved in the experimental buffer (pH 9.2, 50 mM glycine, 200 mM KCl) [34].

Figure 3a–g shows the topographic map (arrows) of the horizontal and vertical growth of collagen fibers. The periodicity of the D-band of a single fiber is independent of the fiber width. The growing collagen fibers gather into a matrix covering almost the entire support surface. In Fig. 3h–l, two fibers disappear after colliding with each other [circle in (I)]. The final support surface is almost completely covered, and almost all defects are annealed or minimized.

3.3.5 Electron Microscopy

The transmission electron microscope used in the experiment is TEM200CX, JEOL, under the acceleration voltage of 80 kV, observing the collagen fibers with D periodicity. TEM sample preparation method: dropped the collagen suspension on a 200-mesh copper grid, and placed a piece of filter paper on the edge of the grid to remove excess water. Then, the fibrils are negatively dyed for 15 s with 1% phosphotungstic acid solution at a pH of 7.4. Milli-Q water should be used to rinse the stained grids, which is then air dried.

The scanning electron microscopy used in the experiment is SEM (JEOL, JSM 6400), at an accelerating voltage of 15 kV. The freeze-dried collagen sample is fixed on the stub and coated with carbon. For each sample, randomly collected at least 5 TEM images with a magnification of 20,000 times, and selected at least 30 different fibers from each image [16].

The morphology of collagen fibers varies greatly under different pH conditions (Figs. 4 and 5). Fibers with typical D-periodicity can be seen in the TEM image. Fibers of varying sizes are formed at low pHs. In the presence of large fibers, many small fibers without

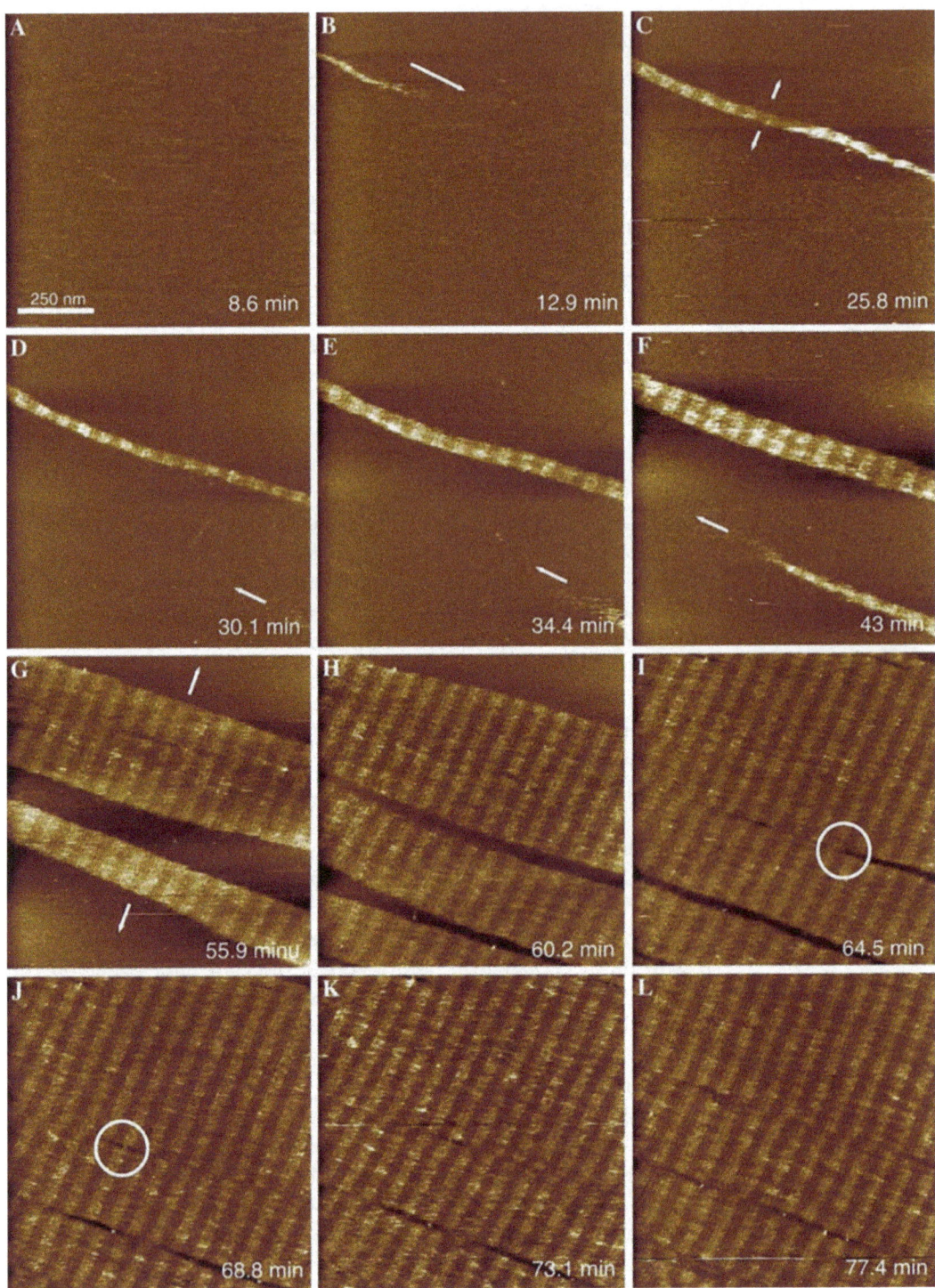

Fig. 3 Observing collagen self-assembly by time-lapse TM-AFM. Topographs were recorded in buffer solution (pH 9.2, 50 mM glycine, 200 mM KCl), at a collagen concentration of 12μg/mL at 27 °C and exhibit a full gray level corresponding to a vertical scale of 5 nm [34]. (Adapted by permission from Elsevier, Copyright 2006)

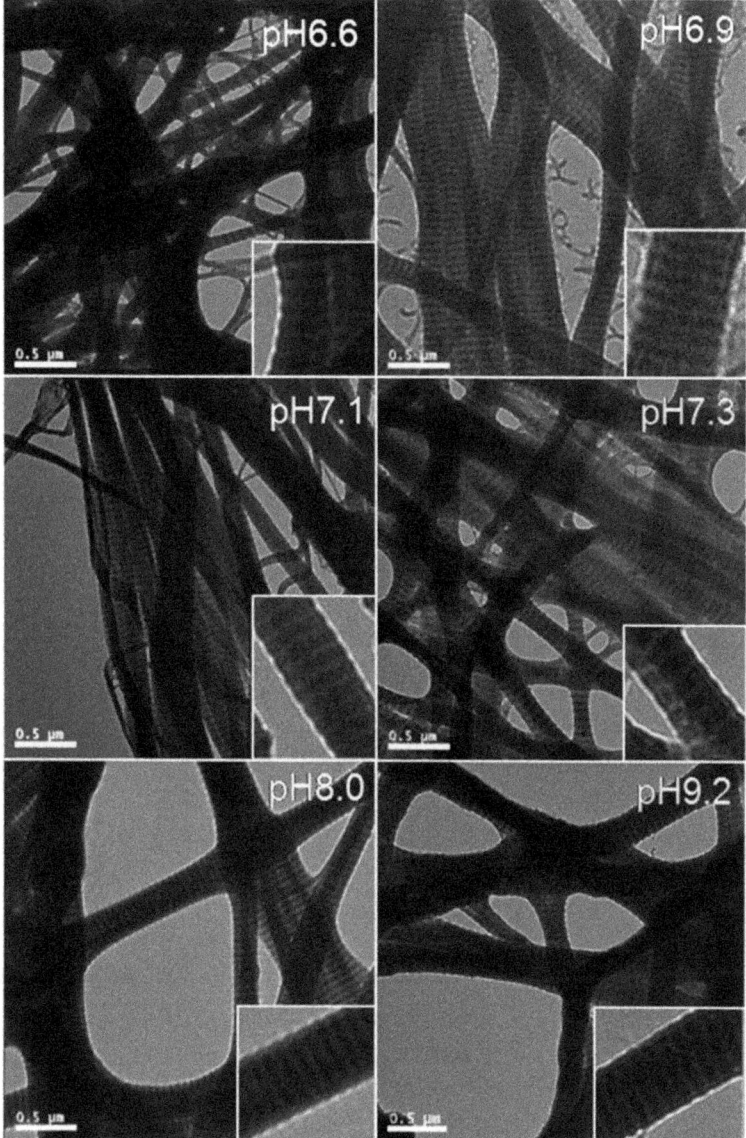

Fig. 4 TEM images of self-assembled collagen fibrils after 3 days of fibrillogenesis [16]. (Adapted by permission from Elsevier, Copyright 2009)

D-periodicity are detected. As the pH value increases, the fibers becomes more interwoven and uniform into a network. When fiber formation is highly inhibited, in the range of pH 6.0 to pH 6.9, the gel is composed of more nonfibrous collagen. The higher the degree of inhibition of fibrillogenesis, the greater the amount of nonfibrous collagen. In the range of pH 7.1 to pH 10.0, there is no significant difference in the SEM results.

Under the condition of pH 6.6, after 3 days of fibrillogenesis, small fibers with diameters of about 85 nm are obtained. Between

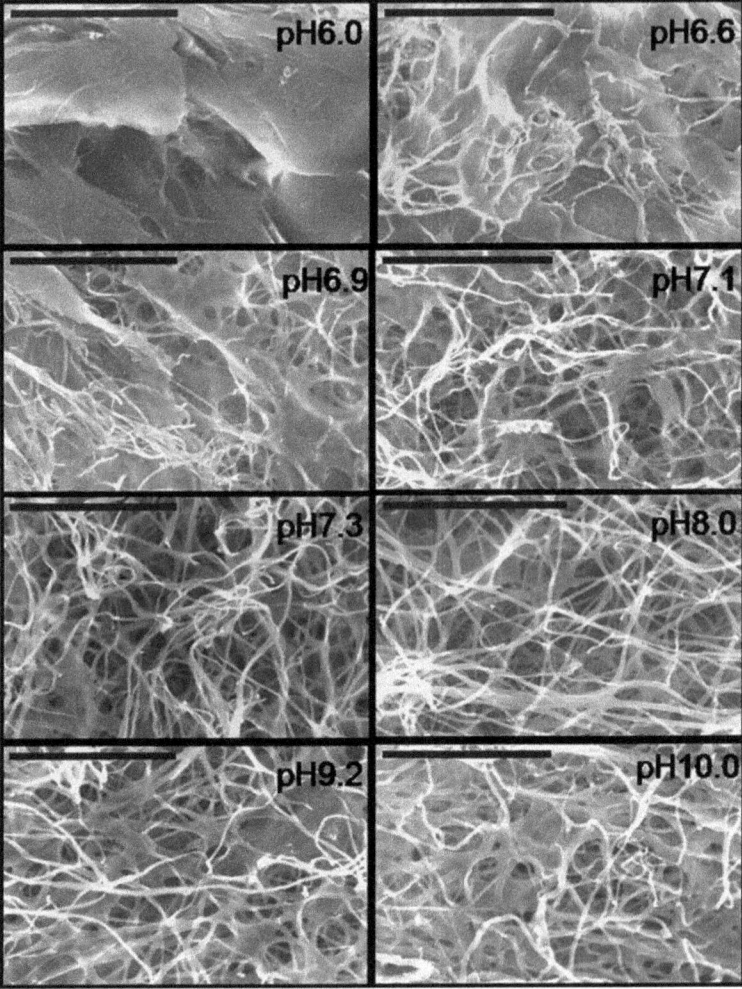

Fig. 5 SEM images of collagen fibrils obtained in different pH conditions. Scale bar is 10μm for all the images [16]. (Adapted by permission from Elsevier, Copyright 2009)

pH 6.9 and pH 8.0, the final fiber diameter is about 200 nm, although there is a significant difference in fiber incidence. Between pH 6.9 and pH 7.3, the fiber diameter increases with time, while under the condition of pH 8.0, the size of the fiber did not change significantly after 1 day is completed.

4   Notes

1. Some of the reagents in the electrophoresis process are volatile and toxic, use proper protective equipment and a fume hood while handling it [24].

2. Dialysate should be changed frequently during dialysis so that dialysis can be faster [24].

3. During the process of centrifugation, the liquid in the centrifuge tube should not be too much, preferably not more than two-thirds [24].

4. In the process of recombinant collagen assembly, the temperature is supposed to be strictly controlled and stable, because the self-assembly process is affected by the temperature [34].

Acknowledgments

This work was supported by the National Natural Science Foundation of China [grant numbers. 51973116, U1832109, 21935002], the Users with Excellence Program of Hefei Science Center CAS [grant number 2019HSC-UE003], the starting grant of ShanghaiTech University, and State Key Laboratory for Modification of Chemical Fibers and Polymer Materials.

References

1. Di Lullo GA, Sweeney SM, Korkko J, Ala-kokko L, San Antonio JD (2002) Mapping the ligand-binding sites and disease-associated mutations on the most abundant protein in the human, type I collagen. J Biol Chem 277 (6):4223–4231

2. Sionkowska A, Skrzynski S, Smiechowski K, Kolodziejczak A (2017) The review of versatile application of collagen. Polym Adv Technol 28 (1):4–9

3. Cen L, Liu W, Cui L, Zhang W, Cao Y (2008) Collagen tissue engineering: development of novel biomaterials and applications. Pediatr Res 63:492–496

4. Song WK, Liu D, Sun L, Li B, Hou H (2019) Physicochemical and biocompatibility properties of type I collagen from the skin of Nile tilapia (*Oreochromis niloticus*) for biomedical applications. Mar Drugs 17(3):137–151

5. Ottani V, Martini D, Franchi M, Ruggeri A, Raspanti M (2002) Hierarchical structures in fibrillar collagens. Micron 33(7-8):587–596

6. Liu Y, Kim Y, Dai L, Li N, Khan SO, Pashley DH, Tay FR (2011) Hierarchical and non-hierarchical mineralisation of collagen. Biomaterials 32(5):1291–1300

7. O'Leary LE, Fallas JA, Bakota EL, Kang MK, Hartgerink JD (2011) Multi-hierarchical self-assembly of a collagen mimetic peptide from triple helix to nanofibre and hydrogel. Nat Chem 3(8):821–828

8. Fratzl P (2008) Collagen: structure and mechanics. Springer US, New York

9. Parenteau-Bareil R, Gauvin R, Berthod F (2010) Collagen-based biomaterials for tissue engineering applications. Materials 3 (3):1863–1887

10. Hafner AE, Gyori NG, Bench CA, Davis L, Saric A (2020) Modelling fibrillogenesis of collagen-mimetic molecules. 119(9):1791-1799

11. Abou Neel EA, Bozec L, Knowles JC, Syed O, Mudera V, Day R, Hyun JK (2013) Collagen--emerging collagen based therapies hit the patient. Adv Drug Deliv Rev 65(4):429–456

12. Gomes S, Leonor IB, Mano JF, Reis RL, Kaplan DL (2012) Natural and genetically engineered proteins for tissue engineering. Prog Polym Sci 37(1):1–17

13. Hulmes DJ, Jesior JC, Miller A, Berthet-Colominas C, Wolff C (1981) Electron microscopy shows periodic structure in collagen fibril cross sections. Proc Natl Acad Sci U S A 78 (6):3567–3571

14. Williams BR, Gelman RA, Poppke DC, Piez KA (1978) Collagen fibril formation. Optimal in vitro conditions and preliminary kinetic results. J Biol Chem 253(18):6578–6585

15. Ling S, Chen W, Fan Y, Zheng K, Jin K, Yu H, Buehler MJ, Kaplan DL (2018) Biopolymer nanofibrils: structure, modeling, preparation, and applications. Prog Polym Sci 85:1–56

16. Li Y, Asadi A, Monroe MR, Douglas E (2009) pH effects on collagen fibrillogenesis in vitro: electrostatic interactions and phosphate binding. Mater Sci Eng C 29(5):1643–1649

17. Myllyharju J, Nokelainen M, Vuorela A, Kivirikko KI (2000) Expression of recombinant human type i-iii collagens in the yeast *Pichia pastoris*. Biochem Soc Trans 28(3):353–357

18. Parkinson J, Kadler KE, Brass A (1995) Simple physical model of collagen fibrillogenesis based on diffusion limited aggregation. J Mol Biol 247(4):823–831

19. Foo CWP, Kaplan DL (2002) Genetic engineering of fibrous proteins: spider dragline silk and collagen. Adv Drug Deliv Rev 54 (8):1131–1143

20. Liu W, Merrett K, Griffith M, Fagerholm P, Dravida S, Heyne B, Scaiano JC, Watsky MA, Shinozaki N, Lagali N, Munger R, Li F (2008) Recombinant human collagen for tissue engineered corneal substitutes. Biomaterials 29 (9):1147–1158

21. Leikin S, Rau DC, Parsegian VA (1994) Direct measurement of forces between self-assembled proteins: temperature-dependent exponential forces between collagen triple helices. Proc Natl Acad Sci U S A 91(1):276–280

22. Jiang F, Horber H, Howard J, Muller DJ (2004) Assembly of collagen into microribbons: effects of pH and electrolytes. J Struct Biol 148(3):268–278

23. Han R, Zwiefka A, Caswell CC, Xu Y, Keene DR, Lukomska E, Zhao Z, Hook M, Lukomski S (2006) Assessment of prokaryotic collagen-like sequences derived from streptococcal Scl1 and Scl2 proteins as a source of recombinant GXY polymers. Appl Microbiol Biotechnol 72 (1):109–115

24. Nokelainen M, Tu H, Vuorela A, Notbohm H, Kivirikko KI, Myllyharju J (2001) High-level production of human type I collagen in the yeast Pichia pastoris. Yeast 18(9):797–806

25. Fratzl P, Weinkamer R (2007) Nature's hierarchical materials. Prog Mater Sci 52 (8):1263–1334

26. Leo L, Bridelli MG, Polverini E (2019) Insight on collagen self-assembly mechanisms by coupling molecular dynamics and UV spectroscopy techniques. Biophys Chem 253:106224

27. Liu X, Zheng C, Luo X, Wang X, Jiang H (2019) Recent advances of collagen-based biomaterials: multi-hierarchical structure, modification and biomedical applications. Mater Sci Eng C 99:1509–1522

28. Barthelat F, Yin Z, Buehler MJ (2016) Structure and mechanics of interfaces in biological materials. Nature Rev Mater 1:16007

29. Abas M, Masry ME, Elgharably H (2020) Collagen in diabetic wound healing. In: Wound healing, tissue repair, and regeneration in diabetes. Elsevier, Amsterdam, pp 393–401

30. Gu L, Shan T, Ma YX, Tay FR, Niu L (2019) Novel biomedical applications of crosslinked collagen. Trends Biotechnol 37(5):464–491

31. Carvalho AM, Marques AP, Silva TH, Reis RL (2018) Evaluation of the potential of collagen from codfish skin as a biomaterial for biomedical applications. Mar Drugs 16(12):495

32. Lee BA, Hudson AR, Shiwarski DJ, Tashman JW, Hinton TJ, Yerneni S, Bliley JM, Campbell PG, Feinberg AW (2019) 3D bioprinting of collagen to rebuild components of the human heart. Science 365:482–487

33. Zhu S, Yuan Q, Yin T, You J, Gu Z, Xiong S, Hu Y (2018) Self-assembly of collagen-based biomaterials: preparation, characterizations and biomedical applications. J Mater Chem B 6(18):2650–2676

34. Cisneros DA, Hung C, Franz CM, Muller DJ (2006) Observing growth steps of collagen self-assembly by time-lapse high-resolution atomic force microscopy. J Struct Biol 154 (3):232–245

35. Yang W, Meyers MA, Ritchie RO (2019) Structural architectures with toughening mechanisms in nature: a review of the materials science of type-I collagenous materials. Prog Mater Sci 103(6):425–483

36. Vuorela A, Myllyharju J, Nissi R, Pihlajaniemi T, Kivirikko KI (1997) Assembly of human prolyl 4-hydroxylase and type III collagen in the yeast pichia pastoris: formation of a stable enzyme tetramer requires coexpression with collagen and assembly of a stable collagen requires coexpression with prolyl 4-hydroxylase. EMBO J 16(22):6702–6712

37. Vuorela A, Myllyharju J, Pihlajaniemi T, Kivirikko KI (1999) Coexpression with collagen markedly increases the half-life of the recombinant human prolyl 4-hydroxylase tetramer in the yeast Pichia pastoris. Matrix Biol 18 (5):519–522

38. Myllyharju J, Lamberg A, Notbohm H, Fietzek PP, Pihlajaniemi T, Kivirikko KI (1997) Expression of wild-type and modified proα chains of human type I procollagen in insect cells leads to the formation of stable $[\alpha 1(I)]$ $2\alpha_2(I)$ collagen heterotrimers and $[\alpha 1(I)]_3$ homotrimers but not $[\alpha 2(I)]_3$ homotrimers. J Biol Chem 272(35):21824–21830

Chapter 9

Recombination and Purification of Elastin-Like Polypeptides

Chenxi Zhao, Yuelong Xiao, Shengjie Ling, Ying Pei, and Jing Ren

Abstract

Elastin, as an extracellular matrix protein, has inherent advantages for biomedical applications. For example, it is highly extensible and biocompatible, biodegradable, and has no immunogenicity. However, directly extracting elastin from biological tissues remains challenging because they usually coexist with other proteins such as collagen. Therefore, an effective strategy to produce elastin is to transfer the elastin's target gene into other expression hosts and synthesize the resultant polypeptides using chemical biology methods. The polypeptides and proteins produced using these methods are usually referred to as elastin-like peptides (ELPs), which have received intensive interests in drug delivery and release, tissue engineering, implanted devices, and so on. Therefore, this chapter introduces the detailed protocol for the preparation of ELPs using genetic recombination, including DNA recombination, expression, and purification. The methods presented here are expected to provide methodological guidance for preparation and application of ELP materials.

Key words Elastin, Elastin-like polypeptide, Recombination

1 Introduction

Elastin is an essential component in the resilient extracellular matrix (ECM) and is abundant in a range of extensible tissues, such as arteries, skin, lungs, and cartilage [1–5]. It accounts for about 28–32% in blood vessels, 50% in elastic ligaments, 30–57% in the aorta, 2–3% in the dry weight of skin, and 3–7% in the lung [6–8]. Biologically, elastin can provide structural integrity and elastic recoil of the biological tissues [9–11]. Chemically, elastin is rich in glycine, proline, alanine, leucine, and valine residues. These residues are assembled into short repeated sequences of three to nine amino acids, Val-Pro-Gly-Xaa-Gly pentapeptides for example, and are further connected into a long molecular chain with a molecular weight of approximately 60–70 kDa, known as tropoelastin [7, 9, 12–14]. The tropoelastin will further assemble into elastin through lysine-mediated cross-linking. Prior to coacervation, polypeptides

Shengjie Ling (ed.), *Fibrous Proteins: Design, Synthesis, and Assembly*, Methods in Molecular Biology, vol. 2347,
https://doi.org/10.1007/978-1-0716-1574-4_9, © Springer Science+Business Media, LLC, part of Springer Nature 2021

appeared as clusters of globular structures with diameters ranging from 25 to 50 nm [15]. From NMR experiments on these tropoelastin fragments, it was proposed that hydrophobic and cross-linking domains of elastin cooperate in protein-folding [16].

Tropoelastin can be isolated and purified from animal tissue, such as fetal and neonatal tissues. However, the purification process is complex, and the amount that can be isolated is quite limited [2, 5, 17]. To this end, recombinant techniques are recommended, which also avoid ethical issues and have better availability. In fact, elastin or elastin-like polypeptides (ELPs) with longer sequences and larger quantities can be produced by gene recombination [3, 18, 19]. Another benefit of using genetic recombination technology to synthesize ELPs is the precise control of the polypeptide sequence and molecular weight can be achieved [11, 12, 20]. For example, chimeric fusion proteins comprising of biological active motifs and ELPs can be synthesized easily through DNA recombination [12, 15, 20, 21]. In addition, ELPs are insoluble when the solution temperature is higher than their phase transition temperature, which enables them to be purified effectively through inverse temperature cycling.

Therefore, this chapter is expected to provide detailed experimental protocols for the synthesis of ELPs through gene recombination techniques, including how to construct plasmids with fused ELP genes, and how to transfer these plasmids to the host for further expression. At last, we will introduce the isolation and purification technologies for extraction target ELPs from the host system.

2   Materials

1. Ultrapure water produced by a Milli-Q ultrapure water purification system.

2. 0.5 mM isopropyl β-D-thiogalactopyranoside (IPTG).

3. Rich LB medium (1% Bacto tryptone, 0.5% NaCl, 1% yeast extract) containing 100μg/mL ampicillin.

4. LB-rich medium supplemented with glucose (1 g/L) and ampicillin (100μg/mL).

5. 1 M phosphoric acid.

6. 1 M sodium hydroxide.

7. PBS buffer (NaCl 137 mM, KCl 2.7 mM, Na_2HPO_4 8 mM, KH_2PO_4 2 mM, pH 7.4).

8. Complete Mini EDTA-free protease inhibitors (Roche).

9. Polyethyleneimine (PEI).

10. Coomassie brilliant blue.

11. Protein standard solution sample buffer solution (containing bromophenol blue indicator).

3 Methods

3.1 Construction of Expression Vector that Fused with ELP Genes

1. The pCR2.1 plasmid provides the ELP gene. The MW $(VPGIG)_{20}C$ sequence (sequence A) is extracted by the EcoRI and HindIII, a process called double digestion.

2. The fragment is ligated into pUC19 plasmid. After recombinant plasmid is transformed into NEB. coli competent cells, pUC19-A is selected by colony PCR and verified by DNA sequencing.

3. Delete the sequence GTTCGCA from pUC19-A by PCR to obtain a reading frame that can synthesize ELP20 correctly and allow for subsequent recursive directed ligation.

4. The PCR product is digested with the DpnI enzyme to eliminate the methylated parental plasmid. The amplified fragment is circularized and used to transform NEB 5a-F′ Iq *E. coli* competent cells.

5. Positive colonies are identified by PCR. Then use DNA sequencing to confirm the sequence (sequence B) of the resultant pUC19-ELP20 plasmid.

6. Sequence A cloned in pUC19-A is double-digested with the enzymes BtgZ1 and BsmF1 to generate sequence C (monomer insert), which corresponds to the DNA sequence of MW $(VPGIG)_{20}C$ minus the MeteTrp N-terminal codons and the Cys C-terminal codon.

7. The clone in the expression vector is as follows: ELP sequences are extracted from pUC19-ELP20 by a double digestion NdeI and BamHI, and then ligated into pET-44a (+) plasmid independently. The new pET-44a (+) plasmid is used to transform NEB5a-F′ Iq (ELP20 containing plasmids) *E. coli* competent cells.

8. Positive clones are screened by PCR. The restriction map confirmed the synthesized plasmid pET-ELP20, and the DNA sequence of the ELP gene is detected by DNA sequencing.

9. Then the pET-ELP20 is transformed into the *E. coli* BLR (DE3) competence cell. The transformed *E. coli* bacteria can be passed down in a petri dish of antibiotics, forming colonies.

   Herein, ELP20 is used as a typical example to briefly explain the general process of recombinant protein preparation method. For ELP40 and ELP60, their preparation processes and primer sequences, PCR programs, and the kit used in experiments, please refer to [20].

3.2 Expression of Target ELPs

1. The ELP gene expression vector is constructed, and the purified plasmid is used for the transformation of BLR (DE3) active cells.

2. With 0.5 mm isopropyl β-D-thiogalactopyranoside (IPTG), the optical density of the cells at 600 nm is induced to reach about 1.2.

3. Select a bacterial colony, in a rotating shaker with a speed of 200 rpm, at 37 °C, in 50 mL rich LB medium containing 100μg/mL ampicillin overnight.

4. The seed culture is inoculated into 0.95 L of LB-rich medium, and glucose (1 g/L) and ampicillin (100μg/mL) are added.

3.3 Purification of Recombinant ELPs

1. After IPTG induction for 5 h, the culture is harvested by centrifugation at 7.5×10^3 g and 4 °C for 15 min, then the supernatant is discarded. The cell pellet is then resuspended in PBS buffer at a wet weight of 10 mL/g supplemented with 1 tablet/10 mL of complete Mini EDTA-free protease inhibitors.

2. The samples are incubated at −80 °C overnight and defrosted slowly by incubation at 4 °C.

3. Total cell lysate is sonicated (125 W, 15 °C, 9 s with 4 s interval) for 15 min.

4. After adding polyethyleneimine (PEI) to the solution at a final concentration of 0.44% (v/v), bacterial DNA will precipitate.

5. Insoluble waste of cells is removed after centrifugation at 1.6×10^3 g under 4 °C for 30 min. The cleared lysate is subjected to three continuous cycles of Inverse Transition Cycling (ITC), containing 'warm spin' and 'cold spin'.

6. ELP is precipitated at 25 °C in the presence of salt (0.5–1.5 mM NaCl) and centrifuged at 1.6×10^3 g and 25 °C for 30 min to precipitation ('warm spin').

7. After removing the supernatant containing the soluble protein and resuspending the pellet in cold PBS buffer, the ELP is redissolved, while the insoluble protein from *E. coli* is removed by centrifugation for 15 min at 1.6×10^3 g and 4 °C ('cold spin').

8. The supernatant containing the ELP is processed through the second and third ITC.

9. It is important to add NaCl to the sample at a final concentration of 0.5 M, 1 M and 1.5 M before each 'warm spin'.

10. The purified ELP is dialyzed with ultrapure water at 4 °C at least 12 h in a dialysis bag with a 1 kDa cutoff.

11. Finally, the samples are lyophilized.

| 3.4 **Characterization** | MALDI-MS (matrix-assisted laser desorption/ionization mass spectrometry) and ESI-MS (electrospray ionization mass spectrometry) are used to determine the molecular weight of purified ELP. |

3.4.1 Mass spectrometry Analyses

MALDI-MS (matrix-assisted laser desorption/ionization mass spectrometry) and ESI-MS (electrospray ionization mass spectrometry) are used to determine the molecular weight of purified ELP.

The ELP samples for MALDI-MS measurement are prepared as follows: lyophilized ELPs obtained after final dialysis in 10 mM ammonium acetate are resuspended in dimethyl sulfoxide and then diluted in water–acetonitrile (1:1, v/v). The standard protein mixture is used for external calibration and all MALDI-MS measurements are acquired in the linear positive mode. The masses measured for ELP20, ELP40, and ELP60 by MALDI-MS are 8908, 17,396, and 25,888 Da, that is, close to the theoretical values [20].

3.4.2 Thermal Turbidimetry

Cary 100 Bio UV-visible spectrophotometer, equipped with a multicell thermoelectric temperature controller, is used to determine T_t (transition temperature) of ELPs (10 and 50µM) in PBS solution by measuring the turbidity at 350 nm from 5 °C to 40 °C at 1 °C/min scan rate.

The thermal behavior of ELPs is studied by turbidimetric method. ELPs showed a lower inverse temperature transition, and the T_t of the shorter ELP20 showed more concentration dependence than ELP40 and ELP60. ELP20 in PBS had a T_t of 26.2 °C at 10 mM concentration and a T_t of 18.6 °C at 50 mM.

3.4.3 Circular Dichroism (CD)

The resultant lyophilized ELPs are resuspended in dimethyl sulfoxide and then diluted in water/acetonitrile (1/1, v/v) and then transferred into a quartz cell with a path length of 10 mm. The spectrum data is recorded from the range of 190 to 250 nm with an interval of 0.5 nm and a scan speed of 50 nm/min under the nitrogen atmosphere at 20 °C.

4 Notes

1. Before starting the formal experiment, preliminary experiments are performed in 250 mL shake flasks with ELP20-producing clones and carbon substrate [20].

2. Samples are collected every hour for measurement of OD_{600nm}. The dry biomass is measured by filtration on cellulose nitrate filter (pore size of 0.45µm) and is dried at 80 °C for 24 h [20].

3. It is important that before each "warm spin", NaCl is added to the sample at a final concentration of 1.5 M, 1 M and 0.5 M for the first, second, and third cycles, respectively [20].

4. For mass spectrometry analyses, the final dialysis step is performed in ammonium acetate (10 mM) [20].

5. During the process of preparing ELP samples for mass spectrometry, in order to avoid aggregation, lyophilized ELPs are first dissolved in dimethyl sulfoxide and then diluted in water–acetonitrile (1:1, v/v) just before mixing with the MALDI matrix or infusion in the ESI source [20].

6. It is really a challenge to prepare ELP samples for mass spectrometry. Their low T_t and rapid aggregation make it difficult to obtain a uniform layer of fine crystals because of the tendency of ELPs to form insoluble aggregates. This heterogeneity leads to low-resolution and low-quality MS signal, especially for ELP40 and ELP60. The difference in mass between experimental and theoretical values of ELP40 and ELP60 may be caused by the adduct of sodium and ammonium cations. ESI-mass spectrometry analyses after deconvolution of ESI spectra raw data using a maximum entropy-based approach (MaxEnt Software, Waters) at ambient temperature confirmed this result [20].

Acknowledgments

This work was supported by the National Natural Science Foundation of China [grant numbers. 51973116, U1832109, 21935002], the Users with Excellence Program of Hefei Science Center CAS [grant number 2019HSC-UE003], the starting grant of ShanghaiTech University, and State Key Laboratory for Modification of Chemical Fibers and Polymer Materials.

References

1. Wen Q, Mithieux SM, Weiss AS (2020) Elastin biomaterials in dermal repair. Trends Biotechnol 38(3):280–291

2. Yeo GC, Mithieux SM, Weiss AS (2018) The elastin matrix in tissue engineering and regeneration. Curr Opin Biomed Eng 6:27–32

3. Mahesh B, Nanjundaswamy GS, Channe Gowda D, Siddaramaiah B (2017) Synthesis of elastin-based polymer and evaluation of its intermolecular interactions with hydroxypropyl methylcellulose. J Appl Polym Sci 134(36):45283

4. Mecham RP, Lange G, Madaras J, Starcher B (1981) Elastin synthesis by ligamentum Nuchae fibroblasts: effects of culture conditions and extracellular matrix on elastin production. J Cell Biol 90:332–337

5. Kozma B, Candiotti K, Poka R, Takacs P (2018) The effects of heat exposure on vaginal smooth muscle cells: elastin and collagen production. Gynecol Obstet Investig 83(3):247–251

6. Partridge SM (1963) Elastin. Adv Protein Chem 17:227–302

7. Li DY, Brooke B, Davis EC, Mecham RP, Sorensen LK, Boak BB, Eichwald E, Keating MT (1998) Elastin is an essential determinant of arterial morphogenesis. Nature 393:276–280

8. Duca L, Floquet N, Alix AJ, Haye B, Debelle L (2004) Elastin as a matrikine. Crit Rev Oncol Hematol 49(3):235–244

9. Mithieux SM, Weiss AS (2005) Elastin. Adv Protein Chem 70:437–461

10. Godwin A, Singh M, Lockhart-Cairns MP, Alanazi YF, Baldock C (2019) The role of fibrillin and microfibril binding proteins in elastin and elastic fibre assembly. Matrix Biol 84(6):17–30

11. Santos M, Serrano-Ducar S, Gonzalez-Valdivieso J, Vallejo R, Girotti A, Cuadrado P, Arias FJ (2018) Genetically engineered elastin-

based biomaterials for biomedical applications. Curr Med Chem 26(40):7117–7146

12. Lescan M, Perl RM, Golombek S, Pilz M, Hann L, Yasmin M, Behring A, Keller T, Nolte A, Gruhn F, Kochba E, Levin Y, Schlensak C, Wendel HP, Avci-Adali M (2018) De Novo synthesis of elastin by exogenous delivery of synthetic modified mRNA into Skin and elastin-deficient cells. Mol Ther Nucleic Acids 11:475–484

13. Cleary EG, Gibson MA (1983) Elastin-associated microfibrils and Microfibrillar proteins. Int Rev Connect Tissue Res 10:97–209

14. Urry DW, Hugel T, Seitz M, Gaub HE, Sheiba L, Dea J, Xu J, Parker T (2002) Elastin: a representative ideal protein elastomer. Philos Trans R Soc Lond Ser B Biol Sci 357 (1418):169–184

15. Bellingham CM, Lillie MA, Gosline JM, Wright GM, Starcher BC, Bailey AJ, Woodhouse KA, Keeley FW (2003) Recombinant human elastin polypeptides self-assemble into biomaterials with elastin-like properties. Biopolymers 70(4):445–455

16. Daamen WF, Veerkamp JH, Van Hest JCM, Van Kuppevelt TH (2007) Elastin as a biomaterial for tissue engineering. Biomaterials 28 (30):4378–4398

17. Bruno PD, Garbay B, Pasqua M, Chevron E, Chinoy ZS, Cullin C, Bathany K, Lecommandoux S, Amedee J, Oliveira H, Garanger E (2019) Production, purification and characterization of an elastin-like polypeptide containing the Ile-Lys-Val-ala-Val (IKVAV) peptide for tissue engineering applications. J Biotechnol 298:35–44

18. Gonzalez I, Alonso M, Rodriguez-Cabello JC (2020) Elastin-based materials: promising candidates for cardiac tissue regeneration. Front Bioeng Biotechnol 8:657

19. Weihermann AC, Lorencini M, Brohem CA, De Carvalho CM (2017) Elastin structure and its involvement in skin photoageing. Int J Cosmet Sci 39(3):241–247

20. Bataille L, Dieryck W, Hocquellet A, Cabanne C, Bathany K, Lecommandoux S, Garbay B, Garanger E (2016) Recombinant production and purification of short hydrophobic elastin-like polypeptides with low transition temperatures. Protein Expr Purif 121:81–87

21. Hernandez B, Crowet JM, Thiery J, Kruglik SG, Belloy N, Baud S, Dauchez M, Debelle L (2020) Structural analysis of Nonapeptides derived from elastin. Biophys J 118 (11):2755–2768

Chapter 10

Methods to Synthesize and Assemble Recombinant Keratins

Wenwen Zhang and Yimin Fan

Abstract

Keratin, as one of the most abundant and underexploited protein sources, is a ubiquitous biological material that commonly exists in epithelial cells. Due to the excellent biocompatibility and biodegradability, keratin is widely used in biomedical applications. Previously, these biomaterials were prepared by dissolving and extracting the keratinous materials. However, the keratins obtained by direct extraction is not pure and contain many by-products. Moreover, natural keratins suffer from limited sequence tenability. In comparison, the recombinant keratin proteins produced by recombinant technology can overcome these drawbacks while maintaining the desired chemical and physical characteristics of natural keratins. Accordingly, this chapter mainly introduces the experimental protocols of the recombination of keratin. As these recombinant keratins are often used for assembly of intermediate filaments (IFs) in vitro, assembly protocols are also introduced in this chapter.

Key words Keratin, Recombinant keratin protein, Self-assembly, Intermediate filament, Characterization

1 Introduction

Keratin is a ubiquitous fibrous protein that mainly exists in hairs, wool, feathers, hooves, claws, horns, and so on. Benefiting from the outstanding biocompatibility, biodegradability, as well as inherent ability to promote cell growth, proliferation, and differentiation, keratin has received intense interest in tissue engineering and drug delivery [1–4]. However, compared with other proteins, keratin has higher stability and lower solubility duo to its complex covalent (disulfide bonds) and noncovalent interactions (electrostatic forces, hydrogen bonds, hydrophobic forces) [5]. Especially, the disulfide bonds provide keratin high resistance to chemical or enzymatic reactions. According to their primary structure, also known as amino acid sequence, keratin can be divided into two main groups, that is, "soft" keratins in epithelial cells and "hard" keratins in epidermal appendages [6]. Compared with the soft keratins, hard keratins have a higher content of cysteine residues—a group that

Shengjie Ling (ed.), *Fibrous Proteins: Design, Synthesis, and Assembly*, Methods in Molecular Biology, vol. 2347,
https://doi.org/10.1007/978-1-0716-1574-4_10, © Springer Science+Business Media, LLC, part of Springer Nature 2021

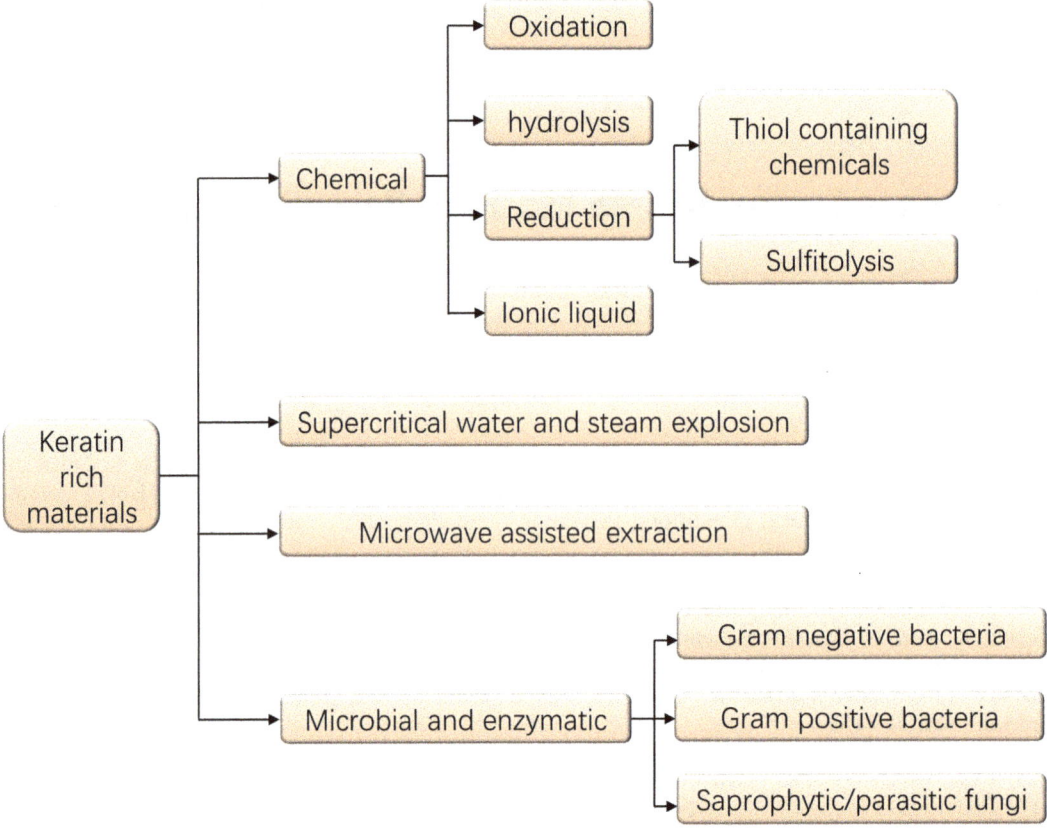

Fig. 1 Various methods to extract keratin from natural sources

can form the intermolecular disulfide bond. Therefore, hard keratins have a tighter packed structure than soft keratins [7].

The primary methods to produce keratin is direct extraction from natural sources [8, 9], such as human hair fibers, wool, feather, are chemical method [10–13], microwave irradiation [14–16], steam explosion [17–19], and microbial and enzymatic [20–22] (Fig. 1). However, several drawbacks have been found for extraction methods. For example, the purity of extracted keratin is low and usually contains many by-products, such as keratin-associated proteins, and IF proteins [23], due to the complex components of keratinous materials. The properties of extracted keratin are also highly dependent on raw material sources [24]. In addition, most extraction methods require the use of toxic chemical reagents and complex processing techniques. These operations significantly weaken the sustainability of keratin and also affect the quality of the extracted keratin. The molecular weight of extracted keratins is usually much lower than the native counterparts as their structure is severely damaged during the extraction process [25–27]. More importantly, the amino acid sequence of the keratin prepared by the extraction method cannot be changed, so its

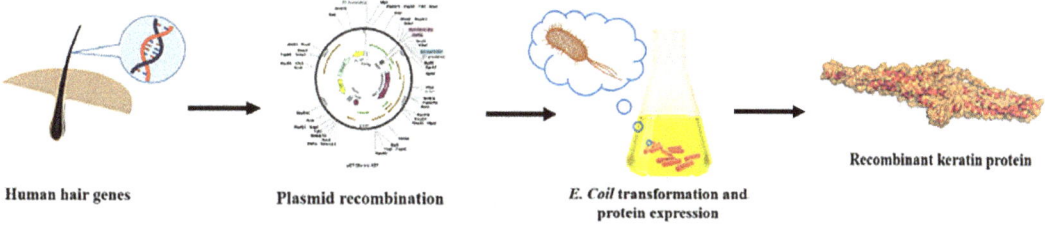

Human hair genes   Plasmid recombination   *E. Coil* transformation and
protein expression   Recombinant keratin protein

Fig. 2 Schematic of the preparation of recombinant keratin proteins. (Adapted by permission from American Chemical Society [28], Copyright 2019)

modification becomes complicated and can only rely on limited exogenous chemistry methods.

The recombinant technique is a rational approach to solve the issues as mentioned above. Indeed, it has become a meaningful way to produce keratins [28, 29]. This technique has been confirmed that can significantly improve the purity and quality of keratin proteins [30, 31]. Many recombinant keratin proteins even maintained their natural conformation and, therefore, can be assembled into IFs in vitro. A series of investigates have also proven that the recombinant method is a practical approach to develop keratin-based biomaterials since the chemical and mechanical properties of these materials can be tuned through sequence design [32, 33].

In this chapter, we first introduce the experimental protocols to recombinant human hair keratin K81 and K37 (Fig. 2), and then introduce the methods to assemble these recombinant proteins into IFs in detail. The methods described in this chapter may provide methodological references for the synthesis of keratin from a scarce source or for desired sequence design. The results of the corresponding characterization techniques are also shown in Figs. 3 and 4.

2   Materials

1. Human hair keratin gene (K37 or K81).

2. TOP10 Ca^{2+}-competent *E. coli.*

3. BL21 (DE3) *E. coli.*

4. Luria–Bertani (LB) medium.

5. Kanamycin.

6. Isopropyl-β-D-thiogalactoside (IPTG).

7. Buffer A (50 mM Tris–HCl, pH 8.0, 150 mM NaCl, 5 mM EDTA, and 20 mM β-mercaptoethanol).

8. Buffer B (50 mM Tris–HCl, pH 8.0, 150 mM NaCl, 5 mM EDTA, 20 mM β-mercaptoethanol, 1 M urea, and 0.5% Triton X-100).

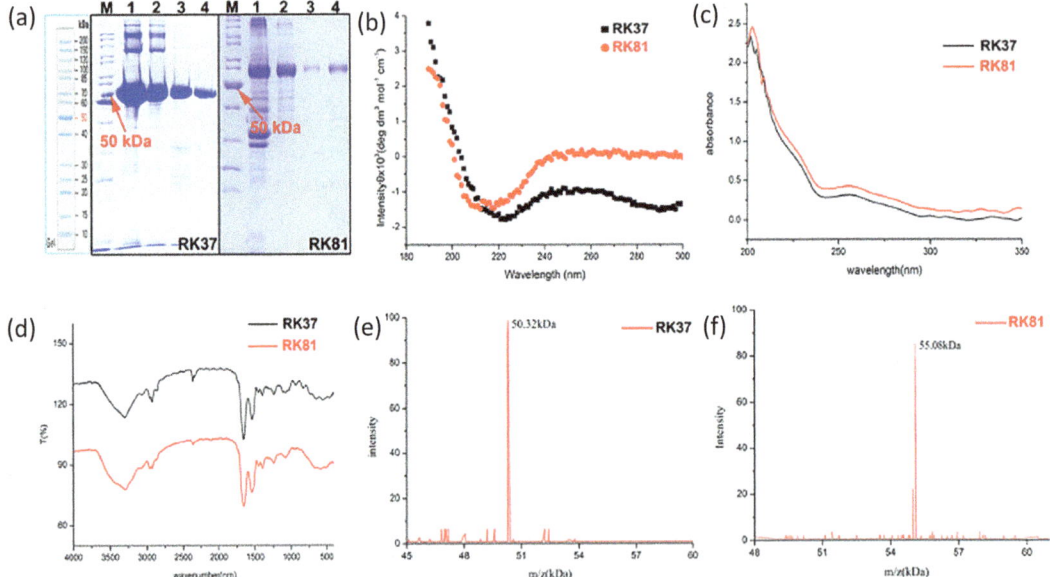

Fig. 3 Characterization of recombinant keratin proteins. (Adapted by permission from American Chemical Society [28], Copyright 2019). Typical Electrophoretic separation patterns (**a**), CD spectra (**b**), UV-vis spectra (**c**), FTIR spectra (**d**), and mass spectra (**e**, **f**) of RK37 and RK81

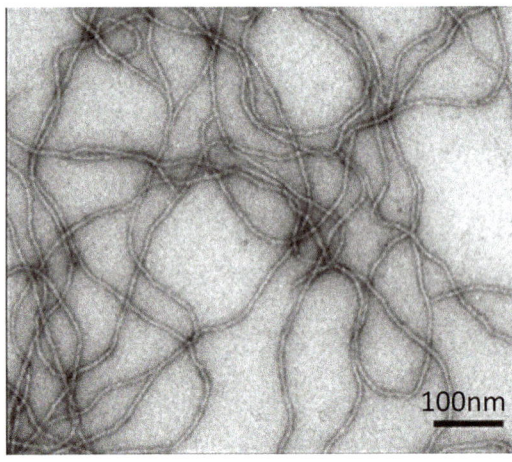

Fig. 4 Electron microscopic analysis of the assembly products of a mixture of K8 and K18. (Adapted by permission from Elsevier [34], Copyright 2012)

9. Buffer C (50 mM Tris–HCl, pH 8.0, 2.5 M NaCl, 8 M urea, 5 mM EDTA, and 20 mM β-mercaptoethanol).

10. DL-dithiothreitol (DTT).

11. Potassium phosphate (K₃PO₄).

12. Guanidine chloride.

13. 10 mM Tris–HCl buffer (pH = 8.5).

14. Sodium dodecyl sulfate(SDS).

15. Ethylenediaminetetraacetic acid (EDTA).

16. Ethylene glycol tetraacetic acid (EGTA).

3   Methods

3.1   Preparation of Recombinant Keratin Proteins

3.1.1   Cloning of Recombinant Keratin Proteins

1. Amplify the human hair keratin gene (K37 or K81) by PCR and clone into the NdeI and XhoI sites in pET-28a (+) by enzyme digestion and ligation (*see* **Note 1**).

2. Amplify the obtained in *Escherichia coli* TOP 10 cells and confirm by plasmid extraction and sequencing.

3.1.2   Expression and Purification of Recombinant Keratin Proteins

1. Transform the pET-28a (+)-K37 and pET-28a (+)-K81 into *E. coli* BL 21(DE3) for protein expression, respectively.

2. Inoculate the transformed cells in 10 mL LB medium at 37 °C with shaking at 250 rpm overnight with 50 μg/mL kanamycin.

3. Dilute the cells in LB medium with a ratio of 1:100 and grow until the OD_{600} reaches 0.6–0.8.

4. Conduct the protein expression with 1 mM IPTG at 37 °C for 4 h.

5. Harvest cells by centrifugation at 5000 × g for 10 min at 4 °C.

6. Resuspend the cell pellets in buffer A with 50 mM Tris–HCl, pH 8.0, 150 mM NaCl, 5 mM EDTA, and 20 mM β-mercaptoethanol.

7. Disrupt the cells by ultrasonication and then centrifuge.

8. Collect the precipitate and resuspend in buffer B containing 50 mM Tris (pH 8.0), 150 mM NaCl, 5 mM EDTA, 20 mM β-mercaptoethanol, 1 M urea, and 0.5% Triton X-100.

9. Centrifuge the sample at 100,000 × g for 1 h at 4 °C.

10. Collect the insoluble fraction.

11. Dissolved in buffer C with 50 mM Tris–HCl, pH 8.0, 2.5 M NaCl, 8 M urea, 5 mM EDTA, and 20 mM β-mercaptoethanol.

12. Centrifuge the sample at 100,000 × g for 1 h at 4 °C.

13. Collect the supernatant and dialyze against 8 M urea and 20 mM β-mercaptoethanol.

14. Collect the solution and store at 4 °C for further use.

3.1.3   Assembly of Recombinant Keratin Proteins

1. Mix the purified keratins with 8 M urea in a 1:1 ratio and dialyze extensively against 10 mM Tris–HCl, pH = 8.5 (~4 h), 2 mM DTT, 6 M urea (*see* **Note 2**).

2. Isolate the heterodimers by elution with a gradient of 0–0.2 M guanidinium chloride in 10 mM Tris–HCl, pH = 8.5 (~4 h), 2 mM DTT, 6 M urea.

3. The day before assembly, dialyze the protein against dialysis buffers containing 4 M, 2 M, 1 M, and 0 M urea (*see* **Note 3**).

4. Dialyze the heterodimers (0.2 mg/mL) against 5 mM Tris–HCl, pH = 8.4, 1 mM EDTA, 0.1 mM EGTA and 1 mM DTT and dilute to a concentration of 45μg/mL with this buffer.

5. Initiate the protein self-assembly by addition of 100 mM NaCl in 40 mM Tris–HCl buffer, pH = 7. (Hours and days are needed to complete the assembly, and size-controllable fibrils could be obtained through time control.)

3.2 Characterization

3.2.1 Gel Electrophoresis Analysis

Dissolving the recombinant keratin in buffer A and then mixed with 4μL of 5× SDS loading buffer and 0.6 M β-mercaptoethanol. Before conducting gel electrophoresis, 20μL of the samples are heated at 100 °C for 10 min in 10–12% gradient Tris–HCl gels at 120 V for 90 min. The gels are subsequently washed three times with deionized water and stained for 30 min with Coomassie Brilliant Blue G-250.

3.2.2 CD Spectroscopy

The CD data is collected with a 1 mm path-length quartz cuvette on a circular dichroism spectrophotometer at 25 °C. The spectra are scanned from 300 to 190 nm at a rate of 100 nm/min and wavelength step of 1 nm. All tests are carried out in triplicate.

3.2.3 UV-Vis Spectra Analysis

The UV-Vis spectra of recombinant keratin is acquired using a spectrophotometer. Before testing, the samples are dissolved in 8 M urea buffer, and the buffer is used as the blank. The spectral data is recorded over a wavelength range of 200–400 nm (*see* **Note 4**).

3.2.4 Fourier Transform Infrared Spectroscopy (FTIR)

The chemical structures of recombinant keratin are analyzed by an FTIR spectrometer. The samples for measurements are conducted by grinding the lyophilized samples with KBr at a ratio of 1:100 and pressing into a thin disk. The data is collected in absorption mode with a resolution of 4 cm^{-1} and a spectral range from 4000 to 400 cm^{-1}.

3.2.5 Mass Spectrometry Analysis

The molecular weights are confirmed by matrix-assisted laser desorption/ionization time-of-flight mass spectrometry (SHIMADZU Corporation, MALDI-7090, Japan). The data are recorded in a linear mode with acceleration voltage of 20 kV.

3.2.6 Electron Microscopy

The morphology of the nanofibrils was examined by a Zeiss 900 transmission electron microscope (Carl Zeiss, Oberkochen, Germany). 5μL samples are deposited on glow-discharged carbon-coated copper grids for 1 min and then washed two times to remove salt. The washed samples are stained with 2% uranyl acetate for 30 s and air-dried for 24 h before observation (*see* **Note 5**).

4   Notes

1. In order to avoid contaminating the sample, it is necessary to check whether the instrument used, including glass vial, cuvette, and beaker, is clean before the experiment.

2. When dialyzing the protein against buffer (10 mM Tris–HCl, pH = 8.5 (~4 h), 2 mM DTT, 6 M urea), the four additional dialysis steps are completed with decreasing amounts of urea equal to 4, 2, 1, and 0 M. Each of the steps is completed at 3 h intervals except for the last step, which is allowed to equilibrate overnight.

3. A freshly prepared solution and protein are preferred for the protocol. When preparing freeze-dried samples, safety precautions should be taken.

4. Since the sample used for UV-Vis analysis is dissolved in urea, the characteristic bands may show a blue shift.

5. For electron microscopy tests, it is essential to ensure that the substrate is clean to avoid contamination of samples during sample drying.

References

1. Rouse JG, Van Dyke ME (2010) A review of keratin-based biomaterials for biomedical applications. Materials 3:999–1014

2. Cheng Z, Chen X, Zhai D, Gao F, Wang B (2018) Development of keratin nanoparticles for controlled gastric mucoadhesion and drug release. J Nanobiotechnol 16:1–13

3. Sawada K, Fujisato T (2012) Keratin protein for biomaterial applications. Sen'i Gakkaishi 68:232–238

4. Wang RM, Li FY, Wang XJ (2010) The application of feather keratin and its derivatives in treatment of potato starch wastewater. J Funct Mater Lett 3:213–216

5. Wang YX, Cao XJ (2012) Extracting keratin from chicken feathers by using a hydrophobic ionic liquid. Process Biochem 47:896–899

6. Coulombe PA, Omary MB (2002) Hard' and 'soft' principles defining the structure, function and regulation of keratin intermediate filaments. Curr Opin Cell Biol 14:110–122

7. Yu J, Yu DW, Checkla DM, Freedberg IM, Bertolino AP (1993) Human hair keratins. J Invest Dermatol 101:56S–59S

8. Shavandi A, Silva TH, Bekhit AA, Bekhit AEA (2017) Keratin: dissolution, extraction and biomedical application. Biomater Sci 5:1699–1735

9. Korniłłowicz-Kowalska T, Bohacz J (2011) Biodegradation of keratin waste: theory and practical aspects. Waste Manag 31:1689–1701

10. Torimoto T, Tsuda T, Okazaki KI, Kuwabata S (2010) New Frontiers in materials science opened by ionic liquids. Adv Mater 22:1196–1221

11. Xu H, Ma Z, Yang Y (2014) Dissolution and regeneration of wool via controlled disintegration and disentanglement of highly crosslinked keratin. J Mater Sci 49:7513–7521

12. Simpson WS, Crawshaw G (2002) Wool: science and technology. Technische Universität Harburg, Hamburg

13. Tsuda Y, Nomura Y (2014) Properties of alkaline-hydrolyzed waterfowl feather keratin. Anim Sci J 85:180–185

14. Yin J, Rastogi S, Terry AE, Popescu C (2007) Self-organization of oligopeptides obtained on dissolution of feather keratins in superheated water. Biomacromolecules 8:800–806

15. Sturm GSJ, Verweij MD, Stankiewicz AI, Stefanidis GD (2014) Microwaves and microreactors: design challenges and remedies. Chem Eng J 243:147–158

16. Zoccola M, Aluigi A, Patrucco A, Vineis C, Forlini F, Locatelli P, Sacchi MC, Tonin C (2012) Microwave-assisted chemical-free hydrolysis of wool keratin. Text Res J 82:2006–2018

17. Zhang Y, Zhao W, Yang R (2015) Steam flash explosion assisted dissolution of keratin from feathers. ACS Sustain Chem Eng 3:2036–2042

18. Yu Z, Zhang B, Yu F, Xu G, Song A (2012) A real explosion: the requirement of steam explosion pretreatment. Bioresour Technol 121:335–341

19. Tonin C, Zoccola M, Aluigi A, Varesano A, Montarsolo A, Vineis C, Zimbardi F (2006) Study on the conversion of wool keratin by steam explosion. Biomacromolecules 7:3499–3504

20. Tork SE, Shahein YE, El-Hakim AE, Abdel-Aty AM, Aly MM (2013) Production and characterization of thermostable metallo-keratinase from newly isolated Bacillus subtilis NRC 3. Int J Biol Macromol 55:169–175

21. Onifade AA, Al-Sane NA, Al-Musallam AA, Al-Zarban S (1998) A review: potentials for biotechnological applications of keratin-degrading microorganisms and their enzymes for nutritional improvement of feathers and other keratins as livestock feed resources. Bioresour Technol 66:1–11

22. Sanchez S, Demain AL (2011) Enzymes and bioconversions of industrial, pharmaceutical, and biotechnological significance. Org Process Res Dev 15:224–230

23. Adav SS, Subbaiaih RS, Kerk SK, Lee AY, Lai HY, Ng KW, Sze SK, Schmidtchen A (2018) Studies on the proteome of human hair - identification of histones and deamidated keratins. Sci Rep 8:1–11

24. Parker RN, Roth KL, Kim C, Mccord JP, Van Dyke ME, Grove TZ (2017) Homo- and heteropolymer self-assembly of recombinant trichocytic keratins. Biopolymers 107:e23037

25. Lee H, Noh K, Lee SC, Kwon I-K, Han D-W, Lee I-S, Hwang Y-S (2014) Human hair keratin and its-based biomaterials for biomedical applications. Tissue Eng Regener Med 11:255–265

26. Nakamura A, Arimoto M, Takeuchi K, Fujii T (2002) A rapid extraction procedure of human hair proteins and identification of phosphorylated species. Biol Pharm Bull 25:569–572

27. Buchanan JH (1977) A cystine-rich protein fraction from oxidized alpha-keratin. Biochem J 167:489–491

28. Gao F, Li W, Deng J, Kan J, Guo T, Wang B, Hao S (2019) Recombinant human hair keratin nanoparticles accelerate dermal wound healing. ACS Appl Mater Interfaces 11:18681–18690

29. Guo T, Li W, Wang J, Luo T, Lou D, Wang B, Hao S (2018) Recombinant human hair keratin proteins for halting bleeding. Artif Cells Nanomed Biotechnol 46:456–461

30. Steinert PM, Steven AC, Roop DR (1983) Structural features of epidermal keratin filaments reassembled in vitro. J Invest Dermatol 81:86s–90s

31. Herrmann H, Wedig T, Porter RM, Lane EB, Aebi U (2002) Characterization of early assembly intermediates of recombinant human keratins. J Struct Biol 137:82–96

32. Dinjaski N, Kaplan DL (2016) Recombinant protein blends: silk beyond natural design. Curr Opin Biotechnol 39:1–7

33. Lin C-Y, Liu JC (2016) Modular protein domains: an engineering approach toward functional biomaterials. Curr Opin Biotechnol 40:56–63

34. Lichtenstern T, Mücke N, Aebi U, Mauermann M, Herrmann H (2012) Complex formation and kinetics of filament assembly exhibited by the simple epithelial keratins K8 and K18. J Struct Biol 177:54–62

Chapter 11

Preparation of Amyloid Fibrils Using Recombinant Technology

Wenwen Zhang and Yimin Fan

Abstract

Amyloid fibrils are widely investigated as they are directly associated with various neurodegenerative diseases. For example, a vast of experimental results have shown that the oligomeric and fibrillar aggregates of the amyloid β-peptide (Aβ) play a critical role in the pathogenesis of Alzheimer's disease (AD). Therefore, the accessibility of certain amounts of pure Aβ peptide is necessary for the studies of the mechanism of neurotoxicity. In this regard, recombinant methods provide the possibility to synthesize the Aβ peptide in vitro and thus promote the investigation of the relationship between peptide structure and pathogenic mechanism. These investigations further provide the fundamental supports for developing potential drugs for AD treatment. In addition to providing support for the study of pathogenic mechanisms, the recombination of Aβ peptides also offers the possibility to utilize these unique protein nanomaterials. For example, Aβ peptides tend to assemble into chiral amyloid fibrils with an ultra-high aspect ratio. These unique nano features, together with the inherent protein characteristics, of amyloid fibrils, allow them to be used in biomedical and environmental fields. Accordingly, herein, we aim to introduce the recombinant protocols for the synthesis of Aβ peptides. The experimental route to assemble these peptides to amyloid fibrils is also summarized in this chapter.

Key words Amyloid fibril, Recombinant amyloid β-peptide, Self-assembly, Characterization

1   Introduction

Amyloid fibrils, as highly ordered insoluble protein aggregations, have a very stable and highly ordered fibrillar structure, which is transformed from the native soluble conformation. The specific misfolded protein fibrils are linked with several neurodegenerative diseases, such as Alzheimer's disease, Parkinson's disease, Huntington disease, Type II diabetes, and prion disorder [1–5]. Therefore, the investigation of the amyloid β-peptide and the corresponding assembly structure—amyloid fibril has the potential to understand the pathogenesis of these diseases. In addition, peptides and amyloid fibers themselves are unique nanomaterials due to their ultra-long aspect ratio and unique chiral structure. These amyloid proteins have been produced different material formats, such as

Shengjie Ling (ed.), *Fibrous Proteins: Design, Synthesis, and Assembly*, Methods in Molecular Biology, vol. 2347,
https://doi.org/10.1007/978-1-0716-1574-4_11, © Springer Science+Business Media, LLC, part of Springer Nature 2021

hydrogels, nanoparticles, porous film, showing promising applications in tissue engineering [6–10], drug delivery [11, 12], water purification [13, 14], biosensor [15, 16], nanodevices [17–19], and energy conversion materials [20, 21].

Chemically, Aβ, originates from proteolytic processing of the amyloid precursor protein (APP), are associated with familial AD and cerebral amyloid angiopathy [22]. And the 40-amino acid alloform (Aβ(1-40)) and the 42-amino acid alloform (Aβ(1-42)) are the most abundant and the 40- and 42-amino acid alloforms aggregate to form fibrils and oligomers in AD [23, 24], The (Aβ (1-42)) aggregates more rapidly and is more neurotoxic than Aβ (1-40) [25]. In addition to the physiological environment changes caused by diseases, a series of external stimuli, such as pH [26], temperature [19], and the presence of salts and denaturing agents [27, 28], can also drive peptide to assemble and aggregate.

The final assembled forms of the Aβ peptides are amyloid fibrils, which generally have a diameter of about 7–10 nm and with length up to a few micrometers [29]. The morphologies of amyloid fibrils vary significantly in the shapes and topologies, although the assembly of Aβ peptides follows the highly similar aggregation dynamics. Another typical feature of amyloid fibrils is that they are composed of continuous cross-β structure [30], which formed by β-strands run perpendicular to the fibril axis. These highly ordered and intensively stacked β-strands are stabilized by high intensive hydrogen bonds, 4.76 Å apart along the length of the fibril [31, 32]. In the most cases, amyloid fibrils are not composed of a single filament (4–5 nm in diameter) [33–35], often termed as protofilament, but are entangled by multiple strands, with a geometry that similar to the twisted yarn, featuring a periodic chiral structure [36–38] (Fig. 1).

There are two technical routes to synthesize Aβ peptides; one is chemical synthesis (including liquid-phase and solid-phase synthesis), and the other is gene recombination technology. Compared with the method of chemical synthesis, the recombinant expression is more favorable due to their advances in low cost, high efficiency,

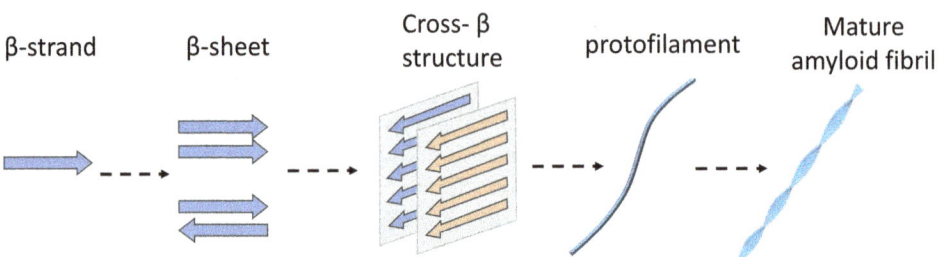

Fig. 1 The formation of amyloid fibrils. A native β-strand undergoes conformation transition into a β-sheet structure which further assemble into a cross-β structure. This structure is then elongated into protofilaments and finally twisted into mature amyloid fibrils

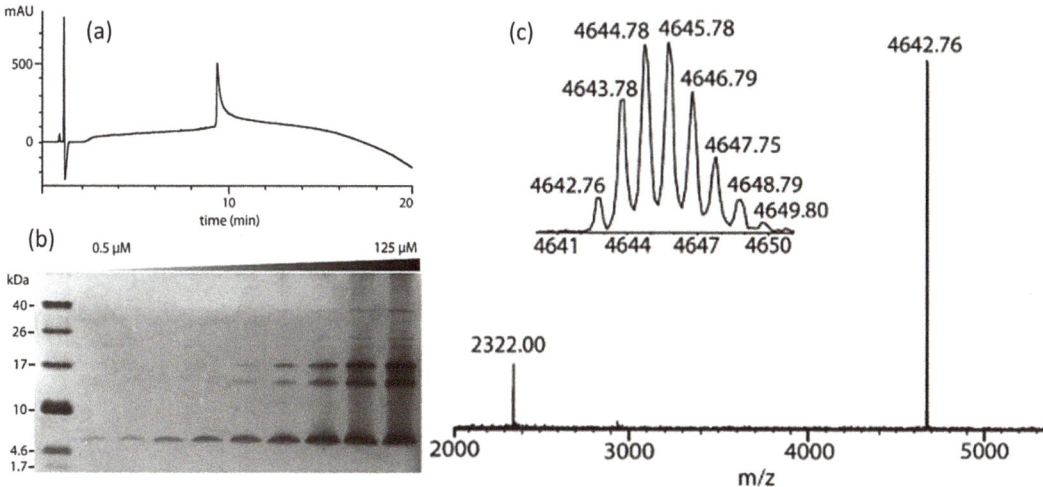

Fig. 2 Characterization of recombinant Aβ(M1-42). (Adapted by permission from American Chemical Society [40], Copyright 2018). Typical analytical HPLC (**a**), electrophoretic separation patterns (**b**) and MALDI mass spectra (**c**) of Aβ(M1-42)

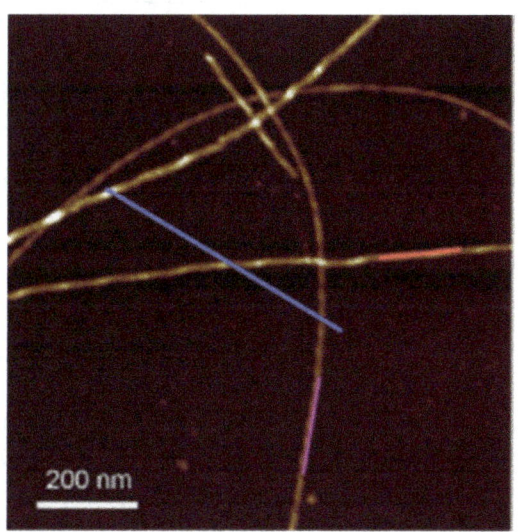

Fig. 3 AFM image of β-lactoglobulin amyloid fibril. (Adapted by permission from John Wiley & Sons [41], Copyright 2014)

high yield, and considerable scalability [39]. Therefore, this chapter will first introduce the methods to synthesize Aβ peptides through gene recombination. We then demonstrate the methods to assemble these peptides into amyloid fibrils. The characterization techniques used in these processes will also be introduced in the last part of this chapter. The results of the corresponding characterization techniques are also shown in Figs. 2 and 3.

2   Materials

1. pET-Sac- Aβ (M1-42).

2. DNA sequences that encode Aβ(M1-42) familial mutants.

3. BL21 DE3 PLysS Star Ca^{2+}-competent *E. coli*.

4. Ethidium bromide.

5. Luria–Bertani (LB) (containing 50 mg/L ampicillin) medium.

6. Buffer A (10 mM Tris–HCl, 1 mM EDTA at pH 8.0).

7. Buffer B (8 M urea, 10 mM Tris–HCl, 1 mM EDTA, pH 8.0).

8. Shrimp alkaline phosphatase (rSAP).

9. Isopropyl β-D-1-thiogalactopyranoside (IPTG).

10. Carbenicillin.

11. Chloramphenicol.

12. Zymo ZR plasmid miniprep kit.

13. Zymoclean Gel DNA Recovery Kit.

3   Methods

3.1   Preparation of Recombinant Aβ (M1-42)

3.1.1   Isolation of pET-Sac-Aβ (M1-42) Plasmid

1. Streak the pET-Sac-Aβ (M1-42) plasmid onto a LB agar-plate containing carbenicillin (50 mg/L) for less than 24 h.

2. Pick the single colonies and inoculate 5 mL of LB broth containing 50 mg/L carbenicillin with constant shaking (225 rpm) at 37 °C overnight.

3. Isolate the pET-Sac-Aβ (M1-42) plasmids.

4. Measure the concentration of the plasmids by a Thermo Scientific NanoDrop instrument (*see* **Notes 1** and **2**).

3.1.2   Expression of Aβ(M1-42)

1. Transform the sequence-verified plasmids into Ca^{2+}-competent *E. coli* cells (BL21 DE3 PLysS Star) by heat shock method.

2. Spread cell cultures on culture media (LB agar plates containing ampicillin (50 mg/L) and chloramphenicol (30 mg/L)), in which single colonies are picked to inoculate 5 mL of culture media for overnight culture.

3. Use all 5 mL of the overnight culture to inoculate 1 L of culture media the next day.

4. Shake the culture at 225 rpm at 37 °C for 3–4 h until the OD_{600} reaches ~0.45.

5. Add the isopropyl β-D-1-thiogalactopyranoside (IPTG) to the culture until the concentration reaches 0.1 mM, and then shake the cells at 225 rpm at 37 °C for 4 h.

6. Harvest the cells by centrifugation at $2800 \times g$ for 25 min at 4 °C.

7. Store the cell pellets at −80 °C.

3.1.3 Cell Lysis

1. Resuspend the cell pellet in 20 mL buffer A containing 10 mM Tris–HCl, 1 mM EDTA at pH 8.0, and sonicate for 2 min on ice (50% duty cycle) until the lysate appeared homogenous (*see* **Note 3**).

2. Centrifuge the lysate at $38,000 \times g$ for 25 min at 4 °C.

3. Remove the supernatant and resuspend the cell pellet in 20 mL buffer A and sonicate for 2 min on ice.

4. Repeat the sonication and centrifugation step three times.

5. Remove the fourth supernatant and resuspend the cell pellet in 15 mL freshly prepared buffer B (8 M urea, 10 mM Tris–HCl, 1 mM EDTA, pH 8.0) and sonicate for 2 min on ice until the solution becomes clear.

3.1.4 Purification of Aβ(M1-42)

1. The purification of Aβ(M1-42) peptides is performed by preparative reverse-phase HPLC equipped with an Agilent ZORBAX 300SB-C8 semipreparative column (9.4 × 250 mm) with a ZORBAX 300SB-C3 preparative guard column (9.4 × 15 mm).

2. Heat the C8 column and the guard column to 80 °C in a water bath. And use the HPLC grade acetonitrile (ACN) and 18 MΩ deionized water, each containing 0.1% trifluoroacetic acid at a flow rate of 5 mL/min as the mobile phase.

3. Split the peptide solution into ~8 mL aliquots and purified in three separate runs (*see* **Note 4**).

4. Load the peptide onto the column by flowing 20% ACN for 10 min and then elute with a gradient of 20–40% ACN over 20 min. (Generally, the fractions containing the monomer elute from 34% to 38% ACN.)

5. Combine the pure fractions and concentrate by rotary evaporation to remove ACN.

6. Lyophilize the samples at −80 °C (*see* **Note 5**).

3.1.5 Assembly of Amyloid Peptide

1. Dissolve a weighed amount of peptide into a buffered solution (20 mM glycine hydrochloride, pH 2.6).

2. Samples are immediately sonicated for 10 min to disassemble preformed nuclei.

3. Centrifuge the solution at $16,100 \times g$ for 5 min to deposit insoluble material.

4. The concentration of the stock solutions is determined by measuring the absorbance at 280 or 220 nm for peptides without aromatic residues.

5. The peptide solutions are incubated at room temperature and checked by CD and TEM at different incubation times (several days to several weeks).

3.2 Characterization

3.2.1 Check the Expression of Aβ (M1-42)

The Analytical reverse-phase HPLC is performed to check whether the expression of Aβ(M1-42) is successful. Dilute the 15 mL solution with 10 mL of buffer A and filtered through a Fisher Brand 0.22μm nonsterile hydrophilic PVDF syringe filter. Inject 40μL of the diluted solution onto an Agilent 1200 instrument equipped with a Phenomenex Aeris PEPTIDE 2.6u XB-C18 column with a Phenomenex SecurityGuard ULTRA cartridges guard column for C18 column with the HPLC grade acetonitrile (ACN) and 18 MΩ deionized water, each containing 0.1% trifluoroacetic acid as the mobile phase. Elute the sample at 1.0 mL/min with a 5–100% acetonitrile gradient at 35 °C for over 20 min.

3.2.2 Assess the Purity of the Peptide

The purity is assessed by analytical reverse-phase HPLC. Inject a 40μL of samples onto the HPLC and elute at 1.0 mL/min with a 5-100% acetonitrile gradient over 20 min, at 35 °C.

3.2.3 Mass Spectrometry

Mass spectrometry is performed using an AB SCIEX TOF/TOF 5800 System. The samples are prepared by dropping ~0.5μL peptide sample onto a MALDI sample support with 0.5μL of 2,5-dihydroxybenzoic acid (DHB). The mixture is allowed to air-dry. All analyses are performed in positive reflector mode, collecting data with a molecular weight range of 2000–8000 Da.

3.2.4 NMR Spectroscopy

NMR spectroscopy is accomplished using a Bruker DRX500 500 MHz spectrometer equipped with a cryogenic probe. And the NMR sample is prepared immediately before the NMR experiment. Dissolving approximately 0.5 mg of NaOH-treated, lyophilized ^{15}N-labeled in 0.6 mL of 50 mM potassium phosphate buffer. The number of points acquired in the direct dimension (^{1}H) is 2048, and the number of increments in the indirect dimension (^{15}N) is 256 experiments (see **Note 6**).

3.2.5 Thioflavin T Aggregation Assays

The measurement is conducted at 37 °C using a FLUOstar Optima plate reader. The samples are prepared by mixing a 70μL aliquot of each Aβ (M1-42) sample with 20μL of 100μM ThT and 10μL of 500 mM glycine–NaOH (pH 8.5). The 100μL resulting solution is then transferred to a 384-well fluorescence plate. Fluorescence is excited at a wavelength of 450 nm and with 485 nm emission filters.

3.2.6 CD Spectroscopy	The CD spectra data is collected from 250 to 190 nm in a 1 mm path length quartz cuvette at 25 °C. The wavelength step for the CD measurements is 1 nm, and the scan rate is 100 nm/min, an integration time of 1 s. The spectra shown were subtracted from the background and averaged over 10 consecutive scans.
3.2.7 Fourier Transform Infrared Spectroscopy (FTIR)	The infrared spectra are recorded using an FTIR spectrometer. The measurements are conducted on samples prepared by grinding the dried specimens with KBr and pressing them to form disks. Spectra are scanned in absorption mode at 25 °C with a spectral resolution of 4 cm^{-1} from 4000 to 500 cm^{-1}, and 256 interferograms are accumulated.
3.2.8 TEM Imaging	The morphology of the amyloid fibrils can be characterized by TEM. Seven microliters of the fibril dilution is absorbed for 2 min on supports with 400-mesh copper grids, and then washed four times using 5 mL of Milli-Q water and stained by 5 mL of 2% uranyl acetate solution. The grids are examined using a JEM-2100 transmission electron microscope (Jeol, Japan) at 80 keV.

4   Notes

1. The concentration of Aβ (M1-42) is determined by absorbance at 280 nm using the extinction coefficient (ε) for tyrosine of 1490 M^{-1} cm^{-1} ($c = A/1490$).

2. When preparing the samples for testing concentration, the pH of the solution should not occur near 5.5 or over 11. When the pH near 5.5, the peptide tends to aggregate, and the solution becomes opaque. If the pH over 11, the tyrosine is mostly deprotonated, which influences the result of UV spectra.

3. To avoid the oxidation of methionine, it is suggested to combine and freeze the purified fractions within 5 h after purification.

4. After the peptide is purified and collected, the column should be washed by injecting 5 mL of filtered buffer B containing 8 M urea, 10 mM Tris–HCl, 1 mM EDTA, pH 8.0 while flushing at 95% ACN for 15 min. The cleaning procedure is necessary to ensure the elution of all peptides and avoid the cross-contamination between runs.

5. Preferably, the ultrasonic treatment is performed under an ice bath.

6. The temperature of the NMR test should be maintained at 5 °C to reduce peptide aggregation.

References

1. Hardy J, Higgins G (1992) Alzheimer's disease: the amyloid cascade hypothesis. Science 256:184–185

2. Eisenberg D, Nelson R, Sawaya MR, Balbirnie M, Sambashivan S, Ivanova MI, Madsen AØ, Riekel C (2006) The structural biology of protein aggregation diseases: fundamental questions and some answers. Acc Chem Res 39:568–575

3. Murphy RM (2002) Peptide aggregation in neurodegenerative disease. Annu Rev Biomed Eng 4:155–174

4. Chiti F, Dobson CM (2006) Protein misfolding, functional amyloid, and human disease. Annu Rev Biochem 75:333–366

5. Dobson CM (1999) Protein misfolding, evolution and disease. Trends Biochem Sci 24:329–332

6. Tena-Solsona M, Alonso-De Castro S, Miravet JF, Escuder B (2014) Co-assembly of tetrapeptides into complex pH-responsive molecular hydrogel networks. J Mater Chem B 2:6192–6197

7. Bolisetty S, Vallooran JJ, Adamcik J, Mezzenga R (2013) Magnetic-responsive hybrids of Fe_3O_4 nanoparticles with β-Lactoglobulin amyloid fibrils and nanoclusters. ACS Nano 7:6146–6155

8. Herbst F-A, Søndergaard MT, Kjeldal H, Stensballe A, Nielsen PH, Dueholm MS (2015) Major proteomic changes associated with amyloid-induced biofilm formation in *pseudomonas aeruginosa* PAO1. J Proteome Res 14:72–81

9. Taglialegna A, Lasa I, Valle J (2016) Amyloid structures as biofilm matrix scaffolds. J Bacteriol 198:2579–2588

10. Gallo Paul M, Rapsinski Glenn J, Wilson RP, Oppong Gertrude O, Sriram U, Goulian M, Buttaro B, Caricchio R, Gallucci S, Tükel Ç (2015) Amyloid-DNA composites of bacterial biofilms stimulate autoimmunity. Immunity 42:1171–1184

11. Larbanoix L, Burtea C, Ansciaux E, Laurent S, Mahieu I, Elst LV, Muller RN (2011) Design and evaluation of a 6-mer amyloid-beta protein derived phage display library for molecular targeting of amyloid plaques in Alzheimer's disease: comparison with two cyclic heptapeptides derived from a randomized phage display library. Peptides 32:1232–1243

12. Mains J, Lamprou DA, McIntosh L, Oswald IDH, Urquhart AJ (2013) Beta-adrenoceptor antagonists affect amyloid nanostructure; amyloid hydrogels as drug delivery vehicles. Chem Commun 49:5082–5084

13. Li D, Furukawa H, Deng H, Liu C, Yaghi OM, Eisenberg DS (2014) Designed amyloid fibers as materials for selective carbon dioxide capture. Proc Natl Acad Sci U S A 111:191–196

14. Bolisetty S, Mezzenga R (2016) Amyloid–carbon hybrid membranes for universal water purification. Nat Nanotechnol 11:365–371

15. Men D, Zhang ZP, Guo YC, Zhu DH, Bi LJ, Deng JY, Cui ZQ, Wei HP, Zhang XE (2010) An auto-biotinylated bifunctional protein nanowire for ultra-sensitive molecular biosensing. Biosens Bioelectron 26:1137–1141

16. Kim S, Kim JH, Lee JS, Chan BP (2015) Beta-sheet-forming, self-assembled peptide nanomaterials towards optical, energy, and healthcare applications. Small 11:3623–3640

17. Zou Q, Liu K, Abbas M, Yan X (2016) Peptide-modulated self-assembly of chromophores toward biomimetic light-harvesting nanoarchitectonics. Adv Mater 28:1031–1043

18. Herland A, Björk P, Nilsson KPR, Olsson JDM, Åsberg P, Konradsson P, Hammarström P, Inganäs O (2005) Electroactive luminescent self-assembled bio-organic nanowires: integration of semiconducting oligoelectrolytes within amyloidogenic proteins. Adv Mater 17:1466–1471

19. Meier C, Lifincev I, Welland ME (2015) Conducting core–shell nanowires by amyloid nanofiber templated polymerization. Biomacromolecules 16:558–563

20. Chaves S, Pera LM, Avila CL, Romero CM, Baigori M, Vieyra FEM, Borsarelli CD, Chehin RN (2016) Towards efficient biocatalysts: photo-immobilization of a lipase on novel lysozyme amyloid-like nanofibrils. RSC Adv 6:8528–8538

21. Ryu J, Kim SW, Kang K, Park CB (2010) Synthesis of diphenylalanine/cobalt oxide hybrid nanowires and their application to energy storage. ACS Nano 4:159–164

22. Wolfe MS, Guénette SY (2007) APP at a glance. J Cell Sci 120:3157–3161

23. Haass C, Selkoe DJ (2007) Soluble protein oligomers in neurodegeneration: lessons from the Alzheimer's amyloid β-peptide. Nat Rev Mol Cell Biol 8:101–112

24. Chiti F, Dobson CM (2017) Protein misfolding, amyloid formation, and human disease: a summary of progress over the last decade. Annu Rev Biochem 86:27–68

25. Jarrett JT, Berger EP, Lansbury PT (1993) The carboxy terminus of the .Beta. Amyloid protein is critical for the seeding of amyloid formation: implications for the pathogenesis of Alzheimer's disease. Biochemistry 32:4693–4697

26. Kobayashi S, Tanaka Y, Kiyono M, Chino M, Chikuma T, Hoshi K, Ikeshima H (2015) Dependence pH and proposed mechanism for aggregation of Alzheimer's disease-related amyloid-β (1-42) protein. J Mol Struct 1094:109–117

27. Abelein A, Jarvet J, Barth A, Gräslund A, Danielsson J (2016) Ionic strength modulation of the free energy landscape of Aβ$_{(40)}$ peptide fibril formation. J Am Chem Soc 138:6893–6902

28. Gladytz A, Abel B, Risselada HJ (2016) Gold-induced fibril growth: the mechanism of surface-facilitated amyloid aggregation. Angew Chem 55:11242–11246

29. Shirahama T, Cohen AS (1967) High-resolution electron microscopic analysis of the amyloid fibril. J Cell Biol 33:679–708

30. Jahn TR, Makin OS, Morris KL, Marshall KE, Pei T, Sikorski P, Serpell LC (2010) The common architecture of cross-β amyloid. J Mol Biol 395:717–727

31. Nelson R, Sawaya MR, Balbirnie M, Madsen A, Riekel C, Grothe R, Eisenberg D (2005) Structure of the cross-β spine of amyloid-like fibrils. Nature 435:773–778

32. Sunde M, Serpell LC, Bartlam M, Fraser PE, Pepys MB, Blake CCF (1997) Common core structure of amyloid fibrils by synchrotron X-ray diffraction. J Mol Biol 273:729–739

33. Kreplak L, Aebi U (2006) From the polymorphism of amyloid fibrils to their assembly mechanism and cytotoxicity. Adv Protein Chem 73:217–233

34. Rochet JC, Lansbury PT (2000) Amyloid fibrillogenesis: themes and variations. Curr Opin Struct Biol 10:60–68

35. Nelson R, Eisenberg D (2006) Structural models of amyloid-like fibrils. Adv Protein Chem 73:235–282

36. Gazit E (2002) Korrekt gefaltete Proteine—ein metastabiler Zustand? Angew Chem 114:267–269

37. Ehud G (2002) The "correctly folded" state of proteins:is it a metastable state? Angew Chem Int Ed 41:257–259

38. Jarrett JT, Lansbury PT Jr (1993) Seeding "one-dimensional crystallization" of amyloid: a pathogenic mechanism in Alzheimer's disease and scrapie? Cell 73:1055–1058

39. HP S, Mortensen KK (2005) Advanced genetic strategies for recombinant protein expression in *Escherichia coli*. J Biotechnol 115:113–128

40. Yoo S, Zhang S, Kreutzer AG, Nowick JS (2018) An efficient method for the expression and purification of Aβ(M1-42). Biochemistry 57:3861–3866

41. Ling S, Li C, Adamcik J, Shao Z, Chen X, Mezzenga R (2014) Modulating materials by orthogonally oriented β-strands: composites of amyloid and silk fibroin fibrils. Adv Mater 26:4569–4574

Chapter 12

Molecular Biology Methods to Construct Recombinant Fibrous Protein

Jicong Zhang

Abstract

Recombinant technologies are often used to synthesize fibrous proteins that are difficult to separate and extract in nature, such as spider silks and elastin. Although the recombination techniques can be diverse, PCR, gel electrophoresis, and seamless cloning, as the basic methods of molecular biology, have been widely used for constructing fibrous proteins' homologous recombinant plasmids. Considering that some readers of this book may not have a molecular biology background, in this chapter, we will introduce these three most used and effective recombination techniques. For PCR, we primarily introduce colony PCR, high-fidelity PCR, and overlap PCR, which are three kinds of the most used methods. In terms of seamless cloning, the detailed protocols of Gibson Assembly and Golden Gate Assembly are introduced. The introduction of this chapter is expected to provide a comprehensive methodological reference for the following chapters to introduce the recombination of specific fibroin proteins.

Key words Polymerase chain reaction, Agarose gel electrophoresis, Seamless cloning

1 Introduction

Many fibrous proteins, such as spider silks and elastin, are not easily obtained from natural products [1–4], so recombinant technologies are usually used to synthesize their recombinant construction from a variety of heterologous hosts, such as *Escherichia coli* (*E. coli*), yeast, tobacco, potato, and goat. Although the recombinant technologies are diverse, polymerase chain reaction (PCR) will be frequently adopted, no matter which recombinant route is selected. PCR is invented in 1984 by the American biochemist Kary Mullis at Cetus Corporation. This technology can amplify a DNA fragment of fibrous protein millions of times in a test tube after a few hours of reaction [5]. Owing to quickly obtain a large number of nucleic acid fragments, it has a pivotal significance in molecular biology research. It has dramatically promoted study in life sciences.

Shengjie Ling (ed.), *Fibrous Proteins: Design, Synthesis, and Assembly*, Methods in Molecular Biology, vol. 2347, https://doi.org/10.1007/978-1-0716-1574-4_12, © Springer Science+Business Media, LLC, part of Springer Nature 2021

There are many types of PCR, in particular, some new types of PCR or methods based on PCR, such as digital droplet PCR (ddPCR), multiplex PCR, first-generation sequencing (Sanger sequencing), and second-generation sequencing, are being developed with actual application requirements. For example, ddPCR has been demonstrated as useful for quantitative measurement of the translocation frequency between chromosome 1 (Chr1) and Chr2 of *Arabidopsis thaliana* [6]. A novel multiplex PCR method for the portable MinION sequencer (Oxford Nanopore Technologies) and the Illumina range of instruments has been successfully used in sequencing Zika and other virus genomes directly from clinical samples [7]. In all PCR methods, colony PCR, high-fidelity PCR, and overlap PCR are three of the most common and widely used techniques in constructing recombinant fibrous proteins. Colony PCR is a convenient high-throughput method for determining the presence or absence of insert DNA in plasmid constructs [8]. High-fidelity PCR, instead, utilizes a DNA polymerase with a low error rate and results in a high degree of accuracy in the replication of the DNA of interest. The fidelity of a polymerase refers to its ability to insert the correct base during PCR. Conversely, the rate of misincorporation is known as a polymerase's error rate. Overlap extension polymerase chain reaction (overlap PCR) is a variant of PCR [9]. It usually is used to create long DNA fragments from shorter ones and insert specific mutations at specific points in a sequence. This method has an advantage over other gene-splicing techniques in not requiring restriction sites.

After PCR amplification, agarose gel electrophoresis, a standard method used for separation and analysis of larger nucleic acids (>100 bases in length) under nondenaturing conditions, is usually performed to separate the target plasmid. By adding the sample with loading buffer to the gel wells and applying a current over the anode and cathode, the negatively charged nucleic acid will migrate to the positive electrode. The DNA and RNA fragments can be separated by length [10]. Analysis requires that the gel contains a DNA stain visible under UV light. The appropriate DNA bonds can be cut and extracted following the protocols of full-fledged gel extraction kits.

Seamless cloning, different from traditional cloning, can insert multiple fragments into a vector without any intervening or unwanted extraneous sequences (such as restriction sites). The "seamless cloning" systems often employ the PCR to amplify the gene of interest, an exonuclease to digest one strand of the insert fragments and vector ends, and either a ligase, recombination event, or in vivo repair to covalently join the insert to the vector through a phosphodiester bond. By combining these attributes of established cloning methods to create a unique mix solution, the seamless cloning enables to perform sequence-independent cloning

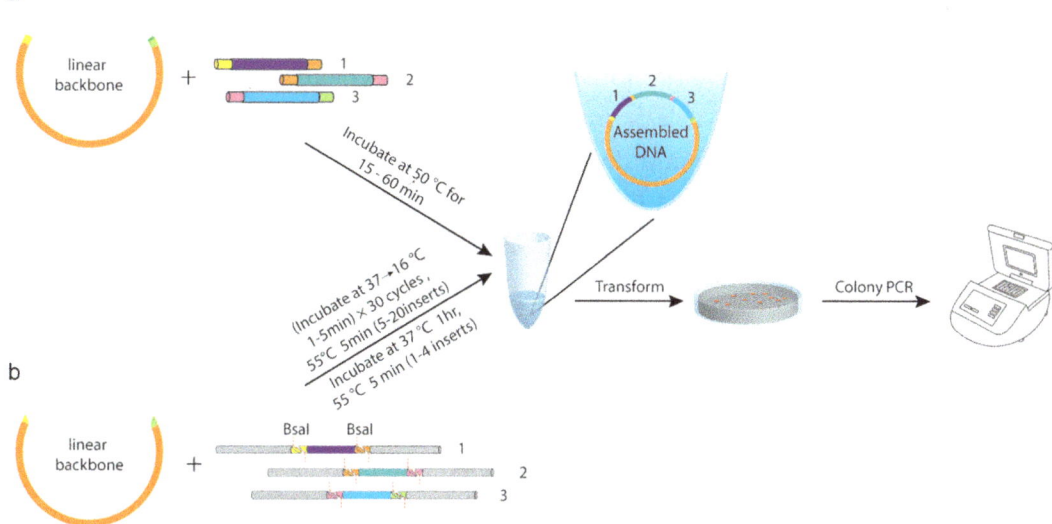

Fig. 1 Scheme of seamless cloning. (**a**) Gibson Assembly. (**b**) Golden Gate Assembly

without retaining undesired nucleotides or scar sequences between vector and insert fragments. There are two different ways of seamless cloning, that is, Gibson Assembly and Golden Gate Assembly.

Gibson Assembly (Fig. 1a), initially described by Daniel G. Gibson (J. Craig Venter Institute), uses a three-enzyme mix to transform the linear DNA fragments to the finished plasmid. The three enzymes are DNA polymerase to fill in the gaps of the annealed single-stranded regions, 5′ exonuclease to generate cohesive terminal ends (overhangs), and a DNA ligase to seal the nicks [11]. Golden Gate Assembly (Fig. 1b) exploits the ability of Type IIS restriction enzymes (such as BsaI, BbsI, BsmBI, and Esp3I) to cleave DNA outside of the recognition sequence. The insert fragments and cloning vectors are designed to place the Type IIS recognition site distal to the cleavage site, such that the Type IIS restriction enzyme can remove the recognition sequence from the assembly [12, 13].

As the above three kinds of molecular biology methods (PCR, agarose gel electrophoresis, and seamless cloning) are widely used in fibroin protein recombination, this chapter will introduce their detailed experimental protocol. For PCR, we mainly introduce the three most used methods—colony PCR, high-fidelity PCR, and overlap PCR. For seamless cloning, the detailed protocols of Gibson Assembly and Golden Gate Assembly are introduced. As these methods are applied to the recombination of all fibrous proteins, this chapter will not specify which protein these methods are aimed at.

2 Materials

2.1 PCR

2.1.1 Colony PCR

1. A plate with restreaked, single colonies.
2. Nuclease-free water.
3. 10 mM dNTPs.
4. 10µM Forward Primer.
5. 10µM Reverse Primer.
6. $10 \times$ Standard Taq Reaction Buffer (100 mM Tris–HCl, 500 mM KCl, 20 mM $MgCl_2$, pH 8.3).
7. Taq DNA Polymerase..
8. Thermal cyclers.

2.1.2 High-Fidelity PCR

1. Samples (cells or DNA).
2. Reaction Buffer (50 mM Tris–HCl, 250 mM KCl, 10 mM $MgCl_2$, pH 8.3).
3. 10µM Forward Primer.
4. 10µM Reverse Primer.
5. High-fidelity DNA Polymerase.
6. 10 mm dNTPs.
7. Nuclease-free water.
8. Thermal cyclers.

2.1.3 Overlap PCR

1. Samples (cells or DNA).
2. Reaction Buffer (50 mM Tris–HCl, 250 mM KCl, 10 mM $MgCl_2$, pH 8.3).
3. 10µM Forward Primer.
4. 10µM Reverse Primer.
5. High-fidelity DNA Polymerase.
6. 10 mm dNTPs.
7. Nuclease-free water.
8. Gel extraction tool kit.
9. Thermal cyclers.

2.2 Agarose Gel Electrophoresis

1. Agarose; $0.5\times$ Tris–Acetate–EDTA (TAE) buffers [diluted from $50 \times$ TAE buffer (242 g Tris, 37.2 g $Na_2EDTA\cdot2H_2O$, 57.1 mL acetic acid pH $= 8.3$)].
2. DNA gel stain.
3. DNA loading buffer.
4. DNA ladder marker.
5. DNA samples.

6. Electrophoresis apparatus.

7. Gel tray.

8. Gel comb.

2.3 Seamless Cloning

2.3.1 Gibson Assembly

1. Linearized plasmid.

2. Insert DNA fragments.

3. Gibson Assembly Master Mix (200μL 1 M Tris–HCl pH = 7.5, 10μL 2 M MgCl$_2$, 16μL 100 mM dNTPs, 20μL 1 M DTT, 0.12 g PEG-8000, 20μL 100 mM NAD, 0.2 U T5 exonuclease, 10 U Phusion polymerase, 1600 U Taq ligase, ddH$_2$O to 1 mL).

4. Thermal cyclers.

5. Deionized H$_2$O.

2.3.2 Golden Gate Assembly

1. The insert DNA fragments.

2. Destination Plasmid (with right and well-designed type II restriction enzyme sites).

3. Golden Gate Buffer (400 mM Tris–HCl, 100 mM MgCl$_2$, 100 mM DTT, 5 mM ATP, pH 7.8).

4. Golden Gate Assembly Mix (20 U/μL Type IIS restriction enzymes, 5 U/μL T4 ligase).

5. Nuclease-free H$_2$O.

6. Thermal cyclers.

---

3 Methods

3.1 Polymerase Chain Reaction (PCR)

3.1.1 Colony PCR

1. Using a sterile loop, pick a small portion of each of these colonies, and suspend individually in 20μL of water. Mix thoroughly as the template solution. Boil the template solution at 100 °C for 5 min will greatly enhance the success of PCR.

2. Calculate the volumes for making up your premix for several reactions (as described above) based on the volumes provided for just one reaction, as listed in Table 1.

   The template DNA usually add 1μL is enough to ensure the success of colony PCR. Primers need to be designed carefully (*see* **Note 1**).

3. Program the thermal cyclers. The detailed steps are given in Table 2 (*see* **Note 2**). When primers with annealing temperatures above 65 °C are used, a two-step thermocycling protocol is possible, as shown in Table 3 (*see* **Note 3**).

   The 30 cycles can change based on *see* **Note 4**.

4. Run the PCR reactions.

5. Analyze a 5μL aliquot of each PCR by agarose gel electrophoresis.

Table 1
Colony PCR reaction mixture

Component	25µL reaction	50µL reaction
10× standard *Taq* reaction buffer	2.5µL	5µL
10 mM dNTPs	0.5µL	1µL
10µM forward primer	0.5µL	1µL
10µM reverse primer	0.5µL	1µL
Template DNA	Variable	Variable
Taq DNA polymerase	0.125µL	0.25µL
Nuclease-free water	to 25µL	to 50µL

Table 2
Procedure of colony PCR

Step	Temperature	Time
Initial denaturation	95 °C	5 min
30 cycles	95 °C 45–68 °C 68 °C	15–30 s 15–60 s 1 min/kb
Final extension	68 °C	5 min
Hold	4–10 °C	

Table 3
Thermocycling conditions for a routine two-step PCR

Step	Temp	Time
Initial denaturation	95 °C	5 min
30 cycles	95 °C 45–68 °C	15–30 s 15–60 s
Final extension	68 °C	5 min
Hold	4–10 °C	

3.1.2 High-Fidelity PCR

1. Reaction setups are given in Table 4. And the specific amounts of template can refer to *see* **Note 5**.

2. Program the thermal cyclers. The details can be found in Table 5.

3. Run the PCR reactions.

Table 4
High-fidelity PCR reaction mixture

Component	25µL Reaction	50µL Reaction
5× Reaction buffer	5µL	10µL
10 mM dNTPs	0.5µL	1µL
10µM forward primer	1.25µL	2.5µL
10µM reverse primer	1.25µL	2.5µL
Template DNA	Variable	Variable
Hot start high-Fidelity DNA polymerase	0.25µL	0.5µL
Nuclease-free water	to 25µL	to 50µL

Table 5
Procedure of high-fidelity PCR

STEP	Temperature	Time
Initial denaturation	98 °C	30 s
25–35 cycles	98 °C *50–72 °C 72 °C	5–10 s 10–30 s 20–30 s/kb
Final extension	72 °C	2 min
Hold	4–10 °C	

The * represent the specific value of this temperature should depend on the primers. And do not exceed the range which there listed in general.

3.1.3 Overlap PCR

1. Design Primers: These primers are like bridges between the two parts you want to assemble together. You will order two primers, which are complements of one another. These primers will each have a 60 °C T_m with one part and a 60 °C T_m with the other part. The "end primers" will not have any complements and will likely only have restriction sites (*see* **Notes 6–11**).

2. High-fidelity PCR amplify the necessary fragments separately.

3. Perform agarose gel electrophoresis. Cut the desired DNA bands under the ultraviolet light. Then extraction the DNA fragments from the agarose gel.

4. The reaction setup based on the high-fidelity PCR, as given in Table 6. And the specific amounts of different fragments are needed (*see* **Note 10**).

Table 6
Overlap PCR reaction mixture

Component	50μL Reaction
5× Reaction buffer	10μL
10 mM dNTPs	1μL
10μM forward primer	2.5μL
10μM reverse primer	2.5μL
Fragment 1 and equimolar to 2 and 3	1μL
Fragment 2 and equimolar to 1 and 3	1μL
Fragment 3 and equimolar to 1 and 2	1μL
…	…
Hot start high-Fidelity DNA polymerase	0.5μL
Nuclease-free water	to 50μL

Table 7
Procedure of overlap PCR

Step	Temperature	Time
Initial denaturation	98 °C	30 s
25–35 cycles	98 °C 50–72 °C 72 °C	5–10 s 10–30 s 20–30 s/kb
Final extension	72 °C	2 min
Hold	4–10 °C	

5. Gently mix the reaction. Collect all liquid to the bottom of the tube by a quick spin if necessary. Overlay the sample with mineral oil if using a PCR machine without a heated lid. Program the thermal cyclers with the steps, as listed in Table 7.

6. Run the PCR reactions.

3.2 Agarose Gel Electrophoresis

3.2.1 Preparing a 100 mL 1% Agarose Gel

1. Weigh 1 g agarose in a 500 mL conical flask. Add 100 mL 0.5 × TAE buffer (*see* **Note 12**). Other concentrations of agarose gel may be needed in some situations (*see* **Note 13**).

2. To dissolve the agarose in the buffer, shake to mix and microwave for a few minutes. Remove the flask occasionally and check whether the agarose has dissolved completely.

3. Let the agarose solution cool down to be touchable. Check the manual of the DNA gel stain and add the appropriate amount to give the desired final concentration.

4. Homogeneous mixing the DNA stain and agarose solution. Pour the solution into the gel tray and remove any air bubbles. Put in the comb.

5. The gel will solidify while cooling down to room temperature. Depending on the initial temperature, it will take ~30 min.

3.2.2 Loading DNA Samples and Electrophoresis

1. Remove the comb carefully and place the gel into the chamber of the electrophoresis apparatus.

2. Add 0.5 × TAE buffer until the gel completely emerges.

3. Load the size marker into a middle well.

4. Mix the DNA samples with loading buffer. The appropriate amount of loading buffer based on the manual recommended.

5. Load the mixture into the other wells while writing down which lanes have which samples.

6. Put the lid onto the buffer chamber and connect it to the power supply (*see* **Note 14**).

7. Run the gel at 170 V for 20–40 min.

8. Stop the run and bring the gel to a UV table to visualize the DNA bands. Take a picture to record the results.

3.3 Seamless Cloning

3.3.1 Gibson Assembly

1. Set up the reaction on ice with the composition ratio, as given in Table 8.

2. Incubate samples in a thermocycler at 50 °C for 15 min when 2 or 3 fragments are being assembled or 60 min when 4–6 fragments are being assembled (*see* **Notes 16–18**).

3. Following incubation, store samples on ice or at −20 °C for subsequent transformation.

4. Transform Competent cells (usually *E. coli*) cells with the assembly reaction, following the transformation protocol.

3.3.2 Golden Gate Assembly

1. Set up assembly reactions with the composition ratio, as given in Table 9. Different insertion options require different composition ratios (*see* **Note 19**).

2. Choose the appropriate assembly protocol, as listed in Table 10. Program the thermal cyclers.

3. Transform competent cells (usually *E. coli*) cells followed by the Golden Gate Assembly reaction, following the transformation protocol.

Table 8
Gibson Assembly reaction mixture

	Recommended amount of fragments used for assembly	
	2–3 fragment assembly	4–6 fragment assembly
Total amount of fragments	0.02–0.5 pmols[a] XμL	0.2–1 pmols XμL
Gibson Assembly master mix (2×)	10μL	10μL
Deionized H_2O	10–XμL	10–XμL
Total volume	20μL	20μL

[a]pmols = (weight in ng) × 1000/(base pairs × 650 Da) (*see* **Note 15**)

Table 9
Golden Gate Assembly reaction mixture

Reagent	Assembly reaction
Destination plasmid, 75 ng/μL	1μL
Inserts	75–100 ng each plasmid
– if precloned	2:1 molar ratio
– if in amplicon form	(insert vector)
Golden Gate buffer (10×)	2μL
Golden Gate Assembly mix	2μL
Nuclease-free H_2O	to 20μL

Table 10
Procedure of Golden Gate Assembly

Insert number	Suggested assembly protocol
For 1–4 inserts	37 °C, 1 h→55 °C, 5 min
For 5–10 inserts	(37 °C, 1 min→16 °C, 1 min) × 30→55 °C, 5 min
For 11–20 inserts	(37 °C, 5 min→16 °C, 5 min) × 30→55 °C, 5 min

4   Notes

1. Oligonucleotide primers are generally 20–40 nucleotides in length and ideally have a GC content of 40–60%. The final concentration of each primer in a reaction maybe 0.05–1μM, typically 0.1–0.5μM.

Table 11
Dosage of different samples

DNA	Amount
DNA genomic	1 ng to 1μg
Plasmid or viral	1 pg to 1 ng

2. An initial denaturation of 30 s at 95 °C is sufficient for most amplicons from pure DNA templates. For difficult templates such as GC-rich sequences, a longer initial denaturation of 2–4 min at 95 °C is recommended prior to PCR cycling to denature the template fully. With colony PCR, an initial 5 min denaturation at 95 °C is recommended.

3. Generally, 25–35 cycles yield sufficient product. Up to 45 cycles may be required to detect low-copy-number targets.

4. The recommended extension temperature is 68 °C. Extension times and the optimal extension temperature may be different; it should refer to the specification of the manufacturer. A final extension of 5 min at 68 °C is recommended.

5. Except for the slightly different in templates, the operation and precautions of high-fidelity PCR are the same as colony PCR. Template: The use of high-quality, purified DNA templates greatly enhances the success of PCR. Recommended amounts of DNA template for a 50μL reaction, as listed in Table 11.

6. This protocol works best for assembling parts higher than 100 bp.

7. Primer: To splice two DNA molecules, special primers are used at the ends that are to be joined. For each molecule, the primer at the end to be joined is constructed such that it has a 5′ overhang complementary to the end of the other molecule. Following annealing when replication occurs, the DNA is extended by a new sequence that is complementary to the molecule it is to be joined to. Once both DNA molecules are extended in such a manner, they are mixed, and a PCR is carried out with only the primers for the far ends.

8. There should be no less than 15 base overlap between two DNA fragments.

9. If the fragment GC content is too high, the GC Buffer should be added.

10. The amount of the two segments should not be added too high. If it is the first PCR product, it should be diluted by 50 times and then add 1μL as a template; If the product is purified, it is advisable to add about 1–10 ng.

11. If you have noticed all the things listed above, but no target bands yet. It may be the annealing temperature of PCR was not appropriate. It is recommended to explore the conditions through gradient PCR or to conduct touchdown PCR.

12. The gel buffer or electrophoresis buffer. Do not use water instead of buffers. Commonly used buffers are TAE and Tris–Borate–EDTA (TBE). TBE provides better-buffering capacity than TAE. The use of a newly prepared buffer during electrophoresis can significantly improve the electrophoresis effect. When repeated use of the electrophoresis buffer many times, the decrease in ionic strength and increase in pH value may lead to the buffer performance decreases, which may cause dim DNA bands and irregular migration of DNA bands.

13. The concentration of agarose gel. The concentration is measured in the weight of agarose over the volume of buffer used (g/mL). The lower the concentration of agarose, the higher the resolution and separation of large molecular bands. Vice versa. For a standard agarose gel electrophoresis, a 0.8% gel gives good separation or resolution of large 5–10 kb DNA fragments, while 2% gel gives a good resolution for small 0.2–1 kb fragments. 1% gels are often used for standard electrophoresis.

14. Do not reverse the orientation of the electrophoresis tank and wrong connect with the power supply cable.

15. pmols = (weight in ng) × 1000/(base pairs × 650 Da).
    The mass of each fragment can be measured using the NanoDrop instrument, absorbance at 260 nm, or estimated from agarose gel electrophoresis followed by ethidium bromide staining.

16. Optimized cloning efficiency is 50–100 ng of the vector with a 2-fold molar excess of each insert. Use five times more inserts if the size is less than 200 bps. The total volume of unpurified PCR fragments in the Gibson Assembly reaction should not exceed 20%.

17. If higher numbers of fragments are assembled, additional Gibson Assembly Master Mix may be required.

18. Extended incubation up to 60 min may help to improve assembly efficiency in some cases.

19. In each cloning step, Golden Gate Cloning can assemble up to nine fragments and only requires homology in type II restriction enzyme sites so that the DNA fragments can be ligated seamlessly.
    Precloned Inserts: Precloning is always an option, and is superior for inserts <250 bp or >3 kb, or those containing repetitive elements that might accumulate errors during PCR

amplification. Note that all sequences that will be part of the assembly must be flanked by correctly oriented Type II restriction sites, facing toward the insert on the top and bottom strands.

Amplicon Inserts: The 5′ flanking bases and type II restriction enzyme recognition site are introduced through PCR primer design upstream and downstream of sequences to be assembled. In all cases, the insert: vector backbone molar ratio = 2:1 is suggested to achieve assembly efficiencies similar to that with precloned inserts.

References

1. Lai Y-T, King NP, Yeates TO (2012) Principles for designing ordered protein assemblies. Trends Cell Biol 22:653–661

2. Bozic S, Doles T, Gradisar H, Jerala R (2013) New designed protein assemblies. Curr Opin Chem Biol 17:940–945

3. Patterson DP (2011) Symmetry assembled supramolecular protein cages: investigating a strategy for constructing new biomaterials. Proquest, Umi Dissertation Publishing, Ann Arbor, Michigan

4. King NP, Lai YT (2013) Practical approaches to designing novel protein assemblies. Curr Opin Struct Biol 23:632–638

5. Whyburn GP, Li Y, Huang Y (2008) Protein and protein assembly based material structures. J Mater Chem 18:3755–3762

6. Beying N, Schmidt C, Pacher M, Houben A, Puchta H (2020) CRISPR–Cas9-mediated induction of heritable chromosomal translocations in Arabidopsis. Nat Plants 6:638–645

7. Saiki RK, Gelfand DH, Stoffel S, Scharf SJ, Higuchi R, Horn GT, Mullis KB, Erlich HA (1988) Primer-directed enzymatic amplification of DNA with a thermostable DNA polymerase. Science 239:487–491

8. Kieleczawa J (2006) DNA sequencing II: optimizing preparation and cleanup (Vol. 2). Jones and Bartlett Publishers, Burlington, Massachusetts

9. Higuchi R, Krummel B, Saiki RK (1988) A general method of in vitro preparation and specific mutagenesis of DNA fragments: study of protein and DNA interactions. Nucleic Acids Res 16:7351–7367

10. Sambrook J, Fritsch EF, Maniatis T (1989) Molecular Cloning: A Laboratory Manual, vol [prepared for Use in the CSH Courses on the Molecular Cloning of Eukaryotic Genes]. Cold Spring Harbor Laboratory Press, Cold Spring Harbor, New York

11. Gibson DG, Young L, Chuang R-Y, Venter JC, Hutchison CA, Smith HO (2009) Enzymatic assembly of DNA molecules up to several hundred kilobases. Nat Methods 6:343–345

12. Engler C, Kandzia R, Marillonnet S (2008) A one pot, one step, precision cloning method with high throughput capability. PLoS One 3: e3647

13. Golden Gate Assembly | NEB. Biolabs, New England.   https://www.neb.com/applications/cloning-and-synthetic-biology/dna-assembly-and-cloning/golden-gate-assembly

Part III

Methods in Assembly of Fibrous Proteins

Chapter 13

Isolation of Nanofibrils from Animal Silks

Ke Zheng

Abstract

The presence of well-organized nanofibrils in animal silks is considered to provide them excellent mechanical and biochemical properties. To direct utilize these unique natural nanomaterials, a variety of physical and/or chemical processes have been developed for directly isolating silk nanofibrils from animal silks. The yield and processability of these techniques as well as the morphologies of resultant silk nanofibrils have apparent differences but also have their own merits. In this chapter, I presented the protocols for isolation silk nanofibrils, including a physical approach of sonication, a chemical approach of salt–formic acid dissolution, as well as three combination approaches, hexafluoroisopropanol liquid exfoliation, urea–guanidine hydrochloride dissolution, and sodium hypochlorite partial dissolution.

Key words Animal silks, Isolation, Nanofibrils

1 Introduction

Animal silks involving spider and silkworm silks have received considerable attention due to their exceptional mechanical and biochemical performance, such as high modulus, strength, extensibility, and low immunogenicity [1–4]. These unique properties of animal silks usually considered as the consequence of their optimized functional adaptation of the structures at each hierarchy [5–7]. In order to replicate these insights of animal silks within artificial materials; therefore, the direct isolation of silk nanofibrils for materials fabrication is a practical option [8–10]. The critical challenges for silk nanofibrils isolation are (1) the infancy of direct extraction of silk nanofibrils due to the complex hierarchical structure and the high crystallinity of native silks, and (2) to obtain silk nanofibrils with versatile processability for matching the distinct requirements in different engineering applications [10–12]. Accordingly, various approaches involving physical and/or chemical isolation techniques have been developed to address these issues.

Shengjie Ling (ed.), *Fibrous Proteins: Design, Synthesis, and Assembly*, Methods in Molecular Biology, vol. 2347,
https://doi.org/10.1007/978-1-0716-1574-4_13, © Springer Science+Business Media, LLC, part of Springer Nature 2021

An initial idea to isolate silk nanofibrils is applying physical processes such as sonication and mechanical squeezing [13]. These physical processes have been successfully applied for the isolation of nanofibrils from biopolymers, including cellulose, chitin, and collagen [14–16]. However, the resultant silk nanofibrils by using physical processes are intertwined in the solvent and lacked processability. Another idea to isolate silk nanofibrils is the chemical dissolution [17, 18]. Chemicals such as sodium hydroxide (NaOH), lithium bromide (LiBr), and calcium chloride–formic acid (CaCl$_2$–FA) that can dissolve silk fibers have been attempted to prepare silk nanofibrils [19–22]. Among these chemicals, NaOH generally degrades silk fibers along their contour length direction, where the degraded fibers have diameters of several micrometers [19]. The CaCl$_2$–FA solution generates silk nanofibrils of 150–200 nm in diameter. However, the subsequent processes require acid tolerance because these nanofibrils are mixed with the acid.

Hexafluoroisopropanol liquid exfoliation [23, 24], urea–guanidine hydrochloride dissolution [25], and sodium hypochlorite partial dissolution approaches [26, 27] have been established to achieve silk nanofibrils. A common feature of these approaches is the integration of physical and chemical techniques to isolate silk nanofibrils. Briefly, these approaches first using a proper chemical to disassemble silk fibers into microscale, and then followed a mechanical treatment (sonication). The silk nanofibrils will disperse in aqueous solution via hydrophobic interaction and/or electrostatic repulsive forces.

In this chapter, I first introduce the protocols to isolate silk nanofibrils by using physical and/or chemical isolation techniques. Then, I describe the details of the morphology of resultant silk nanofibrils at different isolation conditions. Of note, I focus on discussing the *Bombyx mori* (*B. mori*) silk, if not mentioned otherwise.

2   Materials

1. Degummed silk fiber.

2. Calcium chloride (CaCl$_2$).

3. Formic acid (FA, 88 v/v%).

4. Lithium bromide (LiBr).

5. Hexafluoroisopropanol (HFIP).

6. Urea.

7. Guanidine hydrochloride (GuHCl).

8. Sodium chlorate (NaClO, 5% available chlorine).

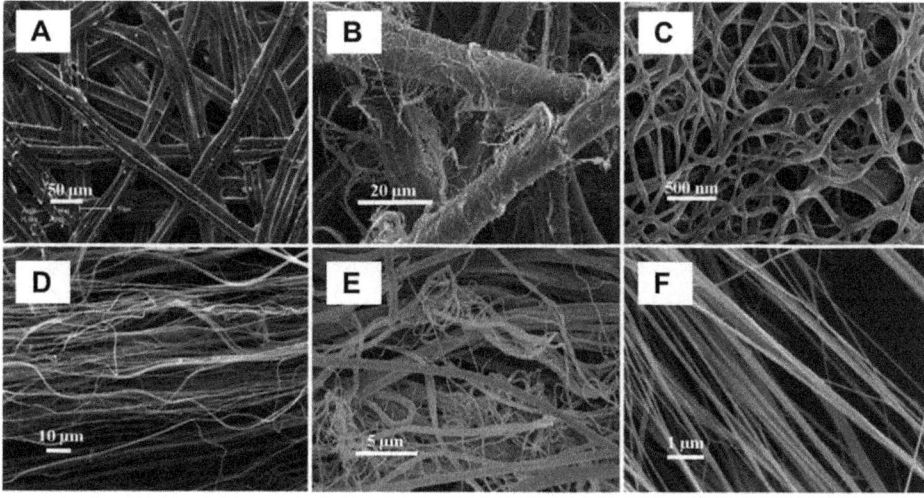

Fig. 1 SEM images of the consecutive sonicated spider and silkworm (*Bombyx mori*) silk fibers. (**a**) Silkworm silk fibers, (**b**) disassembled silkworm silk fibers and (**c**) silkworm silk nanofibers. (**d**) Spider silk fibers, (**e**) disassembled spider silk fibers and (**f**) spider silk nanofibers. (Adapted by permission from American Institute of Physics [13], Copyright 2007)

3   Methods

3.1   Sonication

1. Place 0.05 g of degummed silk fiber in a 250 mL glass beaker, then add 100 mL distilled water (*see* Fig. 1 and **Note 1**).

2. Immerse the sonicating probe into the silk fiber suspension. Do not touch the probe on the sides or the bottom of the glass beaker.

3. <Critical Step>Sonicate the silk fiber suspension at 20 kHz in frequency equipped with a cylindrical titanium alloy probe tip of 2.5 cm in diameter. Of note, use a water bath to avoid the overheating of the suspension. The applying sonication is conducted at different powers and times for different fibers. For example, the silkworm silk requires sonication at 900 W for 30 min to complete disassembly, while the spider silk requires a power of 1000 W for 45 min.

4. Collect the silk nanofibrils at the bottom of the glass beaker.

3.2   Salt–Formic Acid Dissolution

1. Dissolve 2 g calcium chloride ($CaCl_2$) in 100 mL formic acid (FA) to obtain a 2 w/v% $CaCl_2$–FA solution. Of note, lithium bromide (LiBr) could also be used to add to FA. The concentrations of salt–FA solution are in the range of 1–5 w/v% (*see* Fig. 2 and **Note 2**).

2. Place 0.5 g of degummed silk fiber in 100 mL of $CaCl_2$–FA solution.

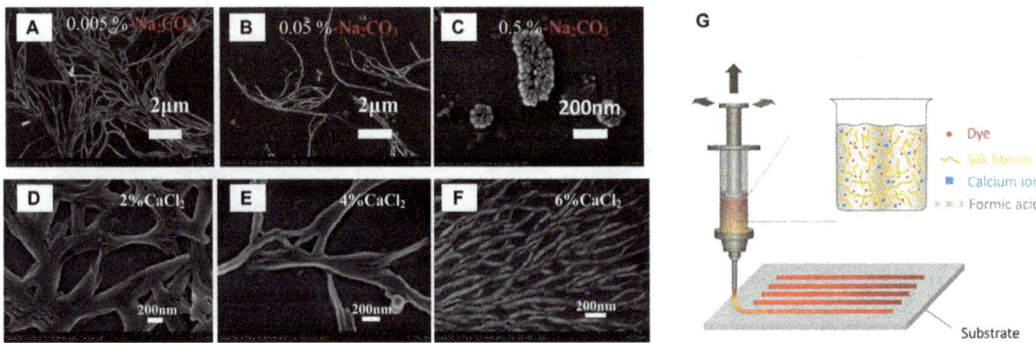

Fig. 2 (**a–f**) Dissolving behavior of the silkworm (*Bombyx mori*) silk nanofibers in CaCl₂–formic acid solution, and (**g**) the printing process of dissolved silk nanofibers. (**a**) 0.005, (**b**) 0.05, and (**c**) 0.5 w/v% Na₂CO₃ degummed silk fibers dissolving in CaCl₂–formic acid solution with the formation of network nanofibers, nanofibers, and micelles, respectively. The length and diameter of the silk nanofibers obtained from the CaCl₂ concentration of (**d**) 2, (**e**) 4, and (**f**) 6 w/v% CaCl₂ in formic acid solution. (**g**) Schematic illustration of the 3D printing process by using a fluorescent silk–CaCl₂—formic acid ink. (Adapted by permission from The Royal Society of Chemistry [17] and [21], Copyright 2014 and 2016)

3. <Critical Step>Incubate the mixture at 25 °C for 3 h. The diameter of as-prepared nanofibrils is ranged from 20 to 200 nm. Of note, higher temperature and/or more incubation time will degrade the silk nanofibrils to micelles and even peptides.

3.3 Hexafluorois-opropanol Liquid Exfoliation

1. Place 0.5 g degummed silk fiber in 15 g hexafluoroisopropanol (HFIP) solution. Make sure the silk fibers are entirely immersed in HFIP solution and maintain the weight ratio of silk to HFIP at 1:30. (*see* Fig. 3 and **Note 3**).

2. Incubate the silk fiber–HFIP mixture at 60 °C for 24 h. Of note, after this step, the silk fiber–HFIP mixture should form a pulp like formation and be split to microfibrils with diameters of 5–50μm and contour length of 50–500μm.

3. Place the silk microfibrils into distilled water with a weight ratio of 1:200, silk to water. Shake the mixture for 1 min and remove the large undispersed microfibrils.

4. <Critical Step>Sonicate the silk fiber suspension at 120μm amplitude, 20 kHz in frequency for 1 h. Use an ice–water bath to avoid the overheating of the mixture (*see* **Note 4**).

5. Centrifuge the mixture at $13,000 \times g$ to remove the precipitates. The transparent silk nanofibrils are obtained in the supernatant, and it will stable for over several months.

3.4 Urea–Guanidine Hydrochloride Dissolution

1. Mix 120 g urea and 95.5 g guanidine hydrochloride (GuHCl) in a 500 mL glass beaker with a molar ratio of 2:1, urea to GuHCl, and incubate the mixture at 90 °C until completely dissolved (*see* Fig. 4 and **Note 5**).

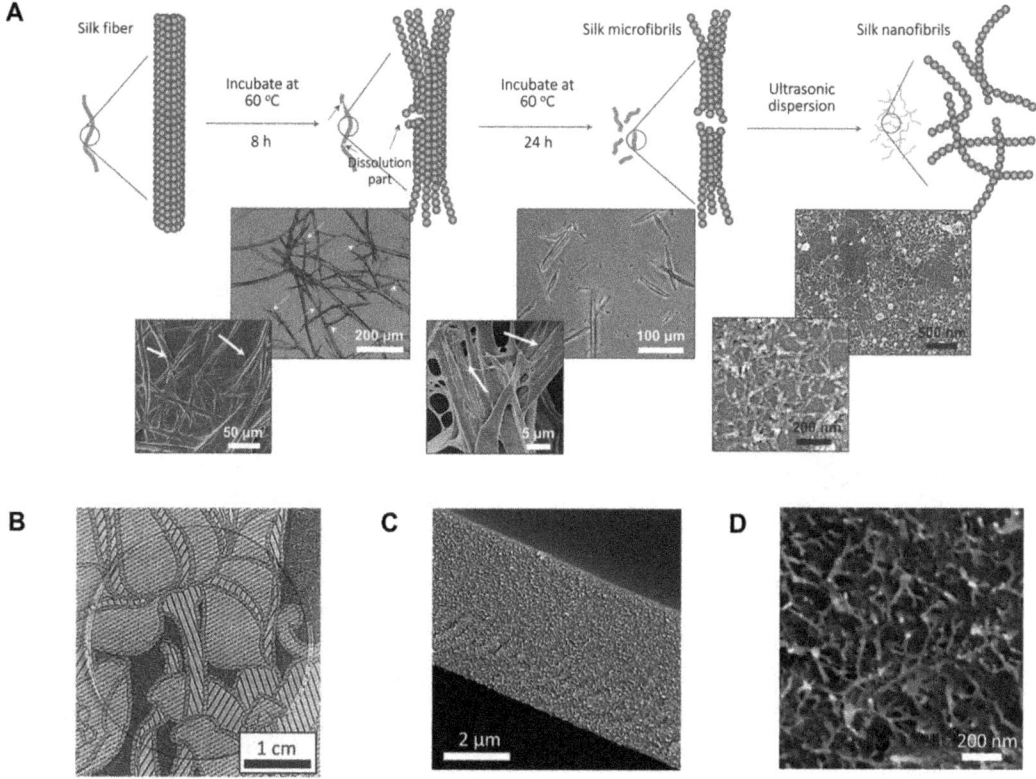

Fig. 3 HFIP liquid exfoliation process of silkworm (*Bombyx mori*) silk fiber. (**a**) Schematic illustration (top row) and photographs (middle and bottom rows) showing the exfoliation steps of silk fibers. Silk microfibers (the first four images in middle and bottom row) and silk nanofibers (the last two images in middle and bottom row) were observed in the representative optical microscopy and SEM images of resultant products in each process. The white arrows indicated the dissolution parts of silk fibers. (**b**) A fully transparency silk nanofiber membranes laid on a colorful cloth under visual light. (**c, d**) The cross-sectional SEM images of silk nanofiber membrane under (**c**) low and (**d**) high magnification. (Adapted by permission from WILEY-VCH Verlag GmbH & Co. KGaA, Weinheim [23], Copyright 2016)

2. Immerse 1 g degummed silk fibers into 100 g urea–GuHCl solution with a weight ratio of 1:100. Then, incubate the mixture at 90 °C with continuously magnetic stirring for 20 h. Of note, the resultant silk fibers will present a gel-like formation after this step.

3. Wash the gel-like silk fibers with distilled water to remove the urea–GuHCl solvent.

4. <Critical Step>Place 1 g of washed silk fiber into 500 mL of distilled water. Sonicate the silk fiber suspension at 40 W, 40 kHz in frequency for 4 h, and using an ice–water bath to keep the temperature at room temperature.

5. Centrifuge the mixture at 13,000 × *g*, and filter the supernatant with a nylon filtration.

6. membrane (pore size: 0.2μm) at 0.1 bar to obtain the silk nanofibril dispersion.

Fig. 4 Urea–guanidine hydrochloride exfoliation of silkworm (*Bombyx mori*) silk fibers. (**a**) Schematic illustration of the exfoliation process of silk fibers. (**b**) Silk nanofiber membranes in different shapes. (**c**) SEM image of the resultant silk nanofibers after urea–guanidine hydrochloride treating at 90 °C. (**d**) SEM image of the rod-like silk nanofibers after urea–guanidine hydrochloride treating at 130 °C. (Adapted by permission from The Royal Society of Chemistry [25], Copyright 2018)

3.5 Sodium Hypochlorite Partial Dissolution

1. <Critical Step>Place 1 g degummed silk fibers into 10 mL sodium chlorate (NaClO, available chlorine: 5%) with a silk fiber to NaClO weight ratio of 1:1. Of note, squeeze the silk fibers for several times by using a tweezers to make sure the NaClO thoroughly infiltrated with silk fibers. Place the mixture in a fume hood for at least 1 h (*see* **Notes 6** and **7**).

2. Homogenize the silk fiber–NaClO solution at 8000 rpm/min for 3 min to achieve pulp-like silk microfibril mixtures.

3. Centrifuge the mixture at 9000 × *g* and wash the silk microfibrils by using distilled water for at least three times to remove residual saline ions.

4. <Critical Step>Place 1 g of silk microfibrils in a 1 L glass beaker with 500 mL distilled water. Sonicate the silk microfibrils at 36 μm amplitude, 20 kHz in frequency for 20 min. Use an ice–water bath to avoid the overheating of the mixture (*see* Fig. 5 and **Note 8**).

5. Centrifuge the mixture at 13,000 × *g* to obtain silk nanofibril aqueous suspensions.

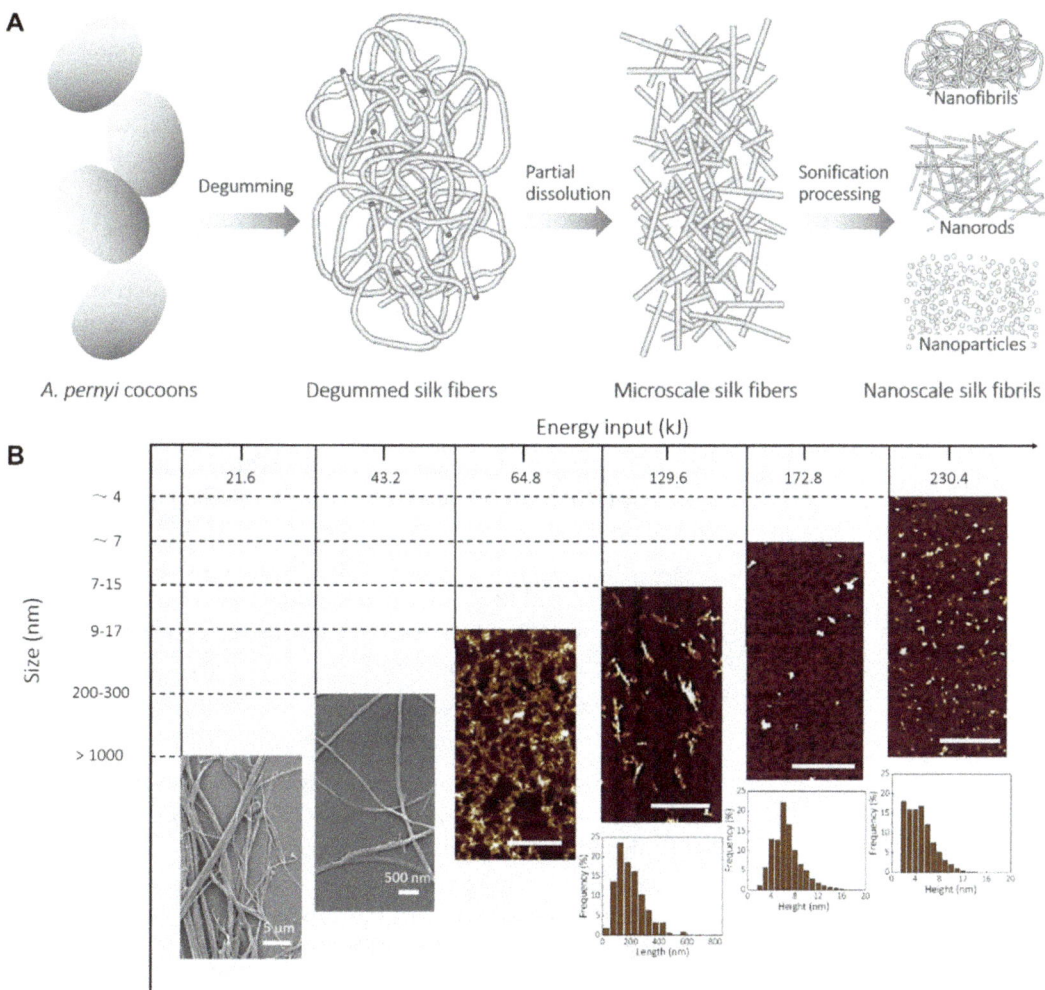

Fig. 5 Sodium hypochlorite partial dissolution of silkworm (*Antheraea pernyi*) silk fibers. (**a**) The downsizing strategy to extract *A. pernyi* silk fibers into nanofibers, nanorods, and nanoparticles. (**b**) The map of energy input versus morphologies of *A. pernyi* silk nanostructures. The bottom statistical columns of each AFM image indicate the average diameter of *A. pernyi* silk nanorods and nanoparticles, respectively. All scale bars are 500 nm unless specifically labeled. (Adapted by permission from WILEY-VCH Verlag GmbH & Co. KGaA, Weinheim [26], Copyright 2018)

4   Notes

1. The sonication process is a technique to physically disassemble native fibers with the capability of processing a variety of natural fibers such as type I collagen, chitin, cellulose, as well as spider and silkworm (*B. mori*) silk fibers. The diameters of resulted nanofibrils are ranged from 25 to 120 nm. Although the

intensive sonication treatment costs a very high energy input, the resultant nanofibrils appeared as mats or aggregates and lacked processability.

2. The salt–formic acid dissolution process generates silk nanofibrils that mixing with the salt–formic acid solvents. The length and diameter of the as-prepared nanofibrils are sensitive to the prior degumming process of silk fibers. For example, by using 0.005 w/v% Na_2CO_3 solution to degum native silk at 100 °C for 1 h will generate the nanofibrils with tens of micrometers in length and 150–200 nm in diameter. These silk nanofibrils mixtures can directly dry to films, as well as apply to spin artificial fibers and used as 3D printing ink.

3. The extracted silk nanofibrils have a diameter of 20 ± 5 nm and a contour length in the range of 300–500 nm. This technique will adapt to both *B. mori* and *Antheraea pernyi* (*A. pernyi*) silks. The liquid exfoliating process gives a yield of silk nanofibrils at around 10% with the sonication parameters of 120μm amplitude, 20 kHz in frequency, and 1 h of time. These silk nanofibrils aqueous dispersion are flexible to generate materials such as ultrafiltration films and optical nanodevices.

4. Applying harsh conditions of sonication will degrade the silk nanofibrils to nanoparticles with the same diameters.

5. Urea–guanidine hydrochloride solution is a kind of green and chemically tailorable solvents that decrease the strength of hydrophobic interactions in silk fibers. The diameters and contour lengths of resultant silk nanofibrils are ranged from 20 to 100 nm and from 0.3 to 10μm, respectively. These silk nanofibrils could be processed into filtration membranes as well as a sustainable separator for supercapacitors.

6. NaClO partial dissolution is an environmentally friendly and scalable technique to isolate native silk fiber (adapt to *B. mori* and *A. pernyi* silks) into microfibrils, nanofibrils, nanorods, and nanoparticles. Of note, the isolated silk nanofibrils have a high modulus of 9.8 ± 3.3 GPa, where almost the same as that of natural silk fibers. The nanofibrils will aggregate into pulps (pH < 5) or redispersed to a uniform dispersion (pH > 7) at different pH values. On the basis of the properties of these silk nanofibrils in tunable sizes, high modulus, as well as excellent redispersibility, they are applied in electronic and environmental fields that including water treatment, recycling organic solvent, paper sensors, and nanofertilizers.

7. If needed, the degummed silk fibers can be pretreated by immersing into formic acid (88 w/v%) solution with a weight ratio of 1:20. This step will disassemble silk fibers, which increases their accessibility with solvents.

8. The morphologies of isolated silk nanostructures are directly related to the amplitude and treating time, that is the energy input of the sonication process. For 1 g of dry silk fibers, increase the energy input of sonication (through the increase of processing time) from 21.6, to 43.2, and to 64.8 kJ, the diameters of resultant silk micro/nanofibrils will decrease from 1–2μm to 200–300 nm and to 13 ± 4 nm, respectively. When the energy input rises to 230.4 kJ, the silk fibers will degrade into nanoparticles with a diameter of 2–7 nm.

References

1. Wang X, Luo H, Zhang R (2018) Innate immune responses in the Chinese oak silkworm, antheraea pernyi. Dev Comp Immunol 83:22–33

2. Guo CC, Li CM, Vu HV, Hanna P, Lechtig A, Qiu YM, Mu X, Ling SJ, Nazarian A, Lin SJ, Kaplan DL (2020) Thermoplastic moulding of regenerated silk. Nat Mater 19:102–108

3. Ling SJ, Kaplan DL, Buehler MJ (2018) Nanofibrils in nature and materials engineering. Nat Rev Mater 3:18016

4. DeBari MK, Abbott RD (2019) Microscopic considerations for optimizing silk biomaterials. Wiley Interdiscip Rev Nanomed Nanobiotechnol 11:e1534

5. Fratzl P, Weinkamer R (2007) Nature's hierarchical materials. Prog Mater Sci 52:1263–1334

6. Wang QJ, Schniepp HC (2018) Strength of recluse spider's silk originates from nanofibrils. ACS Macro Lett 7:1364–1370

7. Zhang WW, Ye C, Zheng K, Zhong JJ, Tang YZ, Fan YM, Buehler MJ, Ling SJ, Kaplan DL (2018) Tensan silk-inspired hierarchical fibers for smart textile applications. ACS Nano 12:6968–6977

8. Vollrath F, Knight DP (2001) Liquid crystalline spinning of spider silk. Nature 410:541–548

9. Jin H-J, Kaplan DL (2003) Mechanism of silk processing in insects and spiders. Nature 424:1057–1061

10. Ling SJ, Chen WS, Fan YM, Zheng K, Jin K, Yu HP, Buehler MJ, Kaplan DL (2018) Biopolymer nanofibrils: structure, modeling, preparation, and applications. Prog Polym Sci 85:1–56

11. Xu GQ, Gong L, Yang Z, Liu XY (2014) What makes spider silk fibers so strong? From molecular-crystallite network to hierarchical network structures. Soft Matter 10:2116–2123

12. Peng ZC, Yang X, Liu C, Dong ZM, Wang F, Wang X, Hu WB, Zhang X, Zhao P, Xia QY (2019) Structural and mechanical properties of silk from different instars of *bombyx mori*. Biomacromolecules 20:1203–1216

13. Zhao HP, Feng XQ, Gao HJ (2007) Ultrasonic technique for extracting nanofibers from nature materials. Appl Phys Lett 90:073112

14. Nechyporchuk O, Belgacem MN, Bras J (2016) Production of cellulose nanofibrils: a review of recent advances. Ind Crop Prod 93:2–25

15. Fan YM, Saito TK, Isogai A (2008) Preparation of chitin nanofibers from squid pen β-chitin by simple mechanical treatment under acid conditions. Biomacromolecules 9:1919–1923

16. Schmidt MM, Dornelles RCP, Mello RO, Kubota EH, Mazutti MA, Kempka AP, Demiate IM (2016) Collagen extraction process. Int Food Res J 23:913–922

17. Zhang F, Lu Q, Ming JF, Dou H, Liu Z, Zuo BQ, Qin MD, Li F, Kaplan DL, Zhang XG (2014) Silk dissolution and regeneration at the nanofibril scale. J Mater Chem B 2:3879–3885

18. Zhang F, Lu Q, Yue XX, Zuo BQ, Qin MD, Li F, Kaplan DL, Zhang XG (2015) Regeneration of high-quality silk fibroin fiber by wet spinning from CaCl$_2$-formic acid solvent. Acta Biomater 12:139–145

19. Mandal BB, Grinberg A, Gil ES, Panilaitis B, Kaplan DL (2012) High-strength silk protein

scaffolds for bone repair. Proc Natl Acad Sci U S A 109:7699–7704

20. Ling SJ, Qin Z, Li CM, Huang WW, Kaplan DL, Buehler MJ (2017) Polymorphic regenerated silk fibers assembled through bioinspired spinning. Nat Commun 8:1387

21. Ling SJ, Zhang Q, Kaplan DL, Omenetto FG, Buehler MJ, Qin Z (2016) Printing of stretchable silk membranes for strain measurements. Lab Chip 16:2459–2466

22. Ling SJ, Wang Q, Zhang D, Zhang YY, Mu X, Kaplan DL, Buehler MJ (2018) Integration of stiff graphene and tough silk for the design and fabrication of versatile electronic materials. Adv Funct Mater 28:1705291

23. Ling SJ, Li CM, Jin K, Kaplan DL, Buehler MJ (2016) Liquid exfoliated natural silk nanofibrils: applications in optical and electrical devices. Adv Mater 28:7783–7790

24. Ling SJ, Jin K, Kaplan DL, Buehler MJ (2016) Ultrathin free-standing *bombyx mori* silk nanofibril membranes. Nano Lett 16:3795–3800

25. Tan XX, Zhao WC, Mu TC (2018) Controllable exfoliation of natural silk fibers into nanofibrils by protein denaturant deep eutectic solvent: nanofibrous strategy for multifunctional membranes. Green Chem 20:3625–3633

26. Zheng K, Zhong JJ, Qi ZM, Ling SJ, Kaplan DL (2018) Isolation of silk mesostructures for electronic and environmental applications. Adv Funct Mater 28:1806380

27. Zheng K, Hu YL, Zhang WW, Yu J, Ling SJ, Fan YM (2019) Oxidizing and nano-dispersing the natural silk fibers. Nanoscale Res Lett 14:250

Chapter 14

Gelation Methods to Assemble Fibrous Proteins

Ning Fan and Ke Zheng

Abstract

Gelation is an efficient way to fabricate fibrous protein materials. Briefly, it is an aggregation process where protein molecules assembly from a random structure into an organized structure such as nanofibrillar networks. According to their mechanisms, the fibrous proteins gelation can be classified into physical gelation and chemical gelation. The physical gelation is formed by the conformational transformation of fibroin proteins, which can be triggered by temperature, concentration, pH, or shear force. On the other hand, the chemical gelation is to cross-link fibrous proteins through chemical and/or enzymatic reactions. In this chapter, we summarize the protocols for preparing fibrous protein hydrogels, including both physical and chemical methods. The mechanisms of these gelation methods are also highlighted.

Key words Fibrous proteins, Physical gelation, Chemical gelation

1 Introduction

Fibrous proteins are ubiquitous building blocks in engineering biomaterials [1–5]. These materials include silk proteins in silk-worm and spider silks, amyloid proteins in Alzheimer associated disease, collagens in tissues, and together with their recombinant peptides/proteins [6–11]. In the laboratory, these fibrous proteins have been widely utilized to construct a variety of biomaterials with tunable functional properties, such as water purification, health-care, and artificial skins [12–17]. In particular, fibrous protein hydrogels have received extensive attention in the fields of drug delivery and release, tissue engineering, software robots, and so on [2, 18–21] Accordingly, how to prepare fibrous protein hydrogels is an important aspect for fibroin protein material fabrication.

Gelation of fibrous proteins can be achieved either physically or chemically. Physical gelation is the result of an aggregation process, where the conformation transition of peptides/proteins occurs [22–26]. For example, the essence of the physical sol-gel transition of silk fibroin is that the conformation of silk fibroin changes from random coil to β-sheet [27]. The β-sheet formed serves as

Shengjie Ling (ed.), *Fibrous Proteins: Design, Synthesis, and Assembly*, Methods in Molecular Biology, vol. 2347, https://doi.org/10.1007/978-1-0716-1574-4_14, © Springer Science+Business Media, LLC, part of Springer Nature 2021

cross-linker to connect the random coil to form a rubber-like network structure. Gelation of type I collagen is another example, which is a process through which collagen partially recovers its triple-helix structure (renaturation), and further in turn forms native-type collagen fibrils [28, 29]. In general, the physical gelation of fibrous proteins can be induced by a series of physical stimuli, such as heating, concentration, pH, and shear force [30–34]. The modalities and kinetics of the conformation transition of fibrous proteins play an important role in channeling the structure and physical properties of the resultant hydrogels.

Chemical gelation of fibroin proteins either can be achieved through the directly cross-linking of the molecules [35, 37] or cross-linking of physical hydrogels through covalent bonds and ionic bonds [23, 37–39]. These chemical reactions can provide chemical handles for the attachment of growth factors, cell-binding domains, and other polymers to the fibrous protein hydrogels [40, 41]. However, the approaches of chemical gelation are varied due to the complex constitute of amino acids in proteins, as well as the diversity of chemical reactions. For example, chemical reactions such as carbodiimide coupling have been well established. Most fibrous proteins contain the primary amines and carboxylic acids in their amino acid side chains (such as aspartic, glutamic acids, and lysine residues) and/or at the C- and N-termini can participate in this reaction [42, 43].

In this chapter, we first provide a critical summary of approaches to prepare fibrous protein hydrogels. Then, we particularly demonstrate the physical and chemical gelation protocols for fibrous proteins of silk fibroin, amyloid β-lactoglobulin, and collagen.

2   Materials

1. 0.3 M Hydrochloric acid (HCl): place 12.44 mL HCl ($c = 12.06$ mol/L) into a 500 mL volumetric flask, fill the flask with deionized water with a final volume of 500 mL.

2. 0.1 M Sodium acetate (NaHAc): weigh 0.82 g NaHAc into a 100 mL volumetric flask, fill the flask with deionized water with a final volume of 100 mL.

3. 20 w/w% Acetate acid (HAc): place 20.4 mL CH_3COOH (98 w/v%) into a 100 mL volumetric flask, fill the flask with deionized water with a final volume of 100 mL.

4. 0.1 M Potassium hydroxide (KOH): weigh 0.56 g KOH into a 100 mL volumetric flask, fill the flask with deionized water with a final volume of 100 mL.

5. 0.1 M Sulfuric acid (H_2SO_4): place 5.43 mL H_2SO_4 (98 w/v%) into a 1 L volumetric flask, fill the flask with deionized water with a final volume of 1 L.

6. 2 M Phosphoric acid (H_3PO_4): place 106.7 mL H_3PO_4 (98 w/v%) into a 1 L volumetric flask, fill the flask with deionized water with a final volume of 1 L.

7. 2 M Lithium chloride (LiCl): weigh 8.49 g LiCl into a 100 mL volumetric flask, fill the flask with deionized water with a final volume of 100 mL.

8. 2 M Sodium chloride (NaCl): weigh 11.69 g NaCl into a 100 mL volumetric flask, fill the flask with deionized water with a final volume of 100 mL.

9. 2 M Potassium chloride (KCl): weigh 14.91 g LiCl into a 100 mL volumetric flask, fill the flask with deionized water with a final volume of 100 mL.

10. 0.52 M Acetate acid (HAc): place 3.18 mL CH_3COOH (98 w/v%) into a 100 mL volumetric flask, fill the flask with deionized water with a final volume of 100 mL.

11. 0.05 M 2-morpholinoethane sulfonic acid (MES): weigh 1.07 g 2-Morpholinoethanesulfonic acid monohydrate into a 100 mL volumetric flask, fill the flask with deionized water with a final volume of 100 mL.

12. 0.1 M Disodium hydrogen phosphate (Na_2HPO_4): weigh 1.42 g 2-Morpholinoethanesulfonic acid monohydrate into a 100 mL volumetric flask, fill the flask with deionized water with a final volume of 100 mL.

13. 20 w/v% Calcium chloride ($CaCl_2$): dissolve 20 g $CaCl_2$ in 100 mL of deionized water.

14. Phosphate buffer saline (PBS): 137 mM NaCl, 2.7 mM KCl, 8 mM Na_2HPO_4, and 2 mM KH_2PO_4, pH 7.4.

15. 4 M Sodium chloride (NaCl): weigh 23.38 g NaCl into a 100 mL volumetric flask, fill the flask with deionized water with a final volume of 100 mL.

3 Methods

3.1 Preparation of Fibrous Protein Solution

3.1.1 Silk Fibroin Extraction

1. Cut silk cocoons into dime-sized pieces and removal of the silkworm. Weigh 5 g of cocoon pieces into a glass beaker.

2. Prepare a large glass beaker, add 2 L of distilled water, and cover with aluminum foil. Heat the glass beaker until boiling.

3. Measure 5 g of $NaHCO_3$ in a small glass beaker. Slowly add the measured $NaHCO_3$ into the water and make it completely dissolve.

4. <Critical Step>Add the cocoon pieces in the boiling water and boil it for 30 min. Occasionally, poke the fibers with a spatula to promote good dispersion of fibroin (Fig. 1).

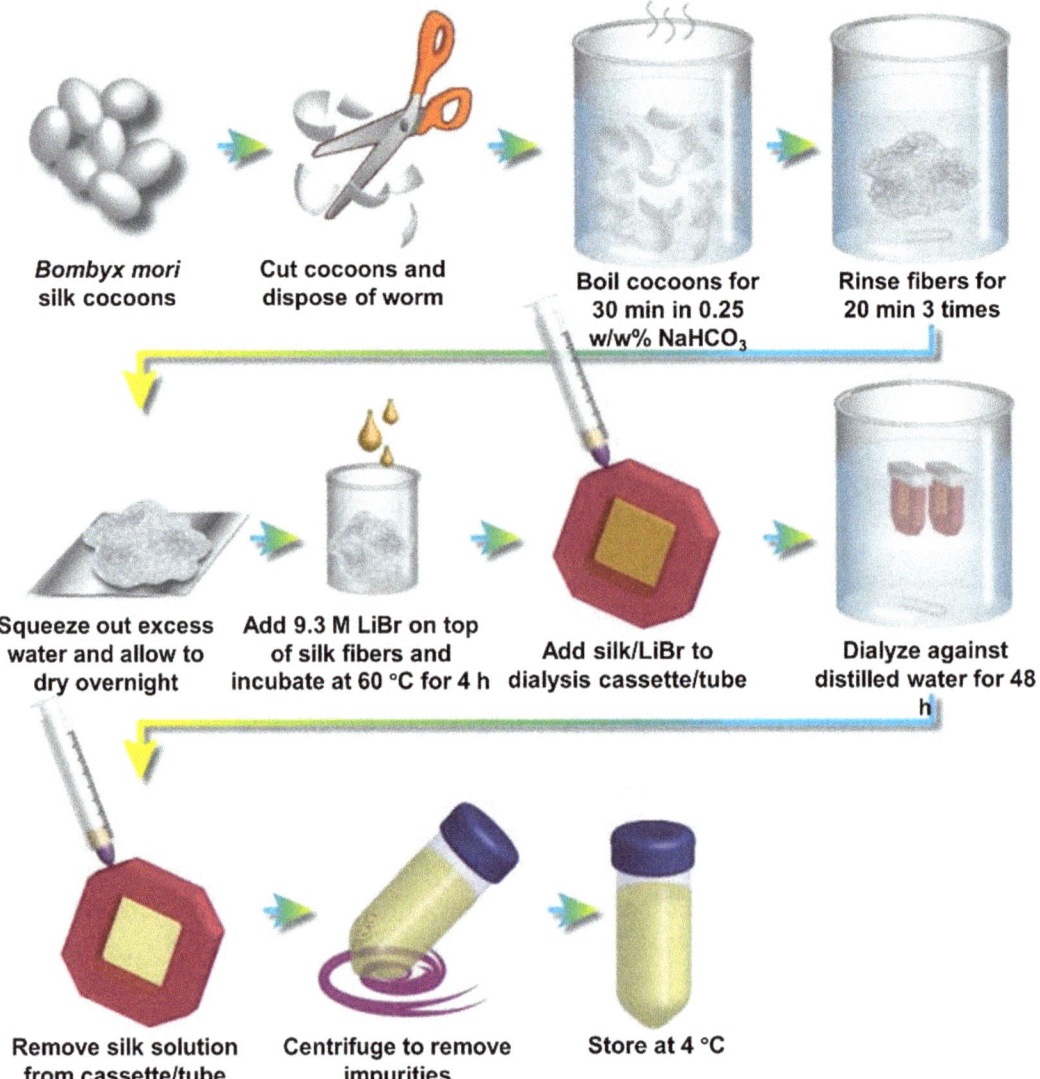

**Bombyx mori
silk cocoons**

**Cut cocoons and
dispose of worm**

**Boil cocoons for
30 min in 0.25
w/w% NaHCO₃**

**Rinse fibers for
20 min 3 times**

**Squeeze out excess
water and allow to
dry overnight**

**Add 9.3 M LiBr on top
of silk fibers and
incubate at 60 °C for 4 h**

**Add silk/LiBr to
dialysis cassette/tube**

**Dialyze against
distilled water for 48
h**

**Remove silk solution
from cassette/tube**

**Centrifuge to remove
impurities**

Store at 4 °C

Fig. 1 Schematic illustration of the silk fibroin extraction strategy. Going from the *Bombyx mori* cocoons to the silk fibroin aqueous solution. (Adapted by permission from Springer Nature [20], Copyright 2011)

5. Separate the silk fibers from the mixture and cool by rinsing in distilled water. Remove excess water out of the fibroin by squeezing and discard the NaHCO₃ solution.

6. Rinse the boiled silk fibers in 2 L distilled water with continuous stirring for 20 min.

7. Repeat **steps 2–6** with the rinsed silk fibers. This step should be processed twice or thrice to ensure the complete removal of sericin. Meanwhile, indicate the total time of boiling to promote reproducibility. Increasing the boiling time will degrade the fibroin.

8. After the removal of sericin, put the silk fibroin on a clean piece of aluminum foil and allow it to dry in a fume hood overnight.

3.1.2 Dissolve Silk Fibroin

Typically, the dissolved silk fibroin solution needed to obtain a 20 w/v% solution which based on the amount of dried silk fibroin. Therefore, the total volume of 9.3 M LiBr is four times that of the dried silk fibroin. That is a ratio of 4:1, LiBr to silk. For example, here we list the protocols to dissolve 5 g of dried silk fibroin.

1. Put 5 g of silk fibroin onto the bottom of a 50 mL glass beaker. Pack silk fibroin tightly and add 20 mL of LiBr solution on top.

2. <Critical Step>Incubate the silk fibroin in an oven at 60 °C for 4 h to let silk fibroin dissolve. The completely dissolved silk fibroin will be highly viscous but appears transparent and amber in color. Black residuals from the silkworm should be removed. The employed 9.3 M LiBr solution should adjust to 50–60 °C before pouring onto the silk fibroin. A lower temperature may case the silk fibroin could not completely dissolve within an appropriate time, while the higher temperature will degrade the silk fibroin (Fig. 1).

3. Hydrate dialysis tube (molecular weight cutoff of 3.5 kDa) in water for 2–3 min. Decant the dissolved silk fibroin (approximately 25 mL) into the dialysis tube. Here, leave enough free volume (3–4 times to the amount of silk fibroin solution) to allow distilled water to exchange into the dialysis tube.

4. Dialysis against 2 L of distilled water per 20 mL of dissolved silk fibroin solution. Use a magnetic stir bar continually stir distilled water to ensure mixing. Change distilled water at 1, 4, 12, 24, 36, and 48 h (at least 6 times within 48 h).

5. Remove the silk fibroin aqueous solution from the dialysis tube and collect it with a 50 mL centrifuge tube.

6. Centrifuge the silk fibroin aqueous solution at $10,000 \times g$, 4 °C for 20 min to remove impurities.

7. Carefully transfer the supernate silk fibroin solution into another centrifuge tube.

8. Repeat **steps 6** and **7**.

9. Measure the weight of a small glass weight disk, and then add 1 mL of the silk solution to the small glass weight disk. Incubate the weight disk at 70 °C to allow the silk fibroin to dry. Once the silk fibroin is dry, measure the weight of the dry silk fibroin and divide it by 1 mL. This will yield the concentration of achieved silk solution.

10. The silk fibroin aqueous solution can either be used directly for gelation or it can be dilute/concentrate to the desired concentration. The silk fibroin aqueous solution should carefully store at 4 °C for no more than 3 days.

*3.1.3 Purification
of β-Lactoglobulin*

1. Dissolve β-Lactoglobulin in deionized water to obtain a solution of 10 w/v%.

2. Adjust the pH of the β-Lactoglobulin solution to 4.6 by HCl.

3. Centrifuge the solution at $15{,}000 \times g$ at 20 °C for 15 min.

4. Adjust the supernatant to pH 2 by HCl and filter through a 0.22 μm Millipore filter membrane.

5. Dialyze the purified protein solution against deionized water by using a dialysis tube (molecular weight cutoff of 6–8 kDa).

6. Adjust the protein solution to pH 2 by HCl and then lyophilized.

3.2 Physical Gelation of Fibrous Proteins

3.2.1 Thermally Inducing

1. Dissolve the β-lactoglobulin powder in deionized water at 25 °C to obtain a concentration of 8 w/v% (or silk fibroin solution with a concentration of 4 w/v%). A lower initial concentration of proteins is usually resulted in the form of jellies, slurries, or pastes and had poor mechanical properties of hydrogels. It might root in the inhomogeneous network formed by the thermally induced assembly of fibrous proteins (*see* **Note 1**).

2. Centrifuged the β-lactoglobulin (silk fibroin) solution at $10{,}800 \times g$ over a period of 1 h at 4 °C, then adjust to the pH of 2 by HCl (pH 7 for silk fibroin).

3. Filter the solution through a 0.22 μm Millipore filter for sterilization.

4. <Critical Step>Incubate the β-lactoglobulin solution in an oven at 80 °C for 10 h (silk fibroin solution at 60 °C for 7 days). To avoid evaporation, the protein solution should store in a hermetically sealed glass tube (*see* Fig. 2 and **Note 2**).

5. Remove the glass tube from the oven and immediately immerse in ice–water mixtures to quench the gelation process.

3.2.2 Alcohol Inducing

1. Dilute silk fibroin aqueous solution to 4 w/v% by using deionized water with gently stir (*see* **Note 3**).

2. <Critical Step>Add appropriate ethanol into the silk solution to obtain the mixed solution of 2 w/v% silk fibroin in 50 v/v% ethanol–water. The volume of ethanol should be considered while making the solution. We suggest dripping ethanol into silk solution slowly with a glass rod dipped. The ethanol can be a defoamer to avoid air bubbles in the solution.

3. Store the silk fibroin–ethanol–water solution in a hermetically sealed glass tube and incubate at 25 °C for 2 h.

4. Quench the gelation of silk fibroin and remove the residual ethanol by removing it into 1 L deionized water.

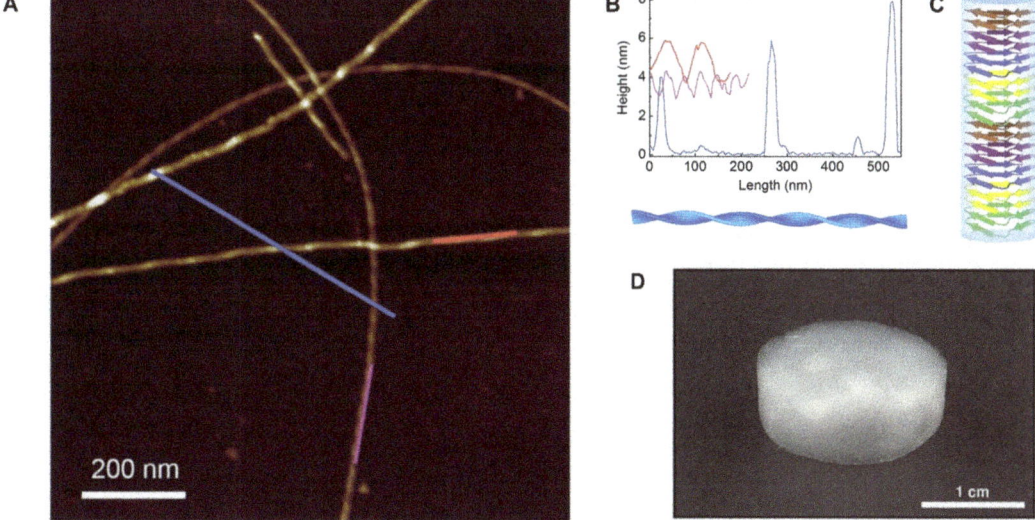

Fig. 2 Morphology and nanostructures of fibrous β-lactoglobulin. (**a**) AFM image of the β-lactoglobulin amyloid fibrils in hydrogels. (**b**) The height profiles of the fibrous β-lactoglobulin. Longitudinal (red and pink curve) and transversal (blue curve) height profiles as indicated by the corresponding traces in (**a**). (**c**) Schematic orientation of β-sheet crystals in amyloid fibrils. (**d**) Photograph of an amyloid aerogel. (Adapted by permission from WILEY-VCH Verlag GmbH & Co. KGaA, Weinheim [24] and [8], Copyright 2014 and 2016)

3.2.3 Squeeze/Shear Force Inducing

1. Draw up 5 mL concentrated silk fibroin solution (approximately 15 w/v%) into a syringe. Here, using a small needle (like 14 gauge) and draw up slowly to avoid including air bubbles (*see* **Note 4**).

2. Place the mandrel and start the mandrel rotation system [44].

3. <Critical Step>Attach a 30-gauge needle (replace a 27 gauge needle if needed) onto the syringe and squeeze the silk fibroin solution onto the mandrel. Keep a generous amount of pressure to make sure the squeezed silk fibroin be uniform, and no beads or discontinuities appeared (*see* **Note 5**).

4. Once the silk fibroin was collected onto the mandrel, drop methanol/ethanol on the top of silk fibroin to further induce β-sheet formation.

5. After the methanol/ethanol has completely dried, immerse the silk fibroin hydrogel into a solution of soap water.

6. Pull off the mandrel tube gently and place the silk fibroin hydrogel into deionized water for 1 h to remove residual soap.

3.2.4 Sonication

1. Adjust fibrous protein (such as silk fibroin) to 4 w/v% by using deionized water.

2. Add 3 mL solution into a 10 mL glass tube (*see* Fig. 3 and **Note 6**).

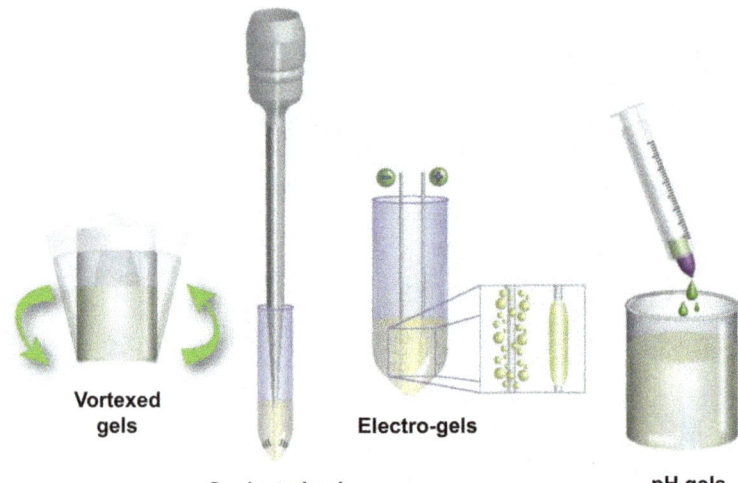

Vortexed gels

Electro-gels

Sonicated gels

pH gels

Fig. 3 Schematic illustration of physical gelation process of fibrous proteins involving vortexing, sonication, electric current, and pH inducing

3. Immerse the sonicating probe into the solution, but not to touch the probe on the sides or the bottom of the glass tube.

4. <Critical Step>Sonicate the solution at 30% amplitude (30 W) for 30 s. Here, use a water bath to avoid the overheating of the solution. Depend on the kinds of protein, the protein concentration can be varied from 1 to 20 w/v%, and sonication time can be changed from 5 s to 1 min at 10–50% amplitude.

5. Incubate the sonicated solutions at 37 °C for gelation.

3.2.5 Vortexing

1. Adjust fibrous protein to 4 w/v% by using deionized water (*see* Fig. 3 and **Note 7**).

2. Place 1.5 mL of protein solution into a tightly sealed glass vial.

3. Secure the vial in an upright position on the vortexer with duct tape.

4. <Critical Step>Vortex the glass vial for 10 min at 4500 × g. Larger volumes and higher concentrations may need more vortex time and higher speed.

5. Incubate the vortexed solution at 37 °C for 24 h to allow for gelation.

3.2.6 Electric Current

1. Place 5 mL of 8 w/v% fibrous protein solution to a 50 mL conical tube. Buffer solution may need for amyloid fibrous protein. For example, the β-insulin solution should be prepared in a buffer solution containing 0.1 M NaAc and 20 v/v% HAc with a pH of 2.74. The conductivity of the β-insulin solution is 5.7 S/m.

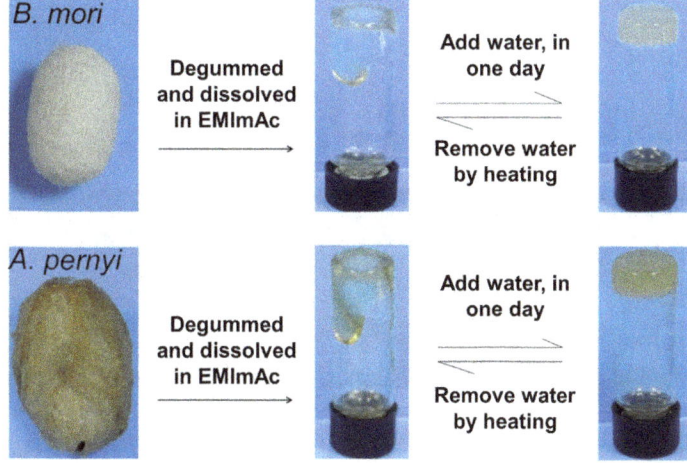

Fig. 4 Schematic representation of the dissolution and ionic liquid/water induced gelation of *B. mori* and *A. pernyi* silk fibroin. (Adapted by permission from American Chemical Society [22], Copyright 2015)

2. Immerse two parallel platinum wire (gold-coated glass coverslips) electrodes into the protein solution.

3. <Critical Step>Apply approximately 25 V_{DC} to the platinum electrodes over a few minutes (*see* Fig. 3 and **Note 8**).

4. Turn off the power supply and collect the hydrogels on the platinum wire (gold-coated glass coverslips) electrode.

3.2.7 pH Inducing

1. Add 1.8 mL of 8 w/v% silk fibroin into a glass vial (*see* Fig. 3 and **Note 9**).

2. <Critical Step>Adjust the pH by 0.3 M HCl (pH 2–5). Keep the volumetric ratio of acid to silk is 1:10. For 1.5 mL of 8 w/v% silk fibroin, 0.2 mL of 0.3 M HCl is required.

3. Seal the glass vial and stir it by inverting.

3.2.8 Ionic Liquid/Water Inducing

1. Dissolve 2 g silk fibroin in 8 g of 1-ethyl-3-methylimidazolium acetate (EMImAc) at 100 °C for at least 4 h to obtain a 20 w/w% silk fibroin–EMImAc solution (*see* Fig. 4 and **Note 10**).

2. Place 3 g of silk fibroin–EMImAc solution together with 3 mL of deionized water in a centrifuge tube. If needed, a conductive component can be added into the mixture to obtain conductive hydrogels (e.g., 0.1 M KOH, 0.1 M H_2SO_4, 2 M H_3PO_4, 2 M LiCl, 2 M NaCl, and 2 M KCl).

3. Centrifuge the mixture at $10,800 \times g$ for 10 min to remove bubbles.

4. <Critical Step>Remove the centrifuged mixture into a sealed glass vial and then incubate at 25 °C for 14 days to complete gelation.

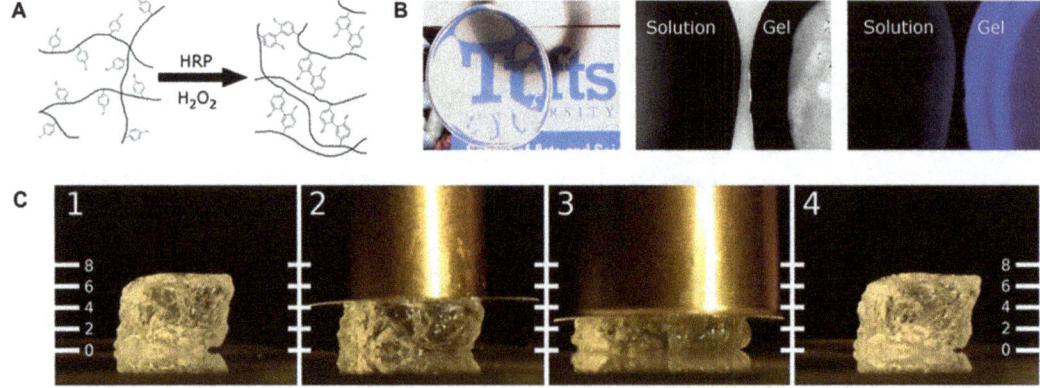

Fig. 5 Horseradish peroxidase cross-linking of silk fibroin hydrogels. (**a**) Schematic representation of the cross-linking of tyrosine residues on silk molecules. (**b**) Images of the resultant silk hydrogels with optically transparency and blue fluorescence when irradiated with UV light. (**c**) Photographs of cross-linked silk hydrogel undergoing ~50% compression, under 50 g (2) and up to 100 g (3) brass weights and showing complete recovery after removal (4). Scale units are in millimeters. (Adapted by permission from WILEY-VCH Verlag GmbH & Co. KGaA, Weinheim [42], Copyright 2014)

3.3 Chemical Gelation of Fibrous Proteins (See Note 11)

3.3.1 Horseradish Peroxidase Cross-Link

1. Dissolve HRP powder in deionized water to form a stock solution with a concentration of 1000 U/mL (*see* Fig. 5 and **Note 12**).

2. Dilute silk fibroin solution to a concentration of 5 w/v%.

3. Add 0.1 mL of HRP solution into 10 mL of 5 w/v% silk fibroin solution. For a larger volume of silk fibroin, maintain a ratio of 10 Units of HRP to 1 mL of silk solution.

4. <Critical Step>Place 100 μL of 165 mM H_2O_2 solution into 10 mL of the mixture. A final concentration of 1.65 mM H_2O_2 in the silk fibroin–HRP solution is required to initiate the gelation.

5. Remove the mixture into a glass vial and mixed by gentle pipetting.

6. Once gelation, wash the preformed hydrogels in distilled water for 12 h.

3.3.2 1-Ethyl-3-(3-Dimethylaminopropyl)-Carbodiimide Cross-Link

1. Dissolve 1 g of type I insoluble collagen in 50 mL 0.52 M HAc solution at 4 °C for 12 h (*see* Fig. 6 and **Note 13**).

2. Add 50 mL of 4 °C deionized water and filter the solution through a 10 μm Millipore filter.

3. De-aerate the collagen solution at a pressure of 0.06 mbar to remove air bubbles. Air-dry the collagen solution at 25 °C to form a thin film.

4. Wash the collagen film with 0.05 M buffer of 2-morpholinoethanesulfonic acid (MES buffer).

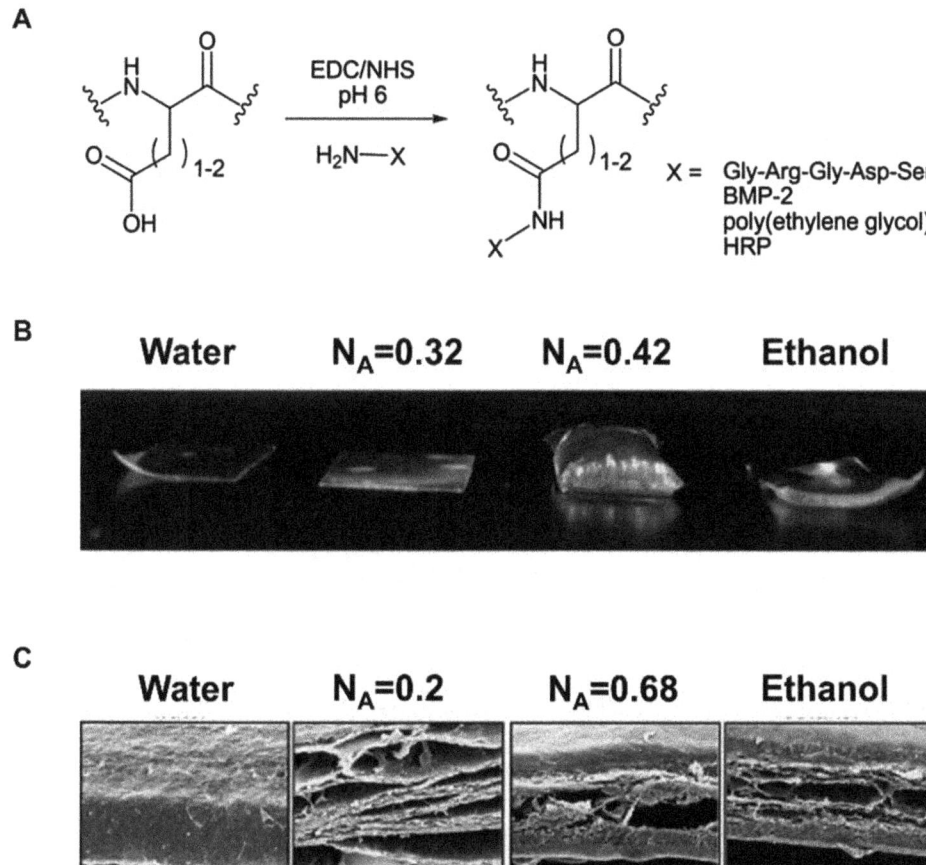

Fig. 6 1-Ethyl-3-(3-dimethylaminopropyl)-carbodiimide cross-linking of fibrous protein hydrogels. (**a**) Schematic mechanism of the carbodiimide coupling reaction on fibrous protein molecules. (**b**) Photographs of the cross-linked collagen hydrogels produced in the ethanol–water solvents. (**c**) The razor-cut surface SEM images of the cross-linked collagen hydrogels. The N_A represents ethanol mole concentration changing from 0 to 1, that is, ethanol–water ratio from 10:0 to 0:10 (v/v). (Adapted by permission from WILEY-VCH Verlag GmbH & Co. KGaA, Weinheim [36], Copyright 2008)

5. <Critical Step>Immerse the washed collagen film in a solution of EDC and N-hydroxysuccinimide (NHS) in MES buffer. To cross-link 1 g of collagen, 1.731 g EDC and 0.415 g NHS in 215 mL MES buffer was used (molar ratio of EDC–NHS–collagen-carboxylic acid groups is 7.0:2.8:1.0).

6. Gently shake the mixture for 4 h, and then wash collagen with 0.1 M Na_2HPO_4 solution for 2 h to quench the reaction.

7. Place the formed collagen hydrogel in distilled water to wash for 30 min. Repeat this step four times.

3.3.3 Metal Ions Cross-Link

1. Place 10 mL of 15 w/v silk fibroin into a 50 mL glass beaker (*see* Fig. 7 and **Note 14**).

2. <Critical Step>Add 2.5 mL of 20 w/v% $CaCl_2$ aqueous solution into the glass beaker to obtain a silk fibroin–$CaCl_2$ mixture

A

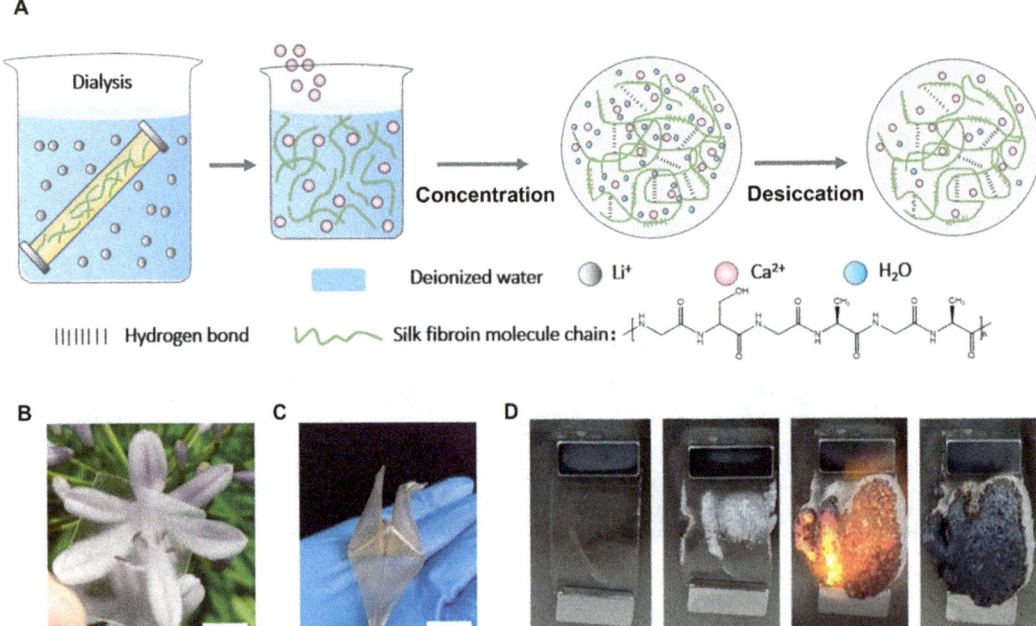

Fig. 7 Metal ions cross-linking of silk fibroin films/hydrogels. (**a**) Schematic of fabrication strategy to obtain silk fibroin/Ca^{2+} films/hydrogels. (**b, c**) Photographs of silk fibroin/Ca^{2+} films/hydrogels with (**b**) optically transparency and (**c**) highly deformability. (**d**) Snapshots of the combustion process indicating the flame resistance of silk fibroin/Ca^{2+} films/hydrogels. (Adapted by permission from American Chemical Society, Weinheim [12], Copyright 2020)

solution at the mass ratio of 75:25 (silk fibroin to $CaCl_2$). The mass ratio of silk fibroin to $CaCl_2$ can range from 85:15, 80:25, 75:25, 70:30 to 65:35.

3. Place the mixture in a fume hood at 25 °C. Let the mixture solution to cure for 72 h.

4. Remove the silk fibroin–$CaCl_2$ to a humidity environment with 60% of relative humidity. After 2 h, the formed silk fibroin–$CaCl_2$ gels will change to a hydro-state.

3.3.4   Glutaraldehyde Cross-Link

1. Dissolve 0.5 g of GA in 100 mL of phosphate buffer solution (PBS, pH = 7.4) to form a 0.5 w/v% GA in PBS solution (*see* **Note 15**).

2. Add 1 g of type I insoluble collagen in 100 mL of the GA/PBS solution.

3. <Critical Step>Incubate the mixture at 25 °C for 2 h to process the cross-link. In order to obtain mechanically enhanced hydrogels, the cross-link time can be increased up to 6 h.

4. Once the cross-link finished, rinse the hydrogels by deionized water for 30 min.

5. Wash the hydrogels twice for 30 min with 4 M NaCl and four times for 30 min with deionized water.

3.3.5 Genipin Cross-Link

1. Place a thermally inducing 8 w/v% β-lactoglobulin hydrogel into a small glass beaker. The initially employed materials could be any fibrous proteins–based hydrogels. The genipin will cross-link the hydrogels with covalent bonds and enhance the interactions between the physical networks of initial hydrogels (*see* **Note 16**).

2. <Critical Step>Cross-link the β-lactoglobulin hydrogel with a 0.5 w/v% genipin at 20 °C for 24 h. The weight ratio of genipin to protein is 1 to 5. The cross-link reaction can be carried out at 4, 20, 37 °C for 6 to 24 h. The weight ratio of genipin to protein can range from 1:10 to 1:2.5.

3. Place the cross-linked hydrogels into 1 L distilled water for 1 h to quench the reaction and remove the residual genipin.

4 Notes

1. The thermally inducing process will adapt to various of fibrous proteins, including β-sheet-based fibrous such as amyloid, silkworm/spider silk, synthetic peptides, recombinant silk-like protein, and other proteins such as collagen and soy protein. The critical of these protocols are the gelation concentration, pH, and temperature, which were minor differ based on the proteins. Here, we specifically established the protocols for two kinds of β-sheet-based fibrous gelation: amyloid β-lactoglobulin and silk fibroin.

2. Generally, the increase of ionic strength (pH, NaCl, etc.), temperature and initial concentration will promote the gelation of β-sheet-based fibrous proteins. For example, 8 w/v% β-lactoglobulin will get gelation in 5 h at 90 °C, as well as 8 w/v% silk fibroin, will get gelation in 5 days at 80 °C. Meanwhile, the shapes of fibrous protein aggregates are varying by the pH and concentration of the protein soliton before thermal treatment. For example, three shapes of rod-like, spherical, and worm-like were obtained by varying the pH of the β-lactoglobulin solution from pH 2.0 to 5.8 and to 7.0, respectively. The strings of silk fibroin aggregates are weakly branched at a lower concentration, but that the degree of branching increases with increasing initial concentration.

3. The assembly of fibrous proteins are not only a thermodynamic process but also a kinetic process. For alcohol inducing processes, alcohols such as methanol and ethanol trigger the gelation of fibrous proteins, rather than the high temperature and concentration in that of thermally inducing process.

4. Squeeze/shear force inducing techniques, including mandrel rotation gel spun, microfluidic spun, and stirring process, will adapt to most β-sheet-based fibrous proteins.

5. A microfluidic spun system was like a micro-sized needle squeeze system, which normally employs a tube-in-tube geometry that allows the central protein solution to be confined by a surrounding buffer solution and permits the easy extraction of gelation material. A stirring process uses a magnetic stirrer bar to stir protein solution for a period of time (from 30 s up to hours), and then collect the gelled flocs. These shear flow processes can induce the cross-β structural feature (where the β-strands arrange perpendicular to the long fibril axis, inconsistent with the parallel-β arrangement in natural silk fibers) into silk fibroin. However, such shearing flow only promoted the gelation of amyloid rather than changed its cross-β-sheet arrangement.

6. Sonication is a simple method to produce hydrogels of fibrous proteins. Since no high temperature or harsh chemicals are required in sonication, this method is widely applying for encapsulating cells into the hydrogels.

7. Vortexing protocol produces fibrous protein hydrogels without needing a probe to contact the solution. This method can encapsulate cells into the hydrogels after vortexing step.

8. Both V_{DC} and V_{AC} can be used to prepare hydrogels. The silk fibroin hydrogel will form at the positive electrode, and the hydrogel is reversible if the polarity of the voltage is switched. In another case, square waveform electric potential with 5–80 V_{AC}, and the frequency of 0.5–100 kHz, was applied to perform β-insulin hydrogels.

9. Fibrous proteins are amphoteric due to their amino acid side chains. Adjust the pH to a proper value (generally around the isoelectric point of proteins) can simply induce weak hydrogels of fibrous protein solution.

10. The ionic liquid–water mixture has applied for physically cross-linking silk fibroin to form hydrogels. As a binary solvent, the ionic liquid–water mixture will enhance to the silk fibroin hydrogel's water retention and freezing tolerance.

11. Chemical gelation of fibrous proteins generally induce enhanced properties into the resulted hydrogels, such as mechanically robust, conductivity, low immunogenicity, and self-healing. There are tons of approaches to cross-link a fibrous protein by using known enzymatic and chemical reactions, due to the variety of protein side chains. Therefore, we focus on exhibit the most common and useful approaches to cross-link fibrous proteins of silk fibroin, amyloid, and collagen.

12. The reaction of horseradish peroxidase (HRP) to a protein in the presence of H_2O_2 is due to the known reaction whereby HRP facilitates cross-linking of the tyrosines in the protein via the formation of free radical species in the presence of hydrogen peroxide.

13. The 1-ethyl-3-(3-dimethylaminopropyl)-carbodiimide (EDC) coupling is a standard method used to react to primary amines with carboxylic acids. This reaction has broadly applied in cross-link proteins as many proteins contain these functional groups in their amino acid sequences.

14. Fibrous proteins such as amyloid and collagen consist in its native state of peptide chains coordinated to metal ions. For example, crystalline insulin is a hexamer, with three insulin dimers coordinated to two Zn^{2+} ions. Collagen fibrous is accompanied by calcified tissue. Therefore, a variety of metal ions such as Ca^{2+}, Zn^{2+}, Li^+, Na^+, and K^+ are applied for cross-link fibrous proteins in aqueous or organic solvents. Herein, we offer a protocol to prepare flame-retardant and conductive gels by using silk fibroin and Ca^{2+} ions.

15. Glutaraldehyde (GA) is a common cross-link reagent that used as a nonspecific coupling agent for amine-containing molecules and polymers. In this manner, glutaraldehyde has been used to improve the mechanical properties of fibrous proteins-based hydrogels.

16. Genipin is a natural cross-linking agent with quite less toxic, biocompatible, and offers very stable cross-linked products. The genipin cross-link approaches will adapt to most fibrous proteins, including silk fibroin, amyloid, and collagen, which have been widely used for tissue engineering, drug delivery, and polymer materials.

References

1. Ling S, Chen W, Fan Y, Zheng K, Jin K, Yu H, Buehler MJ, Kaplan DL (2018) Biopolymer nanofibrils: structure, modeling, preparation, and applications. Prog Polym Sci 85:1–56

2. Huang W, Ebrahimi D, Dinjaski N, Tarakanova A, Buehler MJ, Wong JY, Kaplan DL (2017) Synergistic integration of experimental and simulation approaches for the de novo design of silk-based materials. Acc Chem Res 50:866–876

3. Zheng K, Ling S (2019) De novo design of recombinant spider silk proteins for material applications. Biotechnol J 14:1700753

4. Vepari C, Kaplan DL (2007) Silk as a biomaterial. Prog Polym Sci 32:991–1007

5. Monks JN, Yan B, Hawkins N, Vollrath F, Wang Z (2016) Spider silk: mother nature's bio-superlens. Nano Lett 16:5842–5845

6. Shen Y, Posavec L, Bolisetty S, Hilty FM, Nystrom G, Kohlbrecher J, Hilbe M, Rossi A, Baumgartner J, Zimmermann MB, Mezzenga R (2017) Amyloid fibril systems reduce, stabilize and deliver bioavailable nanosized iron. Nat Nanotechnol 12:642–647

7. Rad-Malekshahi M, Lempsink L, Amidi M, Hennink WE, Mastrobattista E (2016) Biomedical applications of self-assembling peptides. Bioconjug Chem 27:3–18

8. Knowles TP, Mezzenga R (2016) Amyloid fibrils as building blocks for natural and

artificial functional materials. Adv Mater 28:6546–6561

9. Ling S, Kaplan DL, Buehler MJ (2018) Nanofibrils in nature and materials engineering. Nat Rev Mater 3:18016

10. Nileback L, Chouhan D, Jansson R, Widhe M, Mandal BB, Hedhammar M (2017) Silk-silk interactions between silkworm fibroin and recombinant spider silk fusion proteins enable the construction of bioactive materials. ACS Appl Mater Interfaces 9:31634–31644

11. DeSimone E, Schacht K, Pellert A, Scheibel T (2017) Recombinant spider silk-based bioinks. Biofabrication 9:044104

12. Liu Q, Yang S, Ren J, Ling S (2020) Flame-retardant and sustainable silk ionotronic skin for fire alarm systems. ACS Mater Lett 2:712–720

13. Lin N, Cao L, Huang Q, Wang C, Wang Y, Zhou J, Liu X-Y (2016) Functionalization of silk fibroin materials at mesoscale. Adv Funct Mater 26:8885–8902

14. Ling S, Jin K, Kaplan DL, Buehler MJ (2016) Ultrathin free-standing *bombyx mori* silk nanofibril membranes. Nano Lett 16:3795–3800

15. Lefèvre T, Auger M (2016) Spider silk as a blueprint for greener materials: a review. Int Mater Rev 61:127–153

16. Koh L-D, Cheng Y, Teng C-P, Khin Y-W, Loh X-J, Tee S-Y, Low M, Ye E, Yua H-D, Zhang Y-W, Han M-Y (2015) Structures, mechanical properties and applications of silk fibroin materials. Prog Polym Sci 46:86–110

17. Leyton CE, Villemagne VL, Savage S, Pike KE, Ballard KJ, Piguet O, James RB, Rowe CC, Hodges JR (2011) Subtypes of progressive aphasia: application of the international consensus criteria and validation using β-amyloid imaging. Brain 134:3030–3043

18. Ling S, Qin Z, Li C, Huang W, Kaplan DL, Buehler MJ (2017) Polymorphic regenerated silk fibers assembled through bioinspired spinning. Nat Commun 8:1387

19. Huang W, Krishnaji S, Tokareva OR, Kaplan D, Cebe P (2017) Tunable crystallization, degradation, and self-assembly of recombinant protein block copolymers. Polymer 117:107–116

20. Rockwood DN, Preda RC, Yucel T, Wang X, Lovett ML, Kaplan DL (2011) Materials fabrication from *bombyx mori* silk fibroin. Nat Protoc 6:1612–1631

21. Zhang Z, Yang Y, Zhou P, Zhang X, Wang J (2017) Effects of high pressure modification on conformation and gelation properties of myofibrillar protein. Food Chem 217:678–686

22. Zhang C, Chen X, Shao Z (2015) Sol–gel transition of regenerated silk fibroins in ionic liquid/water mixtures. ACS Biomater Sci Eng 2:12–18

23. Pasternack RF, Gibbs EJ, Sibley S, Woodard L, Hutchinson P, Genereux J, Kristian K (2006) Formation kinetics of insulin-based amyloid gels and the effect of added metalloporphyrins. Biophys J 90:1033–1042

24. Ling S, Li C, Adamcik J, Shao Z, Chen X, Mezzenga R (2014) Modulating materials by orthogonally oriented beta-strands: composites of amyloid and silk fibroin fibrils. Adv Mater 26:4569–4574

25. Williams TL, Serpell LC, Urbanc B (2016) Stabilization of native amyloid β-protein oligomers by copper and hydrogen peroxide induced cross-linking of unmodified proteins (CHICUP). Biochim Biophys Acta 1864:249–259

26. Humenik M, Magdeburg M, Scheibel T (2014) Influence of repeat numbers on self-assembly rates of repetitive recombinant spider silk proteins. J Struct Biol 186:431–437

27. Gong Z, Huang L, Yang Y, Chen X, Shao Z (2009) Two distinct beta-sheet fibrils from silk protein. Chem Commun 48:7506–7508

28. Zhang X, Chen X, Yang T, Zhang N, Dong L, Ma S, Liu X, Zhou M, Li B (2014) The effects of different crossing-linking conditions of genipin on type I collagen scaffolds: an in vitro evaluation. Cell Tissue Bank 15:531–541

29. Yunoki S, Ohyabu Y, Hatayama H (2013) Temperature-responsive gelation of type I collagen solutions involving fibril formation and genipin crosslinking as a potential injectable hydrogel. Int J Biomater 2013:1–14

30. Zheng Z, Jing B, Sorci M, Belfort G, Zhu Y (2015) Accelerated insulin aggregation under alternating current electric fields: relevance to amyloid kinetics. Biomicrofluidics 9:044123

31. Jung J, Savin G, Pouzot M, Schmitt C, Mezzenga R (2008) Structure of heat-induced β-lactoglobulin aggregates and their complexes with sodium-dodecyl sulfate. Biomacromolecules 9:2477–2486

32. Martel A, Burghammer M, Davies R, Dicola E, Riekel C (2008) A microfluidic cell for studying the formation of regenerated silk by synchrotron radiation small- and wide-angle X-ray scattering. Biomicrofluidics 2:024104

33. Bolisetty S, Harnau L, Jung JM, Mezzenga R (2012) Gelation, phase behavior, and dynamics of beta-lactoglobulin amyloid fibrils at varying concentrations and ionic strengths. Biomacromolecules 13:3241–3252

34. Pouzot M, Nicolai T, Visschers RW, Weijers M (2005) X-ray and light scattering study of the

structure of large protein aggregates at neutral pH. Food Hydrocoll 19:231–238

35. Seo J-W, Kim H, Kim K, Choi SQ, Lee HJ (2018) Calcium-modified silk as a biocompatible and strong adhesive for epidermal electronics. Adv Funct Mater 28:1800802.1–1800802.13

36. Nam K, Kimura T, Kishida A (2008) Controlling coupling reaction of EDC and NHS for preparation of collagen gels using ethanol/water co-solvents. Macromol Biosci 8:32–37

37. Chronopoulou L, Daniele M, Perez V, Gentili A, Gasperi T, Lupi S et al (2018) A physico-chemical approach to the study of genipin crosslinking of biofabricated peptide hydrogels. Process Biochem 70:110–116

38. Imsombut T, Srisuwan Y, Srihanam P, Baimark Y (2010) Genipin-cross-linked silk fibroin microspheres prepared by the simple water-in-oil emulsion solvent diffusion method. Powder Technol 203:603–608

39. Lian J, Agban Y, Cheong S, Kuchel RP, Raudsepp A, Williams MAK, Rupenthal ID, Henning A, Tilley RD, Holmes G, Prabakar S (2016) ZnO/PVP nanoparticles induce gelation in type I collagen. Eur Polym J 75:399–405

40. Murphy AR, Kaplan DL (2009) Biomedical applications of chemically-modified silk fibroin. J Mater Chem 19:6443–6450

41. Vrana N, Builles N, Kocak H, Gulay P, Justin V, Malbouyres M, Ruggiero F, Damour O, Hasirci V (2012) EDC/NHS cross-linked collagen foams as scaffolds for artificial corneal stroma. J Biomater Sci Polym Ed 18:1527–1545

42. Partlow BP, Hanna CW, Rnjak-Kovacina J, Moreau JE, Applegate MB, Burke KA, Marelli B, Mitropoulos AN, Omenetto FG, Kaplan DL (2014) Highly tunable elastomeric silk biomaterials. Adv Funct Mater 24:4615–4624

43. Elahi M, Guan G, Wang L, King MW (2014) Influence of layer-by-layer polyelectrolyte deposition and EDC/NHS activated heparin immobilization onto silk fibroin fabric. Materials 7:2956–2977

44. Lovett ML, Cannizzaro C, Vunjaknovakovic G, Kaplan DL (2008) Gel spinning of silk tubes for tissue engineering. Biomaterials 29:4650–4657

Chapter 15

Spinning Methods Used for Construction of One- and Two-Dimensional Fibrous Protein Materials

Leitao Cao

Abstract

Natural silk protein fibers have shown a great attraction to the researchers due to the extraordinary mechanical property, biocompatibility, and functional diversity. Unfortunately, the low yield and unevenness have hampered the scale use of the natural silk fibers. Herein, the appearance of the bioinspired artificial spinning strategy offers an effective way to fabricate silk fibers with controllable structures and functionality. This chapter describes an experimental method to prepare silk protein fibers on a large scale and summarizes the method to investigate the effects of the structure–property relationship of the recombinant protein fibers.

Key words Silks, Recombinant protein fibers, Wet spinning, Dry spinning, Electrospinning

1 Introduction

Natural silk fibrous materials have gained substantial attention in the past few years triggered mainly by their availability, superior mechanical property, functional diversity, and flexibility in assembly [1, 2]. Particularly, remarkable features such as hierarchical structure, biocompatibility, and biodegradability make silk fibers a good candidate for the fabrication of functional biomaterials [3–5]. However, the scarcity of the natural silk fibers (such as spider silks) has hampered their usage. Besides, the natural silk fibers are obtained through puling the silk protein from the spigot of the insects such as silkworm or spider; the swing of the insect heads during spinning would inevitably result in the structure unevenness of the silk fibers, which bring negative impact on the application performance of the fiber assemblies [6–9]. Thus, it is meaningful to large-scale fabricate silk fibers with controllable structures and functionality through bioinspired artificial spinning.

Artificial spinning is a process that extruding or pulling the spinning dope from the spinneret, which has been pursued to produce regenerated silk fibers due to its convenience,

Shengjie Ling (ed.), *Fibrous Proteins: Design, Synthesis, and Assembly*, Methods in Molecular Biology, vol. 2347, https://doi.org/10.1007/978-1-0716-1574-4_15, © Springer Science+Business Media, LLC, part of Springer Nature 2021

designability, and scalability [10]. The usually used artificial methods to produce regenerated silk fibers contain wet spinning, dry spinning, and electrostatic spinning; the abovementioned strategy can be summarized as bottom-up methods, which assemblies biopolymers into nano or microfibers [11].

Wet-spinning is a relatively simple procedure to fabricate silk fibers, in which the spinning dope is first extruded through the spinneret and directly immersed into a coagulating bath to remove the solvents and form the solid fiber [12, 13]. Furthermore, the structure of the regenerated fibers can be turned during the spinning process. For example, the orientation of the fibrils can be controlled by regulating the structure of the spinneret and adding a drafting system, which has an obvious effect on the mechanical strength of the regenerated fibers [14, 15]. Besides, to control the morphology of the regenerated fibers, the component of the coagulating bath can be adjusted; in general, a dense silk fiber would be obtained when using the liquid that has a fast diffusion rate and the ability to induce the conformational changes of the protein [13, 16].

Dry-spinning is a strategy that more like the spinning process of natural insects such as silkworms and spiders [17, 18]. During the spinning process, a highly concentrated spinning dope is introduced into the capillary spinneret and pulled out with proper drafting force; at the same time, the solvent evaporates during the drafting process, and the solid fiber is formed [19–21]. The spinning system can also be designed to mimic the silkworm spinning gland; the state of the spinning dope can be adjusted by introducing aqueous buffers with different pH during the spinning process. Besides, the geometry of the spinneret can be designed like a natural spinning apparatus. Thus the shear force flow can be increased by geometric constriction of the spinneret, which is beneficial to form the anisotropic liquid crystals, making the artificial spun silk fibers as strong as the natural silks [14, 22, 23].

Electrospinning technique has gained great attention in recent years due to the potential applications of electrospun nanofibers in nanoscience and nanotechnology [24–26]. At the same time, this technique has been considered as one of the most promising approaches to scalable preparation of nanofibrous materials for its advantages in continuity, low cost, and easy operation [27]. Particularly, the remarkable features of the fiber assemblies (such as high porosity, good interconnectivity, and large specific surface area) have made them suitable to be used in the fields of biomedical, filtration, separation, and sensors [28, 29]. The electrospinning process mainly obeys the following steps: the spinning dope extruded from the metallic spinneret which connected to a high-voltage direct-current power supply, then the viscous spinning dope is stretched by the electric field force and transferred to the whipped spinning jets, the solvent evaporates on the way, and the jet solidifies to form a nonwoven fibrous membrane on the collector [30]. What's more, the electrospinning technique can be applied to fabricate functional protein nanofiber assembly, which can be realized by controlling and adjusting the electrospinning setup (such as the viscous and conductivity of the spinning dope, the evaporate rates of the solvent, and the humidity and temperature during spinning) [31–36].

The successful design of the protein fibers and precise control over the fiber assemblies may benefit to make the protein-based natural materials more suitable to be used in a variety of bio-related application fields. The following protocols introduce the fabrication of silk protein fibers and try to provide a general method with the design and implementation of a variety of silk-based protein assemblies for potential applications.

2 Materials

1. *B. mori* silkworm cocoons.
2. Deionized water (*see* **Note 1**).
3. Sodium carbonate aqueous solution (Na_2CO_3, 1.4 g/L).
4. Sodium bicarbonate aqueous solution ($NaHCO_3$, 5 g/L).
5. Lithium bromide aqueous solution (LiBr, 9.3 M).
6. Polyethyleneglycol (PEG, M_w = 20 kDa).
7. Polyethyleneoxide (PEO, M_w = 900 kDa).
8. Multiwalled carboxylic carbon nanotubes (MWCNT-COOH, with a diameter of 8–15 nm, length of 5–30μm).
9. Sodium dodecyl sulfate (SDS).
10. Alcohol aqueous solution (80 wt%).
11. 2-(*N*-morpholino)ethanesulfonic acid-tris hydroxymethyl aminomethane (MES-Tris).

3 Methods

3.1 Extract the Silk Protein

1. Cut the *B. mori* silkworm cocoons into small pieces.
2. Boil the *B. mori* silkworm cocoon pieces in Na_2CO_3 aqueous solution or $NaHCO_3$ aqueous solution for 30 min; repeat this process twice to remove sericin thoroughly.
3. Wash the degummed silk fibers with distilled water and then dry them in the air at room temperature.
4. Add 13.5 g degummed silk fibers to 50 mL LiBr aqueous solution (9.3 M) and incubate at 60 °C with constant shaking for 1–4 h.
5. Transfer the above solution into a dialysis bag, dialyzed against distilled water for 72 h to remove the salt, then remove the

insoluble impurity with a centrifuge and get the silk protein aqueous solution from the supernatant.

6. Freeze the silk protein aqueous solution at −20 °C for 12 h. Freeze-dry the frozen protein for 24 h to get the silk protein foam.

3.2 Wet Spinning Procedure

3.2.1 Preparation of Silk Fibroin Fibers

1. Transfer the solution prepared from **step 5** in Subheading 3.1 to a dialysis tube, then dialyzed against aqueous PEG at 5 °C for 72 h to concentrate the protein content around 15 wt%.

2. Prepare $(NH_4)_2SO_4$ aqueous solution with a concentrate of 30% (w/v) as the coagulation bath.

3. Load the condensed spinning dope on an extruder and extrude it through a spinning nozzle with a diameter of 0.2 mm.

4. Directly spin into the $(NH_4)_2SO_4$ aqueous coagulant solution at 60 °C; the spinning jet would solidify to a gel-like filament in the coagulant solution.

5. Solidify the surface of the gel-like filament in the air with a temperature of 23 ± 2 °C and relative humidity of about 60% (*see* **Note 2**).

6. Stretch the filaments by using at least two rollers that rotate with different rotation rates (*see* **Note 3**), then wind the drawn filaments onto a spindle (Fig. 1).

7. Immerse the spindle in the $(NH_4)_2SO_4$ aqueous coagulant solution at 25 °C for 24 h to further solidification.

8. Rinse the solidified filaments in deionized water to remove the $(NH_4)_2SO_4$, then dry them in an oven at a temperature of 40 °C.

3.2.2 Fabrication of Silk Fibroin/Ion Fibers

1. Prepare the protein spinning solution through the method mentioned in **step 1** in Subheading 3.2.1.

2. Add $CaCl_2$ into distilled water to obtain the $CaCl_2$ aqueous solution (*see* **Note 4**).

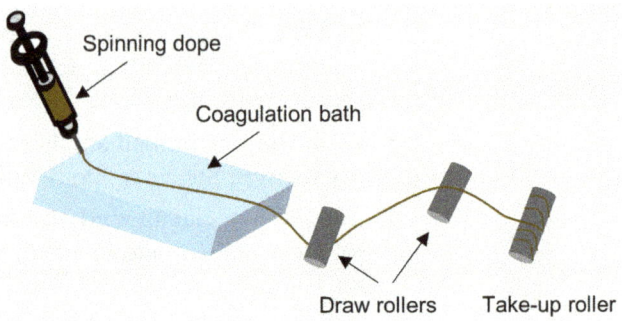

Fig. 1 Schematic of the wet spinning process

3. Add the $CaCl_2$ aqueous solution to the protein spinning solution carefully and stir gently to make homogeneous spinning dope (see **Note 5**).

4. Prepare $(NH_4)_2SO_4$ aqueous solution coagulation bath by referring to **step 2** in Subheading 3.2.1.

5. The spinning and collection procedure is similar to **steps 3–8** in Subheading 3.2.1.

3.2.3 Manufacture of Silk Nanocomposite Fibers

1. The protein spinning solution can be prepared according to **step 1** in Subheading 3.2.1.

2. Add 1 g SDS and 0.5 g MWCNT-COOH into 100 mL deionized water, then prepare the stable MWCNT-COOH dispersion by treating the mixture with an ultrasonic dispersion machine for 30 min (see **Note 6**).

3. Add the MWCNT-COOH dispersion to the protein spinning solution with stirring to prepare the homogenous spinning dope.

4. The spinning procedure is the same as **steps 3–8** in Subheading 3.2.1.

3.3 Dry Spinning Procedure

3.3.1 Fibers Manufactured Through Dry Spinning

1. Immerse the degummed silk fibers (from **step 3** in Subheading 3.1) in HFIP with a weight ratio of 1:20.

2. Incubate the mix silk/HFIP at 60 °C for 7 days to get the spinning dope (see **Note 7**).

3. Transfer the spinning dope to an extruder, then extrude the spinning dope through a spinning nozzle with a diameter of 0.8 mm.

4. Directly extrude the spinning dope onto a PTFE roller (diameter of 10 mm) with a spinning speed of 20 mL h^{-1}, the spinning nozzle contacts with the top of the PTFE roller and repeatedly move from one side to another with a speed of 1 mm s^{-1}, the rolling speed is about 0.785 rad s^{-1} (Fig. 2).

3.3.2 Fibers Fabricated by Dry-Jet Wet Spinning

1. Concentrate the spinning dope prepared from **step 5** in Subheading 3.1 to a weight ratio of 20 wt% by using the forced airflow at 10 °C.

2. Add 0.1 M MES-Tris buffer into the 20 wt% concentrated spinning dope with a volume ratio of 1:2, the aimed PH should be regulated to 4.8.

3. Add 3 M $CaCl_2$ aqueous solution into the mixture and adjust the concentration of Ca^{2+} to 0.3 M.

4. Transfer the concentrated spinning dope to an extruder and extrude it to the air through a spinning nozzle with a diameter of 0.2 mm.

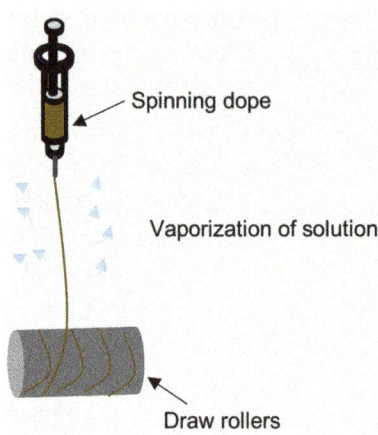

Fig. 2 Sketch of the dry-spinning process

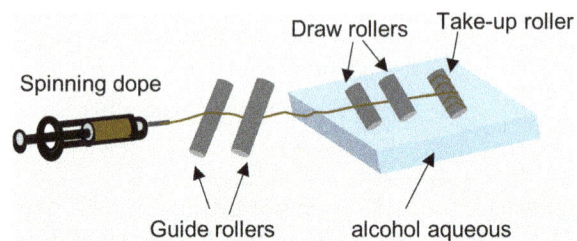

Fig. 3 Illustration showing the process of dry-jet wet spinning

5. The relative humidity and the temperature of the spinning environment can be controlled around 40 ± 5% and 23 ± 2 °C, respectively.

6. Immerse the as-spun fiber into an alcohol aqueous solution and draw the fiber into a demanded length.

7. Immerse the drawn fiber in the alcohol aqueous solution for another 1 h without further extension, then dry the fibers in the air at room temperature (Fig. 3).

3.4   Electrospinning Procedure

3.4.1   Aqueous Solution Electrospinning

1. Transfer the solution prepared from **step 5** in Subheading 3.1 to a dialysis tube, then dialyzed against 10 wt% PEO solution for 5 days to concentrate the protein content around 30 wt%, the concentrated silk solution is prepared.

2. Mix the concentrated silk solution with the PEO solution, then stirring for another 6 h to make the spinning dope (*see* **Note 8**).

3. Transfer the spinning dope into a 10 mL syringe. Assemble a stainless needle (21G) on the syringe.

4. Load the syringe on an electrospinning machine. Use a metallic rotary drum as a collector and connect the drum with an earth ground connection. Fix the tip-to-collector distance to 15 cm. Keep the relative humidity and the temperature of the environment around 20–30% and 20–25 °C, respectively.

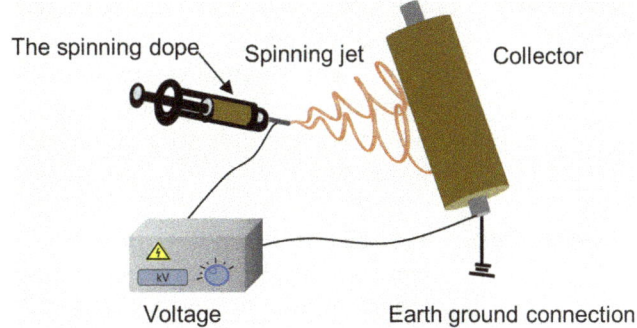

The spinning dope   Spinning jet   Collector

Voltage   Earth ground connection

Fig. 4 A general schematic for the electrospinning process

5. Apply a voltage of 25 kV on the stainless needle. Set the spinning speed and the rolling speed to 1 mL h^{-1} and 50 rpm, respectively (Fig. 4).

6. After spinning, peel off the fiber membrane from the drum and then transfer it to the vacuum oven at 40 °C for 24 h to remove the residual solvent.

3.4.2 Solvent-Based Electrospinning

1. For the FA based spinning dope: add 6 g degummed silk fibers prepared in **step 3** in Subheading 3.1 into a reagent bottle containing 44 g FA. Stir vigorously the mixture for 3 h by a magnetic stirrer to obtain the spinning dope (*see* **Notes 9** and **10**).

2. For the HFIP based spinning dope: add 8.5 g the silk protein foam from **step 6** in Subheading 3.1 into 41.5 g HFIP, stir vigorously for 12 h to get the spinning solution (*see* **Notes 9** and **10**).

3. The assembly of electrospinning equipment is similar to that described in **steps 3** and **4** in Subheading 3.4.1.

4. Apply a voltage of 25 kV on the stainless needle. Set the spinning speed and the rolling speed to 0.5 mL h^{-1} and 50 rpm, respectively (*see* **Note 11**).

5. When finish spinning, peel off the fiber membrane from the drum and then transfer it to the vacuum oven with 40 °C for 24 h to remove the residual solvent.

3.5 Characterization Method

3.5.1 SEM Imaging

SEM characterization can be performed using a field emission scanning microscope operated at an acceleration voltage of 10 kV. To mitigate electrical charging effects, a 2-nm-thick Pd/Pt conductive layer needs to be deposited on the surface of the materials, using a sputter coater.

3.5.2 *Polarized S-FTIR Microspectroscopy*

S-FTIR microspectra are collected in the mid-infrared range of 800–3800 cm^{-1} at a resolution of 4 cm^{-1} with 256 coadded scans. The orientation of individual moieties can be obtained from the angular dependence of the absorbance at wavenumber, which corresponds to a vibration of the molecular group under investigation.

3.5.3 *Mechanical Testing*

For the nonwoven fiber membrane structured electrospun nanofiber assembly, the membrane should be cut to oblong shape with a length of 30 mm, and a width of 3 mm, the thickness of the fiber membrane should also be tested. Then tested on a tensile tester; the clamp distance was 10 mm, and the stretching rate was 10 mm min^{-1}.

For the monofilament, cut the filament into segments with a length of 40 mm, then clip it between two hard-cardboard frames with a length of 20 mm and fix it with adhesives. Mount the frame on the tensile testing machine and cut the side support of the frame away so that the fiber would withstand the tensile stress.

4 Notes

1. Water used is from a Milli-Q ultrapure water purification system.

2. An electric fan can be used to adjust the solidification speed of the gel filament.

3. The number of rollers and the difference in the rotational speeds can be determined by the mechanical property of the solidified filaments. If the filaments were relatively weak, more rollers should be used, and the rotational speeds should be controlled lower; also, the speed difference between neighboring rollers should be minimized to avoid the fracture of the filaments during the draft process. On the contrary, if the filaments were strong enough to be drawn with large distortion, fewer rollers with a faster rotation speed can be directly used to obtain the demanded draft ratio.

4. Here we use $CaCl_2$ as an example. Other ions can also be introduced into the system according to the actual demand.

5. The mass ratio of the Ca^{2+}–silk can be tuned through adjusting the concentration of $CaCl_2$ aqueous solution or by regulating the mass ratio of the $CaCl_2$ aqueous solution to the protein spinning solution.

6. Other materials such as TiO_2 and GO can also be used to fabricate the composite fibers with similar methods according to the actual demand.

7. This procedure should be performed in a chemical hood, and the silk–HFIP mixture should be held in sealed bottles.

8. Pure silk protein solution shows a viscoelastic property, which makes the solution a poor spinnability for electrospinning. Therefore, the easily spun polymer such as PEO is introduced into the spinning system to make the spinning process smoothly.

9. The bottle needs to be sealed during string, and this process should be operated in a chemical hood.

10. The additives (such as carbon nanotubes, graphene oxide, and nanoparticles) can also be introduced into the spinning system in this process to fabricate functional composite electrospun fibers.

11. This process should be operated in a chemical hood.

Acknowledgments

Project funded by China Postdoctoral Science Foundation (No. 2020M681344).

References

1. Yoshioka T, Tsubota T, Tashiro K, Jouraku A, Kameda T (2019) A study of the extraordinarily strong and tough silk produced by bagworms. Nat Commun 10:1469

2. Porter D, Guan J, Vollrath F (2013) Spider silk: super material or thin fibre? Adv Mater 25:1275–1279

3. Hu F, Lin N, Liu XY (2020) Interplay between Light and Functionalized Silk Fibroin and Applications. iScience 23:101035

4. Wang C, Xia K, Zhang Y, Kaplan DL (2019) Silk-based advanced materials for soft electronics. Acc Chem Res 52:2916–2927

5. Vepari C, Kaplan DL (2007) Silk as a biomaterial. Prog Polym Sci 32:991–1007

6. Huang W, Ling S, Li C, Omenetto FG, Kaplan DL (2018) Silkworm silk-based materials and devices generated using bio-nanotechnology. Chem Soc Rev 47:6486–6504

7. Yang N, Zhang W, Ye C, Chen X, Ling S (2019) Nanobiopolymers fabrication and their life cycle assessments. Biotechnol J 14: e1700754

8. Zheng K, Ling S (2019) De novo Design of Recombinant Spider Silk Proteins for material applications. Biotechnol J 14:e1700753

9. Zhang W, Ye C, Zheng K, Zhong J, Tang Y, Fan Y, Buehler MJ, Ling S, Kaplan DL (2018) Tensan silk-inspired hierarchical fibers for smart textile applications. ACS Nano 12:6968–6977

10. Zhang C, Xia L, Deng B, Li C, Wang Y, Li R, Dai F, Liu X, Xu W (2020) Fabrication of a high-toughness polyurethane/fibroin composite without interfacial treatment and its toughening mechanism. ACS Appl Mater Interfaces 12:25409–25418

11. Ling S, Chen W, Fan Y, Zheng K, Jin K, Yu H, Buehler MJ, Kaplan DL (2018) Biopolymer nanofibrils: structure, modeling, preparation, and applications. Prog Polym Sci 85:1–56

12. Andersson M, Jia Q, Abella A, Lee XY, Landreh M, Purhonen P, Hebert H, Tenje M, Robinson CV, Meng Q, Plaza GR, Johansson J, Rising A (2017) Biomimetic spinning of artificial spider silk from a chimeric minispidroin. Nat Chem Biol 13:262–264

13. Madurga R, Ganan-Calvo AM, Plaza GR, Guinea GV, Elices M, Perez-Rigueiro J (2017) Production of high performance bioinspired silk fibers by straining flow spinning. Biomacromolecules 18:1127–1133

14. Li S, Hang Y, Ding Z, Lu Q, Lu G, Chen H, Kaplan DL (2020) Microfluidic silk fibers with aligned hierarchical microstructures. ACS Biomater Sci Eng 6:2847–2854

15. Heim M, Keerl D, Scheibel T (2009) Spider silk: from soluble protein to extraordinary fiber. Angew Chem Int Ed Engl 48:3584–3596

16. Zhou G, Shao Z, Knight DP, Yan J, Chen X (2009) Silk fibers extruded artificially from aqueous solutions of RegeneratedBombyx moriSilk fibroin are tougher than their natural counterparts. Adv Mater 21:366–370

17. Ren J, Wang Y, Yao Y, Wang Y, Fei X, Qi P, Lin S, Kaplan DL, Buehler MJ, Ling S (2019) Biological material interfaces as inspiration for mechanical and optical material designs. Chem Rev 119:12279–12336

18. Wei W, Zhang Y, Zhao Y, Luo J, Shao H, Hu X (2011) Bio-inspired capillary dry spinning of regenerated silk fibroin aqueous solution. Mater Sci Eng C 31:1602–1608

19. Peng Q, Shao H, Hu X, Zhang Y (2015) Role of humidity on the structures and properties of regenerated silk fibers. Prog Nat Sci Mater Int 25:430–436

20. Yue X, Zhang F, Wu H, Ming J, Fan Z, Zuo B (2014) A novel route to prepare dry-spun silk fibers from CaCl2–formic acid solution. Mater Lett 128:175–178

21. Luo J, Zhang L, Peng Q, Sun M, Zhang Y, Shao H, Hu X (2014) Tough silk fibers prepared in air using a biomimetic microfluidic chip. Int J Biol Macromol 66:319–324

22. Ling S, Qin Z, Li C, Huang W, Kaplan DL, Buehler MJ (2017) Polymorphic regenerated silk fibers assembled through bioinspired spinning. Nat Commun 8:1387

23. Gu L, Jiang Y, Hu J (2019) Scalable spider-silk-like Supertough fibers using a Pseudoprotein polymer. Adv Mater 31:e1904311

24. Wang X, Ding B, Yu J, Wang M (2011) Engineering biomimetic superhydrophobic surfaces of electrospun nanomaterials. Nano Today 6:510–530

25. Wang X, Ding B, Sun G, Wang M, Yu J (2013) Electro-spinning/netting: a strategy for the fabrication of three-dimensional polymer nano-fiber/nets. Prog Mater Sci 58:1173–1243

26. Zhang CL, Yu SH (2014) Nanoparticles meet electrospinning: recent advances and future prospects. Chem Soc Rev 43:4423–4448

27. Wang X-X, Yu G-F, Zhang J, Yu M, Ramakrishna S, Long Y-Z (2021) Conductive polymer ultrafine fibers via electrospinning: preparation, physical properties and applications. Prog Mater Sci 115:100704

28. Mishra RK, Mishra P, Verma K, Mondal A, Chaudhary RG, Abolhasani MM, Loganathan S (2018) Electrospinning production of nanofibrous membranes. Environ Chem Lett 17:767–800

29. Xue J, Wu T, Dai Y, Xia Y (2019) Electrospinning and electrospun nanofibers: methods, materials, and applications. Chem Rev 119:5298–5415

30. Jiang SH, Chen YM, Duan GG, Mei CT, Greiner A, Agarwal S (2018) Electrospun nanofiber reinforced composites: a review. Polym Chem 9:2685–2720

31. Qian CY, Xin TW, Xiao WS, Zhu HJ, Zhang Q, Liu LL, Cheng RY, Wang Z, Cui WG, Ge ZL (2020) Vascularized silk electrospun fiber for promoting oral mucosa regeneration. NPG Asia Mater 12:39

32. Chen D, Narayanan N, Federici E, Yang Z, Zuo X, Gao J, Fang F, Deng M, Campanella OH, Jones OG (2020) Electrospinning induced orientation of protein fibrils. Biomacromolecules 21:2772–2785

33. Zhu M, Gu J, He L, Mahar FK, Kim I, Wei K (2019) Fabrication and osteoblastic adhesion behavior of regenerated silk fibroin/PLLA Nanofibrous scaffold by double syringe electrospinning. Fibers Polym 20:1850–1856

34. Zhou CJ, Li Y, Yao SW, He JH (2019) Silkworm-based silk fibers by electrospinning. Results Phys 15:102646

35. Kong N, Wan F, Dai W, Lu Y, Cheng P, Dai J, Li Y-Y, Gong J, Ling S, Yao Y (2020) Bioinspired polypeptide as building blocks for multifunctional material design. Appl Mater Today 20:100683

36. Dai J, Wang YQ, Wu DH, Wan FJ, Lu Y, Kong N, Li XC, Gong JK, Ling SJ, Yao Y (2020) Biointerface mediates cytoskeletal rearrangement of pancreatic cancer cell and modulates its drug sensitivity. Colloid Interface Sci Commun 35:100250

Chapter 16

Three-Dimensional Printing to Build Fibrous Protein Architectures

Huanhuan Qiao and Ke Zheng

Abstract

Fibrous proteins are promising bioinks for three-dimensional printing techniques to fabricate sophisticated structures that find applications in both biomedical engineering and materials science. The critical point of manufacturing these fibrous protein inks is to adjust the cross-linking and rheology properties of proteins that matching the requirements of various printing techniques. In recent years, 3D printing techniques such as extrusion-based printing, droplet-based printing, and light-assisted printing techniques have widely been applied to build sophisticated fibrous protein architectures. In this regard, a series of fibrous protein-based bioinks have been developed, such as bioinks prepared from silk fibroin, collagen, fibrin, gelatin, and recombinant spider silk. In this chapter, we present the protocols to make various fibrous protein inks, as well as how to use these bioinks to print 3D structures via different printing techniques.

Key words Fibrous proteins, Inks, Three-dimensional printing techniques

1 Introduction

Three-dimensional (3D) printing techniques promise to fabricate sophisticated constructions that matching distinct structural and functional needs in biomedical and material engineering [1–5]. Natural fibrous proteins (such as collagen, gelatin, fibrin, and silk fibroin) or recombinant fibrous proteins (such as recombinant spider silks and recombinant elastin) are able to form to hydrogels when their molecules spontaneously self-assemble into fibrous networks [6–9], which is self-supportable. This feature enables these fibrous proteins to be assembled using 3D printing techniques [4, 10, 11]. According to the printing mechanisms, 3D printing can be further divided into extrusion-based printing [12–14], droplet-based printing [15, 16], and light-assisted printing (i.e., laser-based printing and digital light processing printing) [17, 18].

The preparation of bioinks that meets printing requirements is the key to the success of 3D printing of fibrous proteins. These printing requirements for bioinks primarily are rheological stability,

Shengjie Ling (ed.), *Fibrous Proteins: Design, Synthesis, and Assembly*, Methods in Molecular Biology, vol. 2347,
https://doi.org/10.1007/978-1-0716-1574-4_16, © Springer Science+Business Media, LLC, part of Springer Nature 2021

times for fibrous network formation, mechanical properties of the resultant materials, and biocompatibility if they need to be printed with living cells [19–21]. Although some of the natural fibrous proteins, such as silk fibroin, collagen, and fibrin, are able to immobilize their surrounding solvent to form hydrated networks [14, 22, 23], directly printing of natural fibrous proteins remain a challenge. This is mainly due to the following reasons: (1) the thermally inducing gelation of these proteins is difficult to be controlled; (2) most natural fibrous protein hydrogels are weak in mechanical properties [24, 25]. To overcome these issues, other materials, such as polymers, bioactive molecules, growth factors, minerals, and other proteins [26–28], are often mixed with fibrous proteins for ink preparation.

In general, the rheology properties of fibrous proteins determined the printing modalities that they can be used [29]. For example, the extrusion-based techniques require their bioinks viscosities to be $30-6 \times 10^7$ mPa S and often require shear-force sensitivity [30–33]. Droplet-based techniques, instead, require bioinks with viscosities lower than 10 mPa·S [3, 34, 35]. Light-assisted printing techniques allow high-resolution deposition of materials; however, they need pre–cross-linking of the protein at the printing ribbon [17].

In this chapter, we first introduce detailed protocols to fabricate printable bioinks that produced from silk fibroin, collagen, fibrin, gelatin, and recombinant spider silk. We, then, describe the extrusion-based printing, droplet-based printing, and light-assisted printing techniques in detail.

2   Materials

1. Gelatin Type B.
2. Tyrosinase.
3. Polyethylene glycol (PEG, 10,000 molecular weight).
4. Polyethylene glycol (PEG, 400 molecular weight).
5. Laponite XLG.
6. Phosphate buffer saline (PBS): 137 mM NaCl, 2.7 mM KCl, 8 mM Na_2HPO_4, and 2 mM KH_2PO_4, pH 7.4.
7. Fibrin (Sigma-Aldrich).
8. Collagen type I (Sigma-Aldrich).
9. 0.52 M Acetate acid (HAc): place 3.18 mL CH_3COOH (98 w/v%) into a 100 mL volumetric flask, fill the flask with deionized water with a final volume of 100 mL.
10. Pluronic F-127.
11. Gum arabic.

12. Ethanol.

13. 4-(2-hydroxyethyl)-1-piperazineethanesulfonic acid (HEPES)

14. Gelatin Type A.

15. Dulbecco's modified Eagle's medium:

16. Poly-L-lysine hydrobromide (PLL).

17. Tris(hydroxymethyl)aminomethane buffer (Tris–HCl): place 50 mL Tris (0.1 mol/L) and 4.03 mL HCl (1 mol/L) into a 100 mL flat flask, fill the flask with deionized water with a final volume of 100 mL, pH = 7.5.

18. Polyethylene glycol (PEG, 35,000 molecular weight).

3   Methods

3.1   Preparation of Fibrous Protein Ink

3.1.1   Silk Fibroin–Gelatin Ink

1. Add 1.5 g of gelatin in 8 w/v% silk fibroin solution.

2. Keep the suspension at 40 °C with agitation to complete dissolve of gelatin powder in SF solution.

3. <Critical Step>Sonicate the silk fibroin–gelatin solution at 30% amplitude (30 W) for 10 s to pregelation of the mixture (*see* Fig. 1a and **Note 1**).

4. Use an extrusion-based printing system to print the silk fibroin–gelatin ink.

5. Maintain the temperature of the syringe barrel at 28 °C and the printing chamber temperature at 18 °C.

6. Dispense the silk fibroin–gelatin ink through a metal needle with an inner diameter of 250 μm. Use a pneumatic pressure of 200–250 kPa and a deposition speed of 60 mm/min.

3.1.2   Concentrated Silk Fibroin Ink

1. Concentrate 10 mL of 8 w/v% silk fibroin against 1 L of 10 w/v polyethylene glycol (PEG, 10,000 molecular weight) solution for approximately 20 h. The final concentration of the silk fibroin solution will be 20 w/v%.

2. Dissolve 50 g of PEG 400 in 100 mL of deionized water to obtain 50 w/v% PEG 400 solution.

3. <Critical Step>Add 2.5 g of laponite XLG into the 50 w/v% PEG 400 solution. Stir the suspension at 25 °C for 8 h. The suspension was then allowed to settle for an additional 8 h. Here, the laponite–PEG 400 mixture is used as a printing medium. The concentration of PEG 400 can be adjusted from 25 to 100 w/v% if needed.

4. Fill the laponite–PEG 400 mixture into the wells of a multiwell plate and then load it into a print holder.

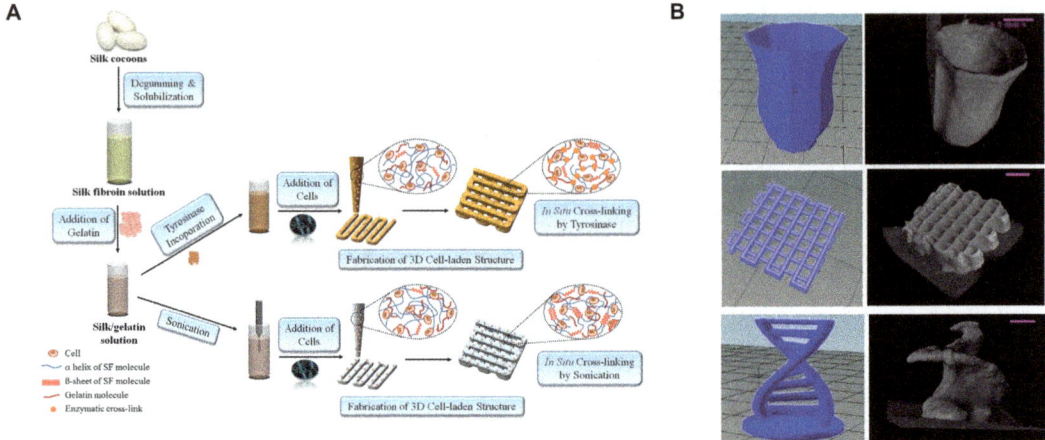

Fig. 1 Schematic illustration of applying gelatin and/or silk fibroin ink for 3D printing. (**a**) The process of in situ cross-linkable silk fibroin–gelatin ink for 3D cell-laden printing. (**b**) 3D freeform printing by using concentrated silk fibroin ink. The left column images showing the design of printed structures, and the right column showing the CT-scan images of printed structures with hollow structure (vase) tissue scaffold and free-standing object (helix). Scale bars represent 1 mm. (Adapted by permission from Elsevier Ltd. [14, 26], Copyright 2015 and 2018)

5. Employ an extrusion-based printing system to print the concentrated silk fibroin ink into the laponite/ PEG 400 medium through a print needle with an inner diameter of 0.51 mm. Pressures are applied to syringes in the range of 50–90 kPA, and printing speed is 2–200 mm/s (*see* Fig. 1b).

3.1.3 Silk Fibroin CaCl$_2$–Formic Acid Ink

1. Dissolve 1 g of CaCl$_2$ in 20 g of formic acid (FA) to obtain a CaCl$_2$–FA solution.

2. Add 3 g of dry silk fibroin into the 20 mL of the prepared CaCl$_2$–FA solution by using a 50 mL glass vial.

3. Shake the silk fibroin–CaCl$_2$–FA solution with a vortexer until there is no visible silk fiber that existed in the glass vial.

4. Remove the silk fibroin–CaCl$_2$–FA solution (SF ink) into another glass vial, and de-aerate at a pressure of 0.06 mbar to remove air bubbles.

5. <Critical Step>Use an extrusion-based printing system to print the silk fibroin CaCl$_2$–formic acid ink. The inner diameter of the print needle is 0.64 mm (*see* **Note 2**).

6. Place the as-printed material into 70% v/v ethanol aqueous solution to cross-link.

3.1.4 Methacrylated Silk Fibroin Ink

1. Dissolve 20 g of dry silk fibroin in 100 mL of 9.3 M LiBr solution to obtain a 20 w/v silk fibroin–LiBr solution.

2. Add 6 mL of glycidyl methacrylate (GMA) into the solution and then stir the mixture with a speed of 300 r/min for 3 h at 60 °C.

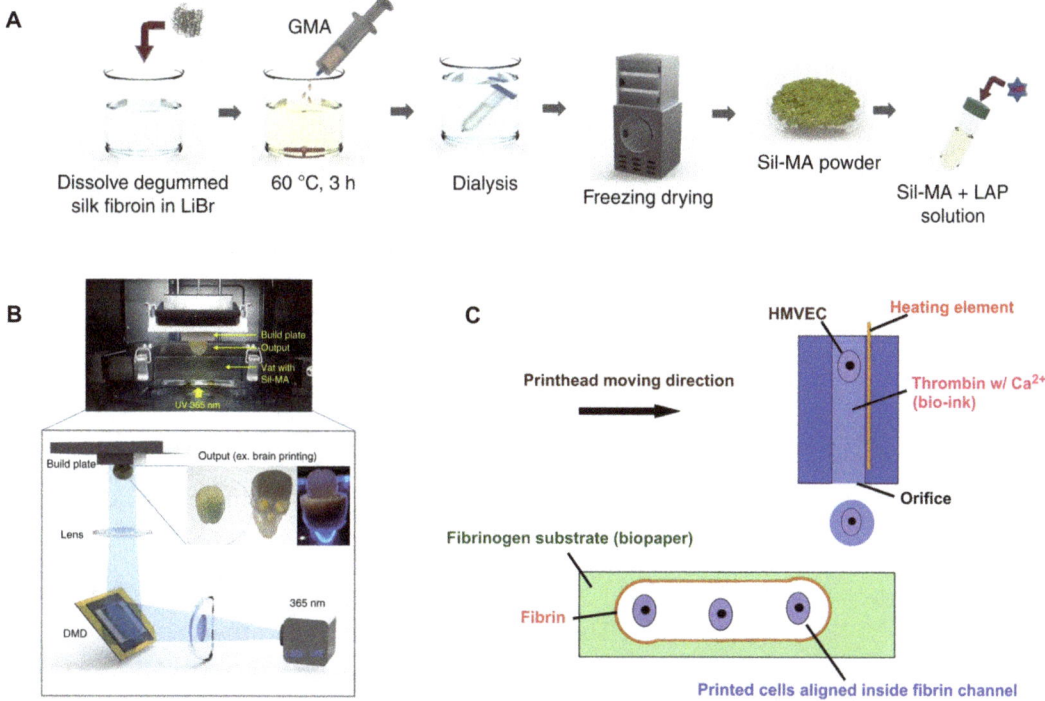

Fig. 2 Schematic illustration of preparing methacrylated silk fibroin ink and the printing process of thrombin cross-linked fibrin ink. (**a**) Fabrication of chemically modified silk fibroin by glycidyl methacrylate as prehydrogel ink. (**b**) Schematic diagram of digital light process printing procedure using methacrylated silk fibroin ink and lithium phenyl(2,4,6-trimethylbenzoyl)phosphinate. (**c**) Mechanism for simultaneous deposition of human microvascular endothelial cells and fibrin ink using a modified droplet-based printer. The printed cells are aligned inside the fibrin substrate. (Adapted by permission from Springer Nature [17], Copyright 2018 and Elsevier Ltd. [16], Copyright 2009)

3. Filter the reacted solution through a miracloth, and then dialysis (molecular weight cutoff 1200–1400 Da) the solution against distilled water using for 4 days.

4. Lyophilize the methacrylated silk fibroin solution at −80 °C for over 48 h to form methacrylated silk fibroin power.

5. Dissolve 3 g of methacrylated silk fibroin powder in 10 mL of deionized water to obtain a 30 w/v% methacrylated silk fibroin solution.

6. <Critical Step>Add 0.02 g of lithium phenyl(2,4,6-trimethyl-benzoyl)phosphinate (LAP) photoinitiator into 10 mL of the mixture (*see* Fig. 2).

7. Print the methacrylated silk fibroin ink by using a digital light processing printing system. The objects are printed layer by layer with a raising of the z-direction. Print 3 layers of base and cure each layer for 4 s.

8. Print the designed object with the ink and buffer solution alternately, the printing thickness is 50 μm, and the curing time is 3 s.

9. Rinse the printed object with deionized water to remove the unreacted solution.

3.1.5 Thrombin–
Fibrin Ink

1. Dissolve 0.6 g of fibrin in 10 mL of deionized water to obtain a 6 w/v% solution.

2. Adjust a 500 unit/mL thrombin solution into phosphate-buffered saline (PBS) solution with 80 mM $CaCl_2$.

3. Filter the fibrin solution and thrombin $CaCl_2$–PBS solution through a 0.22 μm Millipore filter for sterilization.

4. <Critical Step>Print the thrombin–fibrin ink by using a droplet-based printing system. The printing system has a droplet volume of 130 pL, and 50 firing nozzles on the printer head. During the printing process, the actual heating occurs in a 10 μs pulse at 30 °C (*see* Fig. 2c and **Note 3**).

5. Incubate printed samples at 37 °C for 6 min until white printed scaffolds are observed.

3.1.6 Collagen Ink

1. Dissolve 1.2 g of type I collagen in 50 mL 0.52 M HAc solution. The solution should be used within 2 h.

2. Dissolve 2 g of gelatin Type B, 0.25 g of Pluronic F-127, and 0.1 g of gum arabic in a 50% (v/v) ethanol solution at 45 °C with a 200 mL beaker. Adjust the pH of the mixture to 6.25 pH by the addition of 1 M hydrochloric acid (HCl).

3. Stir the mixture overnight, and then centrifuge at $300 \times g$ for 5 min to compact the gelatin microparticles.

4. Wash the gelatin slurry with PBS (pH 7.4) solution for 3 times.

5. Deaerate the uncompacted slurry at a pressure of 0.06 mbar for 15 min to remove air bubbles, and then centrifuge at $2000 \times g$ for 5 min.

6. <Critical Step>Load 5 mL of the collagen ink into a 10 mL, and then printing by using an extrusion-based printing system at 22 °C. Set the slicer with a speed of 23 mm/s, layer height of 40–100 μm, 0–2 perimeters, and 30–80% rectilinear infill. A needle was fitted to the syringe with a length of 0.5 in. and an inner diameter of 150 μm. Of note, print the collagen ink into the gelatin slurry (support bath) (*see* Fig. 3).

7. Incubate the collagen constructs at 37 °C to melt the support bath and release the constructs. Once released, wash the collagen constructs by 50 mM 4-(2-hydroxyethyl)-1-piperazineethanesulfonic acid (HEPES, pH 7.4) solution.

3.1.7 Gelatin Ink

1. Mix 1 g of porcine skin-derived type A gelatin with 10 mL of Dulbecco's modified Eagle's medium™ (DMEM) to obtain a receiving substrate.

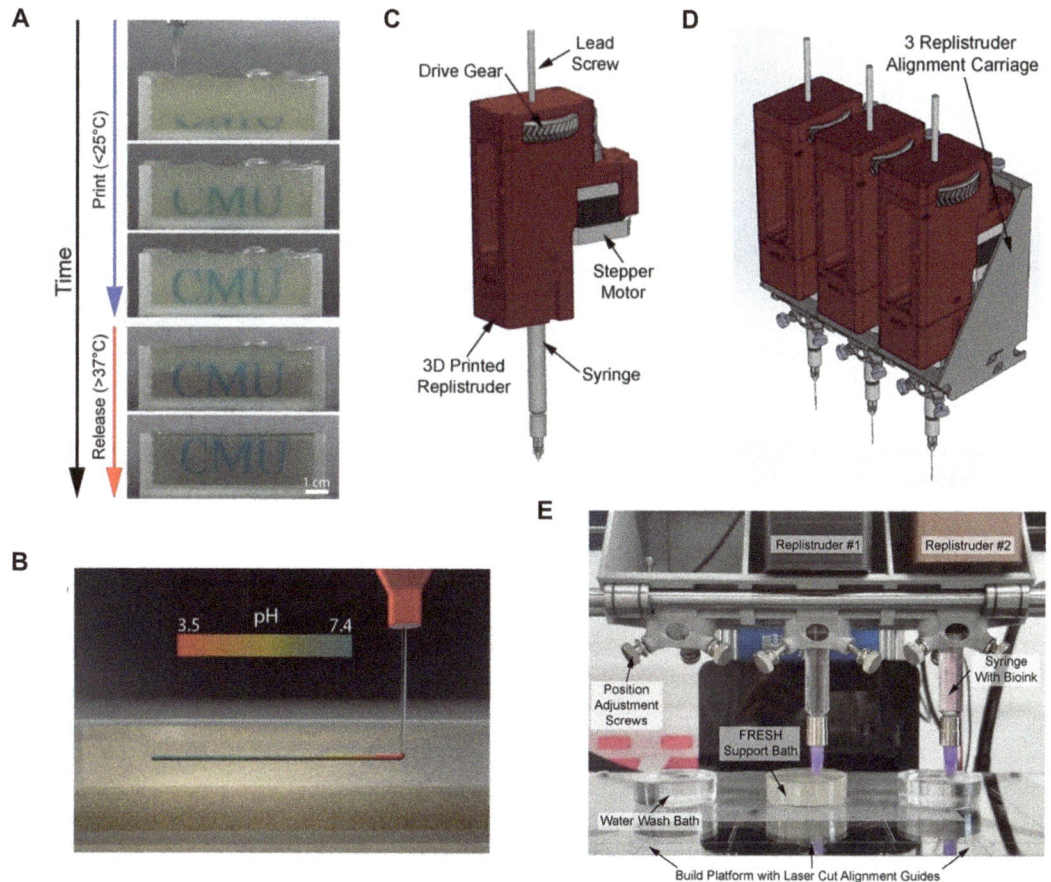

Fig. 3 3D printing of collagen ink using a freeform reversible embedding of suspended hydrogels system. (**a**) Illustration of time-lapse sequence of 3D printing of collagen ink into the support hydrogels. (**b**) Schematic of acidified collagen ink printed into the support hydrogels buffered to pH 7.4. The rapid neutralization of collagen ink causes gelation. (**c–d**) Rendering of the three assembled Replistruder 3 syringe-pump extruder allowing the printing of up to three separate inks. (**e**) The photograph of a multiextruder system with two Replistruder 3 syringe-pump extruder mounted in the carriage. Adjust the needle tip in each extruder aligned to the laser-cut alignment by using the position adjustment screws located around the syringe body and the vertical adjustment screws. (Adapted by permission from American Association for the Advancement of Science [12], Copyright 2019)

2. Dissolve 4 g of gelatin in 20 mL of deionized water to create the print ribbon.

3. Coat 1.5 mL of poly-L-lysine (PLL) hydrobromide onto a 100 mm diameter petri dish for 5 min.

4. Remove the excess PLL from the dish and drying in a laminar flow hood for 1 h.

5. <Critical Step>Add 1 mL of 10 w/v% gelatin solution onto the receiving dish and then spinning at 4000 r/min, 60 °C for 25 s via a spin coater (*see* Fig. 4 and **Note 4**).

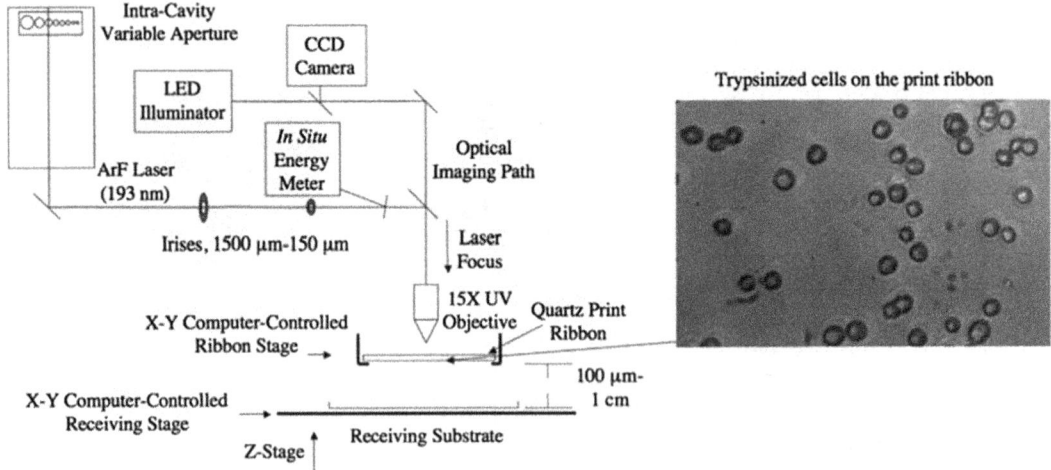

Fig. 4 Schematic drawing of the matrix-assisted pulsed laser evaporation direct-write system used to print cells with gelatin substrate. The inset image highlights the cells on the print substrate as well as the optical clarity of the gelatin. (Adapted by permission from Mary Ann Liebert, Inc. [18], Copyright 2011)

6. <Critical Step>Wash a 50 mm diameter UV transparent quartz flat disk (act as print ribbon) with 70 v/v% ethanol solution. Coat 1.5 mL of PLL and 20 w/v% gelatin onto the ribbon with a spin coater at 2000 r/min, 60 °C for 20 s.

7. Directly print the gelatin ink with a laser-based printing system. Set the laser beam with a diameter of 50 mm to achieve a fluence of 1 J/cm^2. Load the 10 w/v% gelatin-coated receiving dish onto the receiving stage and moved to within 500 mm of the ribbon by using a Z-stage translator.

3.1.8 Recombinant Spider Silk–Gelatin Ink

1. Dissolve 40 mg of lyophilized recombinant spider silk protein eADF4(C16) in 10 mL of 6 M guanidinium thiocyanate (*see* **Note 5**).

2. Dialyze the recombinant spider silk solution against 10 mM Tris(hydroxymethyl)aminomethane (Tris–HCl) buffer (pH 7.5) overnight with a dialysis tube (molecular weight cutoff of 6000–8000 Da). Change the dialysis buffer 3 times during the dialysis process.

3. Concentrate the recombinant spider silk solution to 7 w/v% by dialysis against 25 w/v% PEG 35,000 accompany with 10 mM Tris–HCl buffer solution.

4. Dissolve 10 mg of gelatin Type B in 10 mM Tris–HCl buffer solution for 30 min.

5. <Critical Step>Mix the gelatin Type B solution and recombinant spider silk solution in a weight-to-weight ratio of 200:1, recombinant spider silk to gelatin.

6. Print the recombinant spider silk solution/gelatin ink with an extrusion-based printing system, where the inner diameter of the loading needle is 0.33 mm. During the printing process, maintain the feed rate of 40 mm/s with the load pressure of 2 bar, the dosing distance of 0.12 mm, and the valve opening time of 200 μs.

3.2 Three-Dimensional Printing Techniques

3.2.1 Extrusion-Based Printing

1. Set up an extrusion-based printing system composed of a syringe pump, a standing moving rail and a substrate. The motion of each part is controlled by a stepper motor, where the standing rail, the substrate, and the syringe pump motion in the x-direction, y-direction, and z-direction, respectively. Of note, the motion part in a printing system can be the moving rail or the substrate alone, as long as it can move in x-, y-, and z-direction simultaneously. The printing system needs to be sterilized if the ink involving living cells.

2. Load 10 mL of the fibrous protein ink into the printing chamber. The printing specimens normally are designed by CADian3D (IntelliKorea, Seoul, Korea) with the document of a STereoLithography (STL) file. Load 5 mL of fibrous protein ink into a 10 mL syringe. Be sure to avoid including air bubbles. The printing pattern was converted to printing code (G-code) using open-source software (such as slic3r and Repetier-Host). The injecting speed and the moving speed along each axis are controlled through a computer installed with the programming software (such as Mach3) (*see* Fig. 5 and **Note 6**).

3.2.2 Droplet-Based Printing

1. A droplet-based printing system can be modified form a commercial desktop printer. For example, a HP Desktop 500/550

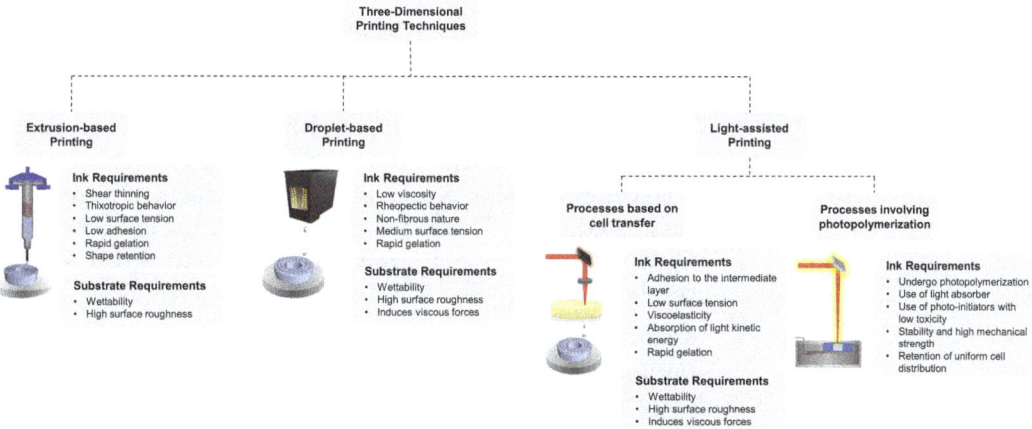

Fig. 5 The technological features of 3D print processes and their ink and substrate requirements. (Adapted by permission from Elsevier Inc. [2], Copyright 2017)

printers with HP 51626a ink cartridge. The printhead of this printer has two rows of 25 orifices, each approximately 50 mm in diameter. The distance between each row of orifices is 170 mm, and the rows are offset by 85 mm. If needed, it can use a UV light to sterilize the modified droplet-based printer overnight and 100% ethanol to clean modified ink cartridges.

2. <Critical Step>The printheads are able to load 85 pL drops with a resolution of 85 mm. Once the print started, it takes 200 µs to launch all 50 chambers, resulting in a print speed of up to 250,000 drops per second. If needed, the print speed can be controlled by digital fabrication—generally, an hourly throughput of structures with 80 mL volume at 85 µm resolution per printhead (*see* Fig. 5).

3.2.3 Laser-Based Printing

1. A laser-based printing system incorporates a pulsed excimer laser operating at a wavelength of 193 nm argon-fluorine (ArF) light source, a computer-assisted design, and a computer-assisted controlling system. The laser beam has a near-Gaussian distribution with a pulse width of 8 ns, and a varied repetition rate ranged from 1 up to 300 Hz.

2. <Critical Step> Once the printing started, the laser beam transmits to the ribbon through an intracavity variable aperture, a series of the lens, two irises to set the spot size, and lastly, through a 15 × objective to focus the beam. The ribbon stage is controlled via software that allows the specified motion of the ribbon in the x- and y-direction, independent of the receiving substrate. Printing patterns are converted into motion, and laser-firing code, thereafter, can be drawn onto the substrate (*see* Fig. 5 and **Note 7**).

3.2.4 Digital Light Processing Printing

1. Set up a digital light processing printing system that consists of three major components: UV Digital Micro-mirror Device™ with a resolution of 30 µm, 365 nm UV-LED with average 30 mW cm^{-2} intensity, and a lens module with two UV-grade biconvex lenses (24 mm diameter). The build area is 35 (L) × 20 (W) × 120 (H) mm with a layer thickness adjustable from 5 to 200 µm. The system was customized by professional manufacturers (NBRTech. Ltd., Chuncheon, Korea, and Illuminated. Ltd., Seongnam, Korea). If the ink involves living cells, the printing system needs to be sterilized with UV light overnight.

2. <Critical Step>Load 10 mL of the fibrous protein ink into the printing chamber. The printing specimens normally are designed by CADian3D (IntelliKorea, Seoul, Korea) with the document of a STereoLithography (STL) file. Then, slice the STL files in the z-direction. Finally, slices are projected by the

projector for every layer to create the designed 3D morphology (*see* Fig. 5 and **Note 8**).

4   Notes

1. A concentration of 500 units of tyrosinase was incorporated in one batch of silk fibroin–gelatin solution for enzymatic gelation.

2. For silk fibroin–$CaCl_2$–formic acid ink, as well as graphene–silk mixture ink, they can be used as a direct writing ink onto the surfaces of paper and skins.

3. Thrombin solution serves as the bioink, while fibrin as a substrate on a microscope coverslip serves as the biopaper.

4. If needed, add 10 mL of DMEM solution in the dish and place it into a 4 °C refrigerator for 5 min. Thereafter, incubate the receiving dish in a standard cell culture incubator (37 °C, 5 v/v % CO_2, and 95% RH) for 20 min. These processes are applied for printing with living cells.

5. The eADF4(C16) protein consists of 16 repeats of module C (sequence: GSSAAAAAAAASGPGGYGPENQGPSGPG GYGPGGP). This recombinant spider silk protein mimics the repetitive core sequence of dragline silk fibroin 4 (ADF4) of the European garden spider *Araneus diadematus*. The protocols to express a recombinant spider silk protein have described in the previous chapter.

6. For extrusion-based 3D printing, the theoretical resolution of the printing pattern is limited by the minimum step angle of the employed stepper motor. Two neighboring bands can be printed as narrow as possible, as long as they do not fuse and become optically indistinguishable. The band thickness is limited by the inner diameter of the syringe tip as a thinner tip significantly increases the shear force, which needs more significant compressing force and can cause ink clogging at the tip. Therefore, in practice, the resolution of a printing pattern mostly depends on the viscosity of the protein ink and the hydrophobicity of the substrate.

7. For laser-based 3D printing, the transferred spot (normally in the diameter of 20–500 μm) can be modulated through adjustments in laser beam diameter, spot size, as well as ink density. The distance between the print ribbon and receiving substrate can be adjust ranged from 1 cm down to 100 μm by using a Z-stage translator.

8. For digital light processing printing, it should be careful to load the fibrous protein ink into the printing chamber. Make sure the fibrous protein does not get crystallization (check the light and pH) before the printing process. Adjust the lens to prevent

overcuring of layers beyond the focal plane. To check the actual resolution of the printing system, print a block in the shape of gradually degrading circle or square patterns ranging from 900 to 70 μm in the planar axis and from 900 to 200 μm in the vertical axis. Calibrate the deviation in both the horizontal and vertical planes.

References

1. Derakhshanfar S, Mbeleck R, Xu K, Zhang X, Zhong W, Xing M (2018) 3D bioprinting for biomedical devices and tissue engineering: a review of recent trends and advances. Bioact Mater 3:144–156

2. Hospodiuk M, Dey M, Sosnoski D, Ozbolat IT (2017) The bioink: a comprehensive review on bioprintable materials. Biotechnol Adv 35:217–239

3. Chia HN, Wu BM (2015) Recent advances in 3D printing of biomaterials. J Biol Eng 9:4

4. Sunna A, Care A, Bergquist PL (eds) (2017) Peptides and peptide-based biomaterials and their biomedical applications. Advances in experimental medicine and biology, vol 1030. Springer, Cham

5. Wang Q, Han G, Yan S, Zhang Q (2019) 3D printing of silk fibroin for biomedical applications. Materials 12:504

6. Zhang C, Chen X, Shao Z (2015) Sol–gel transition of regenerated silk fibroins in ionic liquid/water mixtures. ACS Biomater Sci Eng 2:12–18

7. Bolisetty S, Harnau L, Jung JM, Mezzenga R (2012) Gelation, phase behavior, and dynamics of beta-lactoglobulin amyloid fibrils at varying concentrations and ionic strengths. Biomacromolecules 13:3241–3252

8. Ohyabu Y, Yunoki S, Hatayama H, Teranishi Y (2013) Fabrication of high-density collagen fibril matrix gels by renaturation of triple-helix collagen from gelatin. Int J Biol Macromol 62:296–303

9. Hiew SH, Guerette PA, Zvarec OJ, Phillips M, Zhou F, Su H, Pervushin K, Orner BP, Miserez A (2016) Modular peptides from the thermoplastic squid sucker ring teeth form amyloid-like cross-beta supramolecular networks. Acta Biomater 46:41–54

10. Marques CF, Diogo GS, Pina S, Oliveira JM, Silva TH, Reis RL (2019) Collagen-based bioinks for hard tissue engineering applications: a comprehensive review. J Mater Sci Mater Med 30:32

11. Schacht K, Scheibel T (2011) Controlled hydrogel formation of a recombinant spider silk protein. Biomacromolecules 12:2488–2495

12. Lee A, Hudson AR, Shiwarski DJ, Tashman JW, Hinton TJ, Yerneni S, Bliley JM, Campbell PG, Feinberg AW (2019) 3D bioprinting of collagen to rebuild components of the human heart. Science 365:482–487

13. Yang X, Lu Z, Wu H, Li W, Zheng L, Zhao J (2018) Collagen-alginate as bioink for three-dimensional (3D) cell printing based cartilage tissue engineering. Mater Sci Eng C Mater Biol Appl 83:195–201

14. Rodriguez MJ, Dixon TA, Cohen E, Huang W, Omenetto FG, Kaplan DL (2018) 3D freeform printing of silk fibroin. Acta Biomater 71:379–387

15. Lee W, Lee V, Polio S, Keegan P, Lee JH, Fischer K, Park JK, Yoo SS (2015) On-demand three-dimensional freeform fabrication of multi-layered hydrogel scaffold with fluidic channels. Biotechnol Bioengineering 105:1178–1186

16. Cui X, Boland T (2009) Human microvasculature fabrication using thermal inkjet printing technology. Biomaterials 30:6221–6227

17. Kim SH, Yeon YK, Lee JM, Chao JR, Lee YJ, Seo YB, Sultan MT, Lee OJ, Lee JS, Yoon SI, Hong IS, Khang G, Lee SJ, Yoo JJ, Park CH (2018) Precisely printable and biocompatible silk fibroin bioink for digital light processing 3D printing. Nat Commun 9:1620

18. Schiele NR, Chrisey DB, Corr DT (2011) Gelatin-based laser direct-write technique for the precise spatial patterning of cells. Tissue Eng Part C Methods 17:289–298

19. Loo Y, Lakshmanan A, Ni M, Toh LL, Wang S, Hauser CA (2015) Peptide bioink: self-assembling nanofibrous scaffolds for three-dimensional organotypic cultures. Nano Lett 15:6919–6925

20. Liu CZ, Xia ZD, Han ZW, Hulley PA, Triffitt JT, Czernuszka JT (2008) Novel 3D collagen scaffolds fabricated by indirect printing technique for tissue engineering. J Biomed Mater Res B Appl Biomater 85:519–528

21. Deitch S, Kunkle C, Cui X, Boland T, Dean D (2015) Collagen matrix alignment using inkjet printer technology. MRS Proc 1094:52

22. Yunoki S, Ohyabu Y, Hatayama H (2013) Temperature-responsive gelation of type I collagen solutions involving fibril formation and genipin crosslinking as a potential injectable hydrogel. Int J Biomater 2013:1–14

23. Ling S, Zhang Q, Kaplan DL, Omenetto FG, Buehler MJ, Qin Z (2016) Printing of stretchable silk membranes for strain measurements. Lab Chip 16:2459–2466

24. Inzana JA, Olvera D, Fuller SM, Kelly JP, Graeve OA, Schwarz EM, Kates SL, Awad HA (2015) 3D printing of composite calcium phosphate and collagen scaffolds for bone regeneration. Biomaterials 35:4026–4034

25. DeSimone E, Schacht K, Pellert A, Scheibel T (2017) Recombinant spider silk-based bioinks. Biofabrication 044104:9

26. Das S, Pati F, Choi YJ, Rijal G, Shim JH, Kim SW, Ray AR, Cho DW, Ghosh S (2015) Bioprintable, cell-laden silk fibroin-gelatin hydrogel supporting multilineage differentiation of stem cells for fabrication of three-dimensional tissue constructs. Acta Biomater 11:233–246

27. Lee YB, Polio S, Lee W, Dai G, Menon L, Carroll RS, Yoo SS (2010) Bio-printing of collagen and VEGF-releasing fibrin gel scaffolds for neural stem cell culture. Exp Neurol 223:645–652

28. Ling S, Wang Q, Zhang D, Zhang Y, Mu X, Kaplan DL, Buehler MJ (2018) Integration of stiff graphene and tough silk for the design and fabrication of versatile electronic materials. Adv Funct Mater 28:1705291

29. Bell A, Kofron M, Nistor V (2015) Multiphoton crosslinking for biocompatible 3D printing of type I collagen. Biofabrication 035007:7

30. Xiong S, Zhang X, Lu P, Wu Y, Wang Q, Sun H, Heng BC, Bunpetch V, Zhang SF, Ouyang HW (2017) A gelatin-sulfonated silk composite scaffold based on 3D printing technology enhances skin regeneration by stimulating epidermal growth and dermal neovascularization. Sci Rep 7:4288

31. Shi WL, Sun MY, Hu XQ, Ren B, Cheng J, Li CX, Duan XN, Fu X, Zhang JY, Chen HF, Ao YF (2017) Structurally and functionally optimized silk-fibroin-gelatin scaffold using 3D printing to repair cartilage injury in vitro and in vivo. Adv Mater 29:1701089

32. Rhee S, Puetzer JL, Mason BN, Reinhart-King CA, Bonassar LJ (2016) 3D bioprinting of spatially heterogeneous collagen constructs for cartilage tissue engineering. ACS Biomater Sci Eng 2:1800–1805

33. Kim YB, Lee H, Kim GH (2016) Strategy to achieve highly porous/biocompatible macroscale cell blocks, using a collagen/genipin-bioink and an optimal 3D printing process. ACS Appl Mater Interfaces 8:32230–32240

34. Yeong WY, Chua CK, Leong K, Chandrasekaran M, Lee M (2006) Indirect fabrication of collagen scaffold based on inkjet printing technique. Rapid Prototyp J 12:229–237

35. Xu T, Gregory CA, Molnar P, Cui XF, Jalota S, Bhaduri SB, Sarit B, Boland T (2006) Viability and electrophysiology of neural cell structures generated by the inkjet printing method. Biomaterials 27:3580–3588

Part IV

Methods in Characterization of Fibrous Proteins

Chapter 17

Synchrotron FTIR Microspectroscopy Methods to Understand the Conformation of Single Animal Silk Fibers

Chao Ye, Leitao Cao, and Shengjie Ling

Abstract

Animal silks have received extensive attention in these years due to their unique mechanical properties. The study of the structure–property relationship of animal silks is not only critical for the understanding of the design secrets of natural materials but also can inspire the engineering material designs. Fourier transform infrared spectroscopy (FTIR) has been used to study the secondary structure of animal silk, which is considered to be critical to the mechanical properties of animal silk. However, most of these characterizations are conducted on silk fiber bundles. In this respect, synchrotron FTIR microspectroscopy (S-micro FTIR) has unique advantages in characterizing single animal silks, as S-micro FTIR has significant advantages in ultrahigh brightness and high spatial resolution to characterize samples with small size. Here, we will introduce the methods for using synchrotron FTIR microspectroscopy to analyze the conformation and orientation of single animal silk fibers, which would be an efficient method to elucidate the "structure–property" relationship within animal silks.

Key words Fibrous proteins, Silks, Synchrotron FTIR, Structure–property relationship

1 Introduction

Animal silks, especially silkworm silks and spider silks, have attracted intensive interests in these years due to their excellent mechanical properties [1]. For example, numerous researches have been devoted to understanding the relationship between secondary structure and mechanical properties. To understand the secondary structure of animal silks, a series of techniques such as X-ray diffraction (XRD), Raman microspectroscopy, nuclear magnetic resonance (NMR), infrared spectroscopy, circular dichroism (CD), and fluorescence spectra have been used [2–9]. Among those characterization technologies, infrared spectroscopy is an efficient way to get the spectrum and exhibits its advantages in characterizing protein structure, as infrared absorption accords with the Lambert–Beer law and the contents of various conformations in silk fibroin can be obtained accurately by mathematical processing such

Shengjie Ling (ed.), *Fibrous Proteins: Design, Synthesis, and Assembly*, Methods in Molecular Biology, vol. 2347, https://doi.org/10.1007/978-1-0716-1574-4_17, © Springer Science+Business Media, LLC, part of Springer Nature 2021

as deconvolution and peak fitting. These complementary characterizations have provided numerous information about the secondary
structure of animal silks, and some consensus has been reached. For
instance, animal silk is usually regarded as a semicrystalline polymer
material with highly oriented chain segments along the long axis of
the fiber. More specifically, most animal silks consist of antiparallel
β-sheet nanocrystals and amorphous random coil and/or helix,
where the nanocrystals are dispersed in the amorphous matrix.

However, due to the small cross section of the animal silks and
the larger spot size of those devices, most characterization was
conducted by using silk bundles instead of single silk fibers. Accordingly, there are still many controversies in the study of the secondary structure of animal silks. For example, there are apparent
differences in the characterization of the relative content of each
secondary structure of animal silks. This difference even appears
when the same characterization technology and the same kind of
animal silk are used. An important reason for these controversies is
the variability of animal silk. The difference in the structure of
animal silks exists not only between the species but also in different
parts of the same kind of silk [10, 11]. The amino acid sequence
[12], spinning condition [13], and the differences between animals
and their surrounding environment [14] will make the structure
and mechanical properties of silk different. Therefore, the characterization of specific areas of the single silk fibers will be particularly
important.

In this regard, FTIR microspectroscopy, a technique that
integrated the FTIR spectroscopy and infrared microscopy, has a
unique advantage. It can provide sufficient spatial resolution (diffraction limit) to characterize the single animal silks. However, the
spectra obtained usually have a poor signal-to-noise ratio (SNR)
due to the low brightness of the conventional infrared light source.
These spectra can only be used for qualitative analysis. These technical difficulties have been overcome by involving the synchrotron
light source in the FTIR microspectroscopy system, in which the
synchrotron infrared light has an intensity of two (or even three)
times magnitude higher than that of the conventional infrared
light, ensures enough luminous flux even with small aperture size
[15]. For example, when the infrared light passes through the
aperture dimension of 10 μm, the synchronous light source can
obtain more than 80% of the maximum infrared signal. In contrast,
the infrared signal generated by the globar infrared light source will
be very weak. This evidence indicates that the synchrotron radiation
infrared light source has obvious advantages for the characterization of the small size samples or a specific small area of the
sample [15].

In this chapter, we will provide a practical guide to using
synchrotron FTIR microspectroscopy (S-microFTIR) to investigate the single animal silk fibers. The introduction starts with the

protocols for sample preparations. Then, the methods to collect FTIR spectra and polarized FTIR spectra of single animal silks are discussed. At last, the data processing, such as deconvolution and orientation parameter calculation are described.

2   Materials

1. *Bombyx mori* (*B. mori*) cocoons.

2. *Antheraea pernyi* (*A. pernyi*) cocoons.

3. *Nephila edulis* (*N. edulis*) spiders.

4. Sodium bicarbonate ($NaHCO_3$).

5. Sodium carbonate (Na_2CO_3).

6. Fourier-transform infrared (FTIR) spectrometer with mid-infrared capability.

7. Infrared microscope should be well connected to FTIR spectrometer.

8. Synchrotron radiation source (*see* **Note** 1).

9. Detectors, such as single element liquid nitrogen-cooled mercury-cadmium-telluride (MCT) infrared detectors, linear array MCT detectors.

10. IR transmission windows, such as barium fluoride (BaF_2), calcium fluoride (CaF_2), and diamond.

11. Accessories, such as IR polarizer and stretching device.

3   Methods

3.1   Preparation of Single Animal Silk Fiber

3.1.1   Preparation of Degummed B. mori Silks

B. mori silkworm cocoon silk fibers are degummed by boiling in two 30 min changes of 0.5% (w/w) $NaHCO_3$ solution. The degummed fibers are then washed with distilled water and obtained by drying at room temperature (*see* **Note** 2).

3.1.2   Preparation of Degummed A. pernyi Silks

A. pernyi silkworm cocoon silk fibers are degummed by boiling in two 30 min changes of 0.5% (w/w) Na_2CO_3 solution. The degummed fibers are then washed with distilled water and obtained by drying at room temperature.

3.1.3   Preparation of N. edulis Spider Dragline Silk

N. edulis dragline silk fiber is artificially reeled with a drawing speed of 2 cm/s from spiders under ambient room conditions of 24 °C and 25% relative humidity. The single fibers are selected from these reeled spider fibers.

3.2 Setting Up S-microFTIR Spectrometer and Experimental Parameters

1. Check the status of the S-microFTIR spectrometer and select the light source as synchrotron radiation.

2. Fill the MCT detector with liquid N_2 and wait ~30 min to reach a stable signal.

3. Place a monofilament sample on the stage and adjust the aperture size to limit the infrared light to pass through the sample completely (*see* **Note 3**).

4. Load your experimental parameters, such as wavenumber range, times of scan, resolution, and so on.

5. Record a background spectrum with the parameters used to record the sample spectrum (only the number of scans can be modified but not recommended). The background spectrum should be recorded just before the sample recording.

6. For polarized S-microFTIR of single animal silk, a polarizer placed in the light path is needed. The goal of the polarizer is to filter circularly polarized light into linearly polarized light with different polarization angles relative to fiber axial. At each polarization angle, background spectrum and sample recording are required, as described in Methods 3.2.5 (*see* **Note 4**).

7. To measure the polarized S-microFTIR of single silk fiber under in situ tensile, single fiber can be mounted on a custom-built apparatus or other devices with a modified engineer's screw micrometer to precisely control the extension of a single fiber. Before recording the spectrum, samples are first fixed at particular tensile strain, then background spectrum and sample recording are carried out under each polarization angle as described in Method 3.2.6.

3.3 Spectral Preprocessing

All processing steps described here can be done on any spectral analysis software, including software supplied by spectrometers or by the synchrotron beamline. In most cases, one needs to go through these steps while preprocessing the raw spectrums.

1. Make a copy of the raw data for data processing and analysis.

2. Carry out smoothing with analysis software (*see* **Note 5**).

3. Apply a linear baseline subtraction on the entire spectrum.

4. Save preprocessed spectrums for future analysis. The spectrum files can be saved as .csv, .txt, .xls, and so on.

3.4 Spectral Analysis

3.4.1 Peak Assignment of Biological Samples

As shown in Table 1, peak assignment of FTIR spectrum of biological samples is summarized. Among of which, assignment of FTIR characteristic bands of animal silk protein chain conformation is also summarized in Table 2.

Table 1
Peak assignment of biological samples

Frequency (cm^{-1})	Assignment
~3500	O – H stretch of hydroxyl groups
~3200	N – H stretch (amide A) of proteins
2959	Asymmetric C – H stretch of –CH_3
2934	Asymmetric C – H stretch of –CH_2
2921	Asymmetric C – H stretch of –CH_2 of fatty acids
2898	C – H stretch of C – H methine
2872	Symmetric C – H stretch of –CH_3
2852	Symmetric C – H stretch of –CH_2 of fatty acids
1741	C=O stretch of esters
1715	C=O stretch of esters, RNA/DNA
1695, 1685, 1675	Amide I band components resulting from antiparallel pleated sheets and β-turns of proteins
~1655	Amide I of α-helical structures
~1637	Amide I of β-pleated sheet structures
1575	Asymmetric stretch COO^-
1548	Amide II
1515	Tyrosine band
1468	C – H deformation of CH_2
~1400	Symmetric C=O stretch of COO^-
1310–1240	Amide III band components of proteins
1250–1220	Asymmetric P=O stretch of PO_2^-
1200–900	C – O – C, C – O dominated by ring vibrations of carbohydrates C – O – P, P – O – P
1085	Symmetric P=O stretch of PO_2^- phosphodiesters
720	C – H rocking of CH_2
900–600	Fingerprint region

3.4.2 Content of Secondary Structure Calculating by Peak Fitting

All of the steps in the peak fitting process can be done by PeakFit 4.12 software, Origin, Ominic, and other programs that meet the corresponding functions.

Peak Fitting of Amide I Band or Amide III Band Using PeakFit 4.12 Software

1. Open PeakFit 4.12 software and wait for it to start.

2. Import single spectrum to be analyzed by clicking File-Import option in menu bar and selecting target spectrum (*see* **Note 6**).

Table 2
Assignment of FTIR characteristic bands of animal silk protein chain conformation

Conformational type	Amide I cm^{-1}	Amide II cm^{-1}	Amide III cm^{-1}	Others cm^{-1}
Helix	1650 ~ 1660	1545 ~ 1550	1270 ~ 1275	895, 624
β-sheet	1620 ~ 1630	1515 ~ 1530	1220 ~ 1245	965, 700
Random	1650 ~ 1660	1545	/	664
β-turn	1690 ~ 1700	/	/	/

3. Cut out amide I band or amide III band. Click on "Section and Graphically Process XY Data" button in toolbar to pop up a new window "Selection Data," then input wavenumber range of Amide I band or Amide III band, followed by clicking "Apply New" button in the new pop-up window. (Wavenumber range of amide I is generally 1600 ~ 1700 cm^{-1} and amide III is generally 1200 ~ 1300 cm^{-1}.)

4. Baseline correction of amide I band or amide III band. After Cutting out amide I band or amide III band, click on "Automatically Fit and Subtract Baseline" button in toolbar to pop up a new window "Automatic Baseline Subtraction" and then click on "√" button in the new window, the baseline would be subtracted.

5. Peak fitting of amide I band or amide III band. Click on "Automatically Peak Detection and Fitting, Method II—Second Derivative" button to pop up a new window "AutoPlace and Fit Peaks—Second Derivative Method". In the new window, select several peaks using Gaussian-model, the number and each position of which are according to FTIR characteristic bands of animal silk protein chain conformation in Table 2. After determining the number and position of peaks to be fitted, right-click on at the highest point of each peak to pop up a small window, in which one need chose "Gauss Amp" as peak type and lock peak position.

6. Optimize the peak fitting and save the final fitting data. Click on "Fast Peak Fit with Numerical Update" button in left toolbar to pop up a new window "Peakfit Numerical Fitting". In the new window, click on "Addl Adjust" button to perform numerical fitting once. Repeat Click on "Fast Peak Fit with Numerical Update" button and "Addl Adjust" button until "Iteration" in "Peakfit Numerical Fitting" window reach 7 to acquire the best fitting. Then, one can click on "Review Peak Fit" button in left toolbar to review fitting results and save original data of those peaks. Figure 1 shows the peak fitting result of Amide III band of *A. pernyi* silk. One can also click on

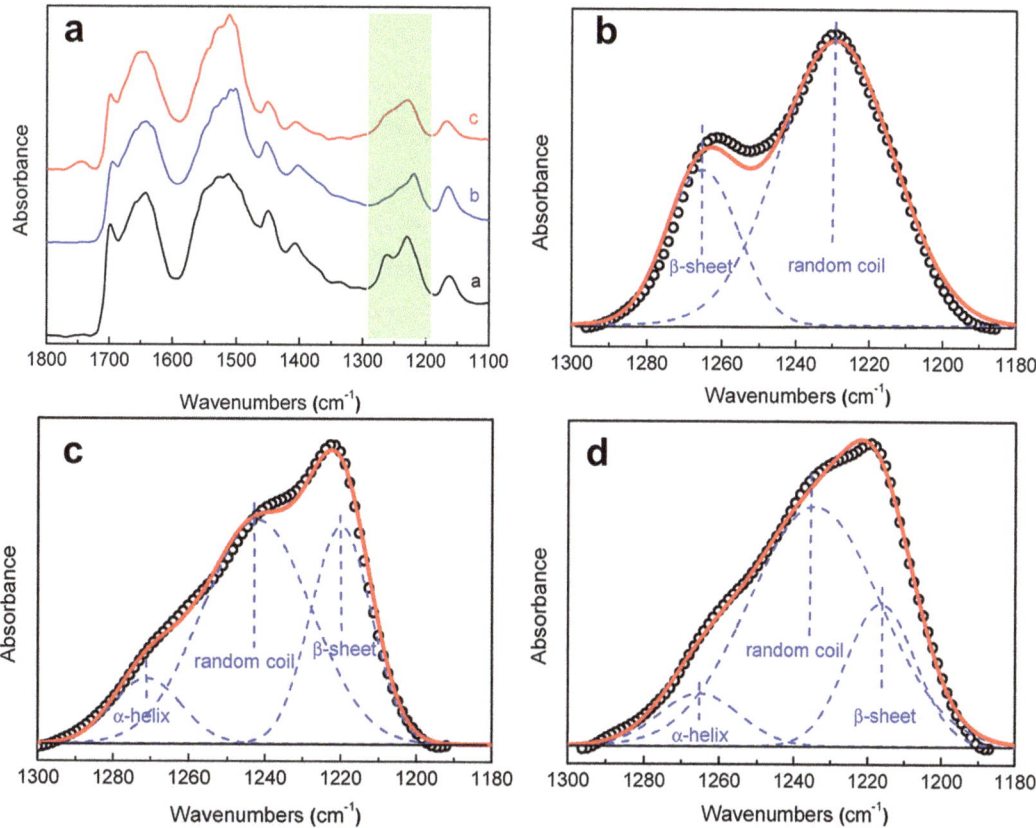

Fig. 1 (**a**) S-microFTIR spectra of single silk fibers: (a) *B. mori*, (b) *A. pernyi*, and (c) *N. edulis*. The marked region is Amide III band. (**b–d**) Deconvolution results of amide III band of single natural silk fibers: (**b**) *B. mori*, (**c**) *A. pernyi*, and (**d**) *N. edulis* (circles, original spectrum; dashed curve, deconvoluted peaks; solid curve, simulated spectrum from summed peaks). (Adapted by permission from American Chemical Society [16], Copyright 2011)

"List Peak Estimates" button to save peak area of each Gaussian model peak [16].

7. After fitting convergence, calculate the area proportion of each conformation represented to acquire content of each secondary conformation.

Peak Fitting of Amide I Band or Amide III Band by Python Program

When there are few of FTIR spectrums, Peak fitting of Amide band using PeakFit software is a good choice. However, it would be very time-consuming when dealing with a large number of FTIR spectra. On this condition, peak fitting using Python programs becomes efficient. Here, an example of Python program to batch process peak fitting of Amide III band of *A. pernyi* silk acquired by Bruker FTIR spectrometer is provided for reference.

```python
1.   #-*- encoding: utf-8 -*-
2.   #Python version = 3.7
3.   #Authored by Chao Ye
4.
5.   #import dependent packages
6.   from matplotlib import pyplot as plt
7.   import numpy as np
8.   from scipy.signal import savgol_filter
9.   from scipy import sparse
10.  from scipy.sparse.linalg import spsolve
11.  from scipy.optimize import curve_fit
12.  from scipy.integrate import quad
13.  import xlwt
14.  from brukeropusreader import read_file #Package for reading opus file
15.  import os
16.
17.  #Define a baseline correction function using the Asymmetric least Square method
18.  def baseline_als(y, lam, p, niter=10):
19.      global z
20.      L = len(y)
21.      D = sparse.csc_matrix(np.diff(np.eye(L), 2))
22.      w = np.ones(L)
23.      for i in range(niter):
24.          W = sparse.spdiags(w, 0, L, L)
25.          Z = W + lam * D.dot(D.transpose())
26.          z = spsolve(Z, w * y)
27.          w = p * (y > z) + (1 - p) * (y < z)
28.      return z
29.
30.
31.  #Define a function composed of three Gaussian distribution functions,
32.  #each of which correspond to a secondary conformation in A. pernyi silk.
33.  def func3(x, a1, a2, a3, m2,s1, s2, s3):
34.      global gussan_1
35.      global gussan_2
36.      global gussan_3
37.      gussan_1 = a1 * np.exp(-((x - 1224) / s1) ** 2)
38.      gussan_2 = a2 * np.exp(-((x - m2) / s2) ** 2)
39.      gussan_3 = a3 * np.exp(-((x - 1270) / s3) ** 2)
40.      return gussan_1 + gussan_2 + gussan_3
41.
42.  #Select file path, the files are original files acquired Bruker FTIR spectrometer
43.  path="C:/Users/"
44.  txtfiles = os.listdir(path)
45.  print(txtfiles[:20])
46.  list1=[]
47.  list2=[]
48.  list3=[]
49.  list4=[]
50.  list5=[]
51.  list6=[]
52.  list7=[]
53.  list8=[]
```

```python
54.
55.  count = 0
56.  for i in range(len(txtfiles)):#Loop through each file
57.        #read file
58.        data=read_file(path+"/"+txtfiles[i])
59.        #Get wavenumber range of amide Ⅲ and Absorbance value corresponding to wavenumber
60.        axis_x=np.array(data.get_range("AB")[701:728])
61.        axis_y=np.array(data["AB"][701:728])
62.
63.        zs = savgol_filter(axis_y, 9, 3)  # Smoothing, window size 9, polynomial order 3
64.        cell = baseline_als(axis_y, 10000000, 0.001)# Acquire baseline
65.        sample = axis_y - cell# Subtract baseline
66.        #Fitting data with curve_fit Toolkit
67.        popt, pcov = curve_fit(func3, axis_x, sample,bounds=([0,0,0,1235,0,10,0],
68.            [np.inf,np.inf,np.inf,1250,14,np.inf,14]),loss="huber", maxfev=5000000)
69.        #Calculate the area of each Gaussian peak after fitting by integration
70.        a,a1=quad(lambda x:popt[0] * np.exp(-((x - 1224) / popt[4]) ** 2),1200,1300)
71.        b,b1=quad(lambda x:popt[1] * np.exp(-((x - popt[3]) / popt[5]) ** 2),1200,1300)
72.        c,c1=quad(lambda x:popt[2] * np.exp(-((x - 1270) / popt[6]) ** 2),1200,1300)
73.        #Calculate the relative content of each conformation.
74.        b_sheet=a/(a+b+c)
75.        random=b/(a+b+c)
76.        b_turn=c/(a+b+c)
77.        #Save data
78.        list1.append(txtfiles[i])
79.        list8.append(a)
80.        list2.append(b)
81.        list3.append(c)
82.        list4.append(b_sheet)
83.        list5.append(random)
84.        list6.append(b_turn)
85.        #Plot curves and review fitting results
86.        plt.plot(axis_x, sample, label='sample')
87.        plt.plot(axis_x, gussan_1, label='β-sheet')
88.        plt.plot(axis_x, gussan_2, label='Random coil')
89.        plt.plot(axis_x, gussan_3, label='β-turn')
90.        plt.plot(axis_x, func3(axis_x, *popt), color='green', label='fit', linestyle='--')
91.        plt.legend(bbox_to_anchor=(1, 1), loc='upper left', borderaxespad=0.)
92.        plt.savefig('D:/fig/{}'.format(txtfiles[i])+".jpg")
93.        plt.clf()
94.        count = count + 1
95.        print("\r Current progress: {:.2f}%".format(count * 100 / len(txtfiles), end=""))
96.  #Data summary
97.  list7.append(list1)
98.  list7.append(list8)
99.  list7.append(list2)
100. list7.append(list3)
101. list7.append(list4)
102. list7.append(list5)
103. list7.append(list6)
```

```
104. #Write the data into excel file
105. f = xlwt.Workbook()  # Create a workbook
106. sheet1 = f.add_sheet(u'sheet1', cell_overwrite_ok=True)
107. for y in range (len(list7)):
108.          j = 0
109.          ii = list7[y]
110.          for iii in ii:
111.                  sheet1.write(j,y,iii) #Loop write data
112.                  j=j+1
113.
114. f.save("D:/" + "data" + ".xls")  # Save file
115. print("Data saved successfully")
```

3.4.3 Calculations of Absorbance Profile and Molecular Order Parameter

The orientation of specific moieties can be obtained from the polarizing angular dependence (the angle between the polarized light and the fiber axis) of the absorbance $A(\nu)$ at certain wavenumber ν, which corresponds to a certain vibration of the molecular group under investigation (Fig. 2). Generally, the angular dependence of the absorbance can be determined by the following function [17–21].

$$A(\nu,\Omega) = -\log_{10}\{\left(10^{-A_{\max}(\nu)}\cos^2(\Omega-\Omega_0) + 10^{-A_{\min}(\nu)}\sin^2(\Omega-\Omega_0)\right\}$$ (1)

where $A(\nu, \Omega)$ is the peak intensity of a specific band, Ω is the polarization angle, Ω_0 is the polarizing angle at maximum absorption, and $A_{\max}$ and $A_{\min}$ are polarizing angles correspond to the maximum and minimum absorbance, respectively. The molecular order parameter (S^{mol}) of the corresponding secondary structural component can be calculated as follows [21].

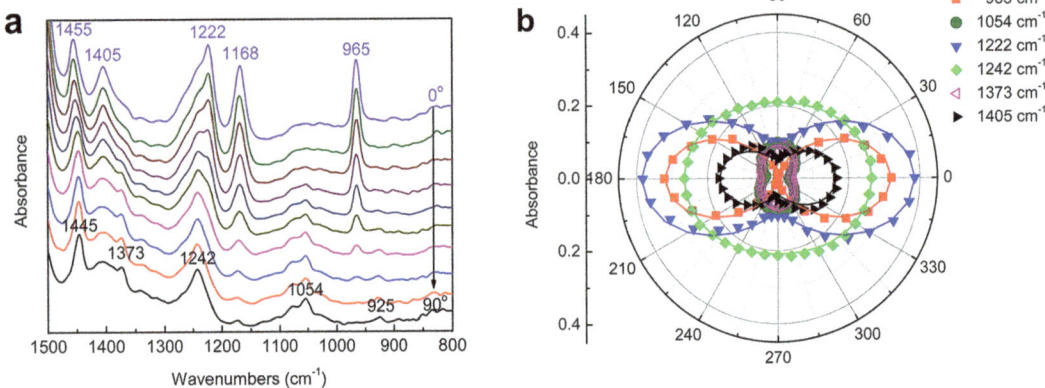

Fig. 2 (**a**) S-microFTIR spectra of single *A. pernyi* silk fibers with different polarization angle from 0° to 90°. (**b**) Polar plot of the absorbance of the characteristic peaks in S-microFTIR spectra in the 1500–800 cm^{-1} region from single *A. pernyi* silk fibers. The symbols represent individual experimental data points, while the curves are fitted using Eq. 1. (Adapted by permission from American Chemical Society [17], Copyright 2013)

$$S^{mol} = \frac{A_{max}(\nu) - A_{min}(\nu)}{A_{max}(\nu) + 2A_{min}(\nu)} \qquad (2)$$

4 Notes

1. In synchrotron radiation facilities, the infrared beamline stations usually equipped the synchrotron source, in which the brightness is 100 ~ 1000 times higher than that of a globar source. Thus, the spatial resolution of the synchrotron FTIR imaging system can reach the infrared diffraction limitation.

2. Sericin has little contribution to the mechanical properties of *B. mori* silk and *A. pernyi* silk, and it is not uniformly distributed on silk. Thus, a degumming process is needed before the test. During the degumming process, it is necessary to stir the silk cocoon for better degumming. Although the effect of the degumming process on the secondary structure of silk protein is not fully understood, previous results of wide-angle X-ray scattering have shown that the degumming process does not obviously change the content of β-sheet conformation in the filament.

3. Single silk fibers should be firmly fixed on the stage to prevent sample slippage during testing. For example, mount a sample on a hard-cardboard frame with a hollow in the middle and fix two edges with cyanoacrylate. After the cyanoacrylate is dried, mount the frame onto the stage. To obtain the high-quality infrared spectrum of silk monofilament, it is necessary to ensure that there is enough infrared light to pass through the aperture; however, strong scattering and diffraction of infrared light through the filament should also be avoided. This contradiction is mainly between the size of the aperture and the shape of the fiber. If the aperture is too small, there is little infrared light that can go through the silk monofilament, and the sufficient infrared light intensity will be very low. On the contrary, if the aperture is too large (even larger than the diameter of the silk monofilament), the infrared light will have a strong diffraction effect when it passes through the silk monofilament (the animal silk is generally cylindrical structure, and the diameter is equivalent to the wavelength of the infrared light, about 5–50 μm), resulting in the distortion of the spectrum. Therefore, it is necessary to determine the appropriate aperture size through the preliminary experiment.

4. It should be noted that the angle displayed on the polarizer does not necessarily correspond to the angle between the polarized light and the fiber axis. This depends on the direction

of samples placed on stage and insertion angle of polarizer in S-microFTIR spectrometer. The recommended way is to place the sample in the direction, where you can see monofilament horizontally or vertically displayed on the screen. After determining the direction of the sample, according to the principle of orthogonal extinction, an analyzer with the direction of zero angles vertical to fiber axial direction can be used to set the correct angle of the polarizer, which causes polarized infrared light parallel to the fiber axis.

5. Five or nine-point smoothing is recommended, as thirteen-point smoothing will cause distortion of the spectrum.

6. Generally, files that can be processed by PeakFit software are text file compiled by ASCII code, such as .csv, .txt, .prn, and .bat text files.

Acknowledgments

This work was supported by the National Natural Science Foundation of China [grant numbers. 51973116, U1832109, 21935002], the Users with Excellence Program of Hefei Science Center CAS [grant number 2019HSC-UE003], China Postdoctoral Science Foundation [grant number 2020M681344], the starting grant of ShanghaiTech University, and State Key Laboratory for Modification of Chemical Fibers and Polymer Materials.

References

1. Zhang W, Ye C, Zheng K, Zhong J, Tang Y, Fan Y, Buehler MJ, Ling S, Kaplan DL (2018) Tensan silk-inspired hierarchical fibers for smart textile applications. ACS Nano 12 (7):6968–6977. https://doi.org/10.1021/acsnano.8b02430

2. Sapede D, Seydel T, Forsyth VT, Koza MM, Schweins R, Vollrath F, Riekel C (2005) Nanofibrillar structure and molecular mobility in spider dragline silk. Macromolecules 38:8447–8453

3. Trancik JE, Czernuszka JT, Bell FI, Viney C (2006) Nanostructural features of a spider dragline silk as revealed by electron and X-ray diffraction studies. Polymer 47:5633–5642

4. Rousseau M-E, Hernández Cruz D, West MM, Hitchcock AP, Pézolet M (2007) Nephila clavipes spider dragline silk microstructure studied by scanning transmission X-ray microscopy. J Am Chem Soc 129:3897–3905

5. Seydel T, Kölln K, Krasnov I, Diddens I, Hauptmann N, Helms G, Ogurreck M, Kang S-G, Koza MM, Müller M (2007) Silkworm silk under tensile strain investigated by synchrotron X-ray diffraction and neutron spectroscopy. Macromolecules 40:1035–1042

6. Yang Z, Liivak O, Seidel A, LaVerde G, Zax DB, Jelinski LW (2000) Supercontraction and backbone dynamics in spider silk: 13C and 2H NMR studies. J Am Chem Soc 122:9019–9025

7. van Beek JD, Hess S, Vollrath F, Meier BH (2002) The molecular structure of spider dragline silk: folding and orientation of the protein backbone. Proc Natl Acad Sci 99:10266–10271

8. Rousseau M-E, Beaulieu L, Lefèvre T, Paradis J, Asakura T, Pézolet M (2006) Characterization by Raman microspectroscopy of the strain-induced conformational transition in fibroin fibers from the silkworm Samia cynthia ricini. Biomacromolecules 7:2512–2521

9. Sirichaisit J, Brookes VL, Young RJ, Vollrath F (2003) Analysis of structure/property relationships in silkworm (Bombyx mori) and spider

dragline (Nephila edulis) silks using Raman spectroscopy. Biomacromolecules 4:387–394

10. Blackledge TA, Cardullo RA, Hayashi CY (2005) Polarized light microscopy, variability in spider silk diameters, and the mechanical characterization of spider silk. Invertebr Biol 124:165–173

11. Vollrath F, Madsen B, Shao Z (2001) The effect of spinning conditions on the mechanics of a spider's dragline silk. Proc R Soc Lond Ser B Biol Sci 268:2339–2346

12. Hayashi CY, Lewis RV (2000) Molecular architecture and evolution of a modular spider silk protein gene. Science 287:1477–1479

13. Mi R, Shao ZZ, Vollrath F (2019) Creating artificial rhino horns from horse hair. Sci Rep 9:16233

14. Vollrath F, Knight DP (1999) Structure and function of the silk production pathway in the spider Nephila edulis. Int J Biol Macromol 24:243–249

15. Dumas P, Tobin MJ (2003) A bright source for infrared microspectroscopy: synchrotron radiation. Spectrosc Eur 15:17–23

16. Ling S, Qi Z, Knight DP, Shao Z, Chen X (2011) Synchrotron FTIR microspectroscopy of single natural silk fibers. Biomacromolecules 12:3344–3349

17. Ling S, Qi Z, Knight DP, Huang Y, Huang L, Zhou H, Shao Z, Chen X (2013) Insight into the structure of single Antheraea pernyi silkworm fibers using synchrotron FTIR microspectroscopy. Biomacromolecules 14:1885–1892

18. Papadopoulos P, Sölter J, Kremer F (2007) Structure-property relationships in major ampullate spider silk as deduced from polarized FTIR spectroscopy. Eur Phys J E Soft Matter 24:193–199

19. Papadopoulos P, Ene R, Weidner I, Kremer F (2009) Similarities in the structural organization of major and minor ampullate spider silk. Macromol Rapid Commun 30:851–857

20. Papadopoulos P, Sölter J, Kremer F (2009) Hierarchies in the structural organization of spider silk—a quantitative model. Colloid Polym Sci 287:231–236

21. Boulet-Audet M, Lefèvre T, Buffeteau T, Pézolet M (2008) Attenuated total reflection infrared spectroscopy: an efficient technique to quantitatively determine the orientation and conformation of proteins in single silk fibers. Appl Spectrosc 62:956–962

Chapter 18

Using FTIR Imaging to Investigate Silk Fibroin-Based Materials

Jiajia Zhong, Xiaojie Zhou, Chao Ye, Wenjie Yu, and Yuzhao Tang

Abstract

The secondary structures of silk fibroin (SF) are critical in the determination of the mechanical properties of the animal silks. Different characterization techniques, such as X-ray diffraction, nuclear magnetic resonance, Raman spectroscopy, and Fourier transform infrared (FTIR) technique, have been applied to study the secondary structure of animal silks. Among these techniques, FTIR is most widely used as it is sensitive to all secondary structures of proteins. Especially with the development of FTIR imaging, it is now possible to image the secondary structures of proteins at the micrometer scale, so as to understand the spatial distribution of proteins and the interaction of proteins with other materials at specific locations of interest. In this chapter, we present the methods and protocols of FTIR imaging to silk protein-based materials. We primarily introduce how to set up the instruments and accessories, as well as how to choose the appropriate imaging methods and sample preparation methods according to sample morphologies. The critical protocols for data analysis are also introduced in the last section.

Key words FTIR imaging, Silk protein, Multivariate imaging

1 Introduction

Silk fibroin (SF), derived from mulberry *Bombyx mori* (*B. mori*) silkworm cocoons, has been widely used for different task-specific applications, such as biomedicine, optics and energy and associated environmental field. In particular, in the last 20 years, silk fibroin has received numerous interests as a biomaterial due to its excellent mechanical properties, biocompatibility, and tunable degradability in vivo. Chemically, SF is primarily composed of a heavy chain (Mw ~390 KDa) and light chain (Mw ~26 KDa). Among them, the heavy chain is hydrophobic and contains blocks of GAGAGS repeat—a motif that is known to form anisotropic β-sheet-rich nanocrystals. In contrast, the light chain is more hydrophilic and relatively elastic.

Practically, in order to improve material properties and/or to achieve material functionalization, SF often blends with other

Shengjie Ling (ed.), *Fibrous Proteins: Design, Synthesis, and Assembly*, Methods in Molecular Biology, vol. 2347,
https://doi.org/10.1007/978-1-0716-1574-4_18, © Springer Science+Business Media, LLC, part of Springer Nature 2021

polymers, such as sodium alginate, chitosan, polyethylene oxide, and polyvinyl alcohol, or inorganic nanomaterials, such as hydroxyapatite and carbon nanomaterials. Due to the integration of the advantages of both SF and other materials, these SF-based materials have been fabricated into different materials, such as membranes, hydrogels, fibers, foams, and nanoparticles, to match the various functional needs in textiles, tissue engineering, and drug delivery.

For a composite, its phase behavior and the chemical structure of each component in different phases or specific locations are critical in determining the performance and function of the materials. The traditional methods to investigate polymer phase behavior include differential scanning calorimetry, scanning electron microscope, and dynamic thermomechanical analysis. These methods have provided sufficient morphologic information related to the phase behavior of SF composites, however, all of these methods are chemically insensitive, which thus are unable to provide the chemical structure information of each component. In particular, for SF component, these methods cannot give the conformational information in each phase.

In this aspect, FTIR related technique has a unique advantage. FTIR spectroscopy technique is the earliest and most common method to analyze the conformation of proteins because of its simplicity and nondestructive. As with other vibrational spectroscopy techniques, such as circular dichroism spectroscopy [1, 2], and Raman spectroscopy [3–5], FTIR spectroscopy is sensitive to the molecular vibrations of components. Samples in FTIR measurement can be solid, liquid, or even the gas. The IR vibrational frequencies of one given group depend on the bond strength and are also sensitive to the surrounding group environment.

For SF, the most critical spectral regions measured are typically the amide I region (1700–1620 cm^{-1}), amide II region (1550–1515 cm^{-1}), and amide III region (1275–1220 cm^{-1}). Various conformations of silk protein, such as β-sheet, random coil, or helix have different peak positions and intensities. Therefore, specific peak positions and peak intensities of certain conformation can be identified by peak splitting or deconvolution to amide bands, so as to study the conformational changes of silk protein [6, 7]. FTIR imaging technique provides a powerful toolkit to disclosure the conformation of SF blends in microscale spatial resolution and thereby can be used to characterization the structural properties of components in a specific phase [8–10]. Therefore, this chapter provides critical methods of using FTIR imaging to study the structure of SF–polymer blends. We first detail the materials and equipment needed in the FTIR imaging system. Then we describe the steps to prepare the instruments and accessories. We also give some ideas about choosing the appropriate FTIR imaging methods and different sample preparation methods to get a high-quality FTIR imaging map. Finally, the analysis protocols of the imaging data are also discussed.

2 Materials

1. Fourier-transform infrared (FTIR) spectrometer with mid-infrared capability.

2. Infrared microscope which should be well connected to FTIR spectrometer.

3. Infrared source, such as a globar source or synchrotron radiation source (*see* **Note 1**).

4. Detectors, such as single element liquid nitrogen-cooled mercury–cadmium–telluride (MCT) infrared detectors, linear array MCT detectors, and focal plane array (FPA) detectors (*see* **Note 2**).

5. Accessories, such micro attenuated total reflection (ATR), IR polarizer, microfluidic devices, and diamond anvil cell (DAC).

6. IR transmission windows, such as barium fluoride (BaF_2), calcium fluoride (CaF_2), or diamond (*see* **Note 3**).

7. IR reflection windows, such as gold-coated slides and Low emissivity glasses.

8. Freezing microtome.

9. Reagents: kalium bromatum (KBr), paraffin oil, heavy water (D_2O), optimal cutting temperature compound (OCT), and liquid nitrogen.

10. Silk fibroin powder, silk fibroin solution, silk film, and single silk fiber.

3 Methods

3.1 Preparation of Instruments

3.1.1 Preparation of FTIR Spectrometer and IR Microscope

1. Check that the FTIR spectrometer and IR microscope are placed in the room with stable temperature and humidity. Generally, the room temperature is about 17 ~ 27 °C, and the room humidity is about 30–50%.

2. Check that the FTIR spectrometer and IR microscope are purged with dry air or nitrogen (*see* **Note 4**).

3. Fill the MCT detector with liquid nitrogen, and wait about 30 min to get the stable signal.

4. Power on the IR microscope, auto sample stage and computer, and open the software to check the correct instrument parameter settings.

5. Select the appropriate size and shape of the aperture according to sample shape and size.

6. Adjust the positions of the sample stage and IR condenser of the microscope to make samples clear in the visual field and IR signal intensity maximum.

<table>
</table>

3.1.2 Preparation of Accessories

Micro ATR

1. Clean the ATR crystal (*see* **Note 5**).

2. Insert the ATR slide into the IR objective of IR microscope (*see* **Note 6**).

3. Mount the ATR sample holder on the microscope auto sample stage. There is a pressure sensor on the ATR sample holder.

4. Connect the ATR signal cable to a microscope. When the ATR crystal touches the sample, the indicator light on the microscope turns green. While pressure exceeds the set value, the indicator light turns red.

5. Check that the IR signal is optimized after installing the micro ATR accessory.

IR polarizer

1. Carefully take out the IR polarizer (*see* **Note 7**).

2. Rotate the appropriate polarization angle and insert the IR polarizer to the microscope. In the transmission imaging mode, the IR polarizer can be inserted in the light path before IR objective and after condenser, while in the trans-flection imaging mode, the IR polarizer can only be inserted in the light path before IR objective.

3. Check that the IR signal is optimized after installing the IR polarizer. Usually, the height and position of the condenser should be adjusted after inserting the IR polarizer in order to optimize the signal.

Microfluidic devices

FTIR imaging of SF aqueous solution remains challenging because of the strong infrared absorption of water. To meet this challenge, microfluidic devices have emerged as a method to control water thickness. In addition, FTIR imaging combined with microfluidic devices can investigate the kinetics of chemical reactions or protein folding. The microfluidic device contains a microfluidic chip in which the windows of the chip are IR transparent materials, injector, injection pump, peristaltic pump, and controller. The preparation steps of microfluidic devices are listed below:

1. Assemble the microfluidic chip and mount the chip between the lid and base. Usually, the base can control the temperature of the chip.

2. Connect the water inlet tube between the microfluidic chip and injection pump, while connecting the water outlet tube between the microfluidic chip and the peristaltic pump.

3. Put the microfluidic chip on the auto sample stage and power on the controller, and then the aqueous protein solution can be

injected into the channels of the microfluidic chip at a controlled speed.

Diamond anvil cell

Most of the pressure-induced elastic or conformational changes in proteins occur under pressure of 500–600 MPa, and this range of pressure is significantly high. FTIR imaging equipment, coupled with the DAC accessory, makes the investigation of the protein secondary structure under extremely high pressure. DAC contains two metal bases inlaid with diamond windows, and the two bases are fixed by few screws. The DAC can be pressurized by tightening the nuts on the screws evenly and gradually. There is also a certain thickness of the gasket between the two diamond windows, and a small hole is opened on the gasket to fill the sample. The preparation of DAC accessory and the sampling steps are listed below:

1. Clean the diamond windows and gasket with ethanol, and wait a few minutes for ethanol to evaporate.

2. If the sample is an aqueous protein solution, just fill the hole with the sample solution. If the sample is a protein powder, grind the solid powder to a particle size of less than 2.5μm, mix the protein powder with H_2O or D_2O, and then fill the hole with the mixed solution (*see* **Note 8**).

3. Add several grains of quartz sand with particle size less than 4μm, and operate under the microscope (*see* **Note 9**).

4. Close the two metal bases carefully and tighten the nuts on the screws evenly and gradually, then put the DAC onto the sample stage.

3.2 Imaging Modes and Sample Preparation Methods

3.2.1 Transmission Mode

In the mode of transmission imaging, the infrared beam passes through the sample and support window, and the beam energy is measured by the detector, then one transmission spectrum can be obtained. This method is suitable for samples with a smooth surface and thin thickness (5–10μm). Generally, the supporting window can be BaF_2, CaF_2, and diamond, which are transparent or partly transparent in the mid-infrared region. Preparation methods of SF materials with different morphologies are listed below:

1. SF powder:Firstly, grind a quantity of the protein powder with KBr powder, usually, the mass ratio is 1:150. Ensure that the crystal size of the mixture must less than 2.5μm in aim to eliminate scattering effects from large crystals. Secondly, press the mixture powder into a thin and transparent pellet by presser. If the pellet is too thick, some absorption bands will show full absorption, while if the pellet is too thin, the absorbance and the signal-noise (S-N) ratio may be too low. The absorbance between 0.5 and 1.4 is appropriate. Thirdly, put the pellet on the infrared window and the measurement can begin.

2. SF films: The most used method to prepare thin protein film is film casting. Firstly, the solid protein materials are dissolved into an aqueous solution. Secondly, one drop of the protein solution cast onto an IR window and allowed to dry at air conditions. Ensure that the thickness of the protein is around 5–20μm, and the film is dry enough before the FTIR imaging because water molecules have strong absorption at 3000–3750 cm^{-1} and 1600–1700 cm^{-1}. The later absorption band is overlapped with the absorption of the amide I band. Therefore, it is necessary to dry the sample enough. The second method to prepare thin protein film is using a microtome to cut a thin film from solid protein materials. Frozen section technology can make thin sections effectively. Firstly, the protein materials are placed in an embedding agent (usually is OCT) and then the mixtures are placed at −20 °C for 30 min. Secondly, transfer the cured sample and OCT to the sample holder and cut out a thin slice (5–20μm) with a freezing microtome. Thirdly, carefully transfer the slice to an IR window (*see* **Note 10**).

3. SF aqueous solution: the critical point to prepare the solution sample is to control the liquid thickness. Since the water absorption intensity at 1650 cm^{-1} and 3600–3200 cm^{-1} will be saturation when the thickness of the water layer is larger than 15μm and 2μm, respectively. One method is using a microfluidic device since the channel depth can be controlled accurately. Another general method is to use D_2O instead of H_2O. The absorption bands of D_2O are mainly at 1200 cm^{-1} and 2500 cm^{-1}, so the amide I band will not be affected by the absorption bands of D_2O.

4. Single SF fiber: usually, the SF fiber is cylindrical in morphology. For example, the diameters of spider silk and *Bombyx mori* silk fibers are approximately 2–50μm, which is equivalent to the IR wavelength. Therefore, a typical Mie scattering will happen when the IR beam passing through the fibers. The baseline of the spectra obtained will be distorted. One method to address this problem is using the accessory of micro compression cells which can flatten the fiber and eliminate the scattering. But it's vital to ensure that the compressing has no significant effect on the IR absorption properties of the SF fibers. The second method is using paraffin oil to wet the samples to decrease the scattering. Although paraffin oil has absorption band at 2950–2800 cm^{-1} and 1450 cm^{-1}, these two bands are not overlapped with the amide I and amide III bands of protein.

3.2.2
Trans-flection Mode

In the trans-flection mode, the incident infrared beam passes through the sample and then reflected from the substrate. The reflected light energy is collected by the detector, and the reflectivity spectrum is obtained subsequently. This method is often used to characterize the sample surface. The sample preparation methods are mostly the same as those described in transmission mode, but the IR windows are different. The most used window materials are gold-coated slides or low emissivity glasses, which have high IR light reflectivity. Under the same instrument setting conditions, the light intensity in the trans-flection mode is half of that in the transmission mode, and the reflection spectrum is usually distorted to a certain extent due to the influence of light scattering. For the samples suitable for both modes, the transmission mode is preferred.

3.2.3 Micro-ATR Imaging Mode

Samples that are hard to make thin slices by routine methods, such as soft, sticky, or hard materials, can be used in micro-ATR imaging mode. In a typical optical arrangement for a micro-ATR objective, the incident beam will be internally reflected. High refractive index ATR crystal can improve the number aperture of the imaging system, and the spatial resolution of ATR imaging can be improve according to the Rayleigh criterion [11]. Usually, the refractive index of Ge crystal is about 4, and the spatial resolution can be reached 8μm at wavelength approximately 6μm [12]. The penetration depth of the evanescent wave which generated on the surface of the ATR crystal is about 1–2μm. So there is no need to control the thickness of the sample precisely. SF aqueous solution and solid SF materials are suitable for this mode.

1. SF aqueous solution: firstly, fill an open liquid cell with a protein solution, and put the liquid cell on ATR sample holder. Secondly, adjust the height of the sample stage and let the ATR crystal fully contact the liquid surface.

2. Solid SF sample: Put the solid protein material on ATR sample holder and adjust the height of the sample stage. Let the sample contact with the ATR crystal directly. Ensure that the sample and crystal contact firmly. Since the evanescent wave into the sample is improved with more intimate contact. In addition, firmly press of the sample also can remove more trapped air, which can cause spectra distortion [13].

3.3 Data Recording

Once the instruments and the samples have been prepared and imaging mode has been selected, then the data recording can be started.

1. According to the sample morphologies, select an aperture with appropriate size and shape, adjust the positions of sample stage, and infrared condenser of the microscope to make the sample in visual field clear and the IR signal intensity maximum (*see* **Note 11**).

2. Check that the instrument parameters are set correctly. Especially, choose the right accessories (ATR or others), IR source

(IR or external source), and imaging mode (transmission mode, reflection mode or micro-ATR mode). Besides, set the appropriate scan times, scan resolution (*see* **Note 12**).

3. Move the auto sample stage, and find the blank area without sample, take the background spectroscopy.

4. Select an imaging area by rectangle tool, set the right step sizes in X and Y directions. Usually, the step size should be smaller than the aperture size, so that every pixel can be taken the IR information repeatedly.

5. Collect the map and save the data.

3.4 Data Processing

One imaging map contains hundreds of spectra according to the number of pixels in the mapping area. In order to obtain accurate protein conformation information, the data processing of every single spectrum should be done. The software used to process data has already been reviewed in other review [14].

3.4.1 Pre-processing

The aim of pre-processing is to improve the spectra quality and accuracy, and this method is primarily for the spectra that do not collect in an optimized condition. Pre-process can be divided into smoothing, baseline correction, normalization, and others.

If the mixture of SF powder and KBr powder are not well ground, the light scattering will happen and can cause the baseline distortion. In trans-flection mode and micro-ATR mode, the spectra are more affected by light scattering. In addition, interference fringes will appear when double window plates are used to support the sample. That baseline distortion can be corrected in the software such as Omnic and Opus. In another situation, the diameters of the spider silk fibers and *Bombyx mori* silk fibers are about 5–20μm, which can cause a typical Mie scattering. A 'mode based' baseline correction method can be adopted. A model is assumed to explain all sample effects, and rigorous theory is used to recover spectra. Extended multiplicative scattering correction (EMSC), resonant Mie scattering correction (RMieSC), and rubber band baseline correction can be used in this method [14–17].

Smoothing is aimed to de-noise the spectra, and it may be carried out with Savitzky–Golay (SG) smoothing. The more smoothing data points, the worse resolution of the spectrum (Generally take 9 points to smooth). In fact, if the S-N ratio of the spectrum is low when collecting the spectrum, the spectral resolution can be reduced (such as set to 8 cm^{-1}), and then the S-N ratio can be improved.

The purpose of normalized spectra is to eliminate the difference absorbance caused by sample thickness. It is convenient for the comparison of the peak intensity and peak area between different samples after normalization. Conventional normalization methods of biomaterial spectra include amide I/II normalization and vector normalization. However, amide I/II normalization is not suitable for SF materials. Because amide I/II absorption bands reflect protein conformation and configuration information, and those bands are variations between samples. The peak that does not change with sample difference should be the normalized peak. Vector normalization is usually applied to the normalization of second derivative spectra.

In some cases, reagents such as OCT and paraffin oil will inevitably mix with the sample. Therefore, the absorption peaks of these reagents may be included in the sample spectra, and it's necessary to eliminate the effect by differential spectra.

3.4.2 Univariate Imaging

Univariate imaging, also known as chemical imaging or functional group imaging, is the most basic and commonly used method in FTIR imaging. In this method, the peak intensity or peak area of the characteristic absorption peak is integrated at each pixel. Then two-dimensional (2D) or three-dimensional (3D) infrared spectral imaging map is assembled according to the position of each pixel. Chemical imaging can directly show the distribution of specific components in the imaging area. The peak assignments of animal silks in the infrared region have been summarized in the previous review [13].

In order to identify the peak position accurately and eliminate the influence of baseline distortion, the calculation of second derivative spectra is also used in chemical imaging. However, the second derivative spectra cause a significant loss in signal-to-noise ratio. Therefore, smoothing must be done in the pre-processing.

Peak-fitting of amide I or amide III is usually conducted to evaluate the content of protein conformation. Reconstruction of the content of conformation at each pixel can also obtain a 2D or 3D imaging map to identify the conformation distribution. In the peak-fitting process, firstly, calculate the second derivatives or deconvolution spectra, then the peak positions and peak number can be identified. Secondly, input parameters to the software, such as Peakfit 4.12 (SPSS Inc.), using iterative least-squares progress to estimate the width, height, and shape of identified peaks. Thirdly, after the final iteration, the integrated area of each identified peak and the percentage of this peak area relative to the total band area is calculated [13, 18].

3.4.3 Multivariate Imaging

Although the univariate imaging method is simple and easy to operate, as mentioned above, univariate imaging is usually according to the intensity or area of one specific absorption peak. Therefore, when the difference of the single spectrum between pixels is small, the result obtained by the univariate imaging method is not reliable. Therefore, Lasch et al. [19, 20] developed multivariate

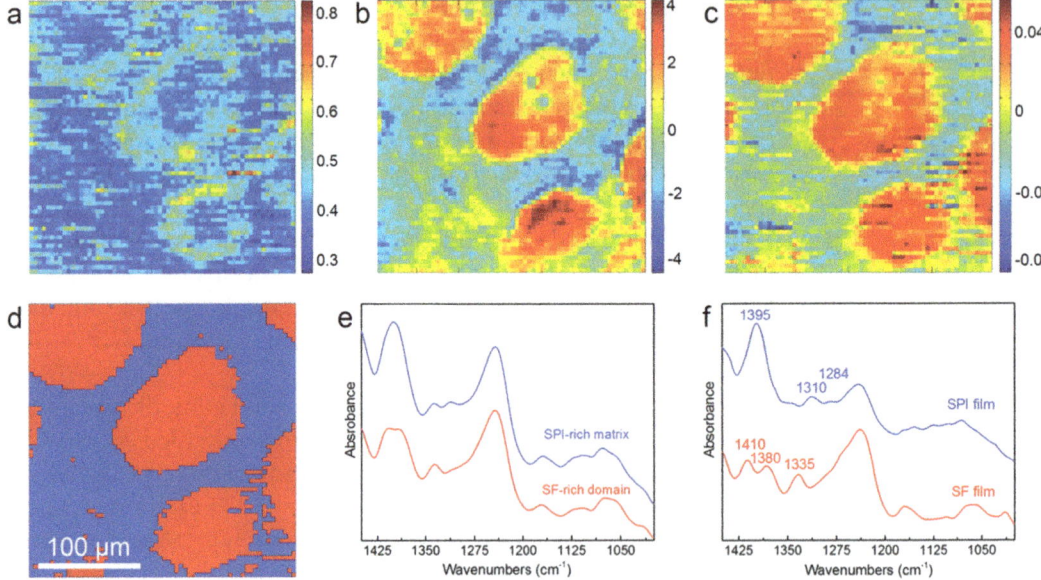

Fig. 1 FTIR images of silk fibroin–soy protein isolate (SF/SPI) blend film obtained by FPA detector and FTIR spectra analysis. (**a**) Amide III/Amide II image. (**b**) PCA image, calculated from Resonance Mie scattering multiplicative scattering correction spectra. (**c**) PCA image, calculated from first derivative spectra. (**d**) Clustering analysis image, calculated from first derivative spectra. (**e**) The mean spectra of each cluster in image (**d**), in which the red and blue plots are the mean spectra of the red domain and blue matrix, respectively. (**f**) The single spectra extracted from the pristine SF and SPI film in (**b**). (Adapted by permission from The Royal Society of Chemistry [8], Copyright 2015)

infrared spectral imaging methods based on multivariate pattern recognition methods, mainly including principal component analysis (PCA) imaging and clustering imaging (Fig. 1). These multivariable imaging methods can not only select a specific absorption peak for analysis but also can use the whole spectral range or specific wavenumber interval for analysis, so those multivariate imaging methods can overcome the shortcomings of univariate imaging to a certain extent.

PCA is a method to reduce the dimension of the original variables. The new variables called principal components (PCs) can reflect most of the information provided by the original variables. The number of PCs is determined according to the cumulative contribution rate of the PCs, usually, the cumulative contribution rate should reach to 85–95%. After determining the number of PCs, the loading and score of each principal component can be further calculated. In a loading curve, the X-axis is the wavenumbers, and Y-axis is the coefficient. According to the absolute value of Y-axis, the correlation of this PC on a specific wavenumber is determined. In PCA imaging method, first several PCs of each pixel spectrum are resembled a 2D or 3D imaging map. The difference of each pixel can be distinguished by color difference from the reconstruction imaging map (Fig. 1b–d) [8]. PCA is an unsupervised statistical method, which aims at dimensionality reduction rather than classification.

Clustering is used to classify data into different classes or clusters. Therefore, objects in the same cluster have great similarities, while objects in different clusters have great differences. The most commonly used clustering method in FTIR imaging is hierarchical clustering (HCA) [19]. Firstly, the single spectrum at each pixel is regarded as a separate category, and then spectra with the highest similarity (according to the shortest distance, such as Euclidean distance or Markov distance) are merged into one cluster at the same hierarchy. Then the spectra are merged and classified successively until all the spectra are merged into the top layer of the hierarchy, or a termination condition is reached. After clustering, the spectra at the same hierarchy are labeled as the same color, and then the colors are arranged according to the spatial position of the pixel. A 2D pseudocolor map is reconstructed, which can reflect the similarity and difference of each pixel. HCA is a supervised statistical method that the number of hierarchies should be determined at first. However, it is impossible to know the number of hierarchies in advance. Therefore, we must evaluate the clustering results repeatedly so as to determine the appropriate number of hierarchies.

4   Notes

1. In general laboratories, commercial infrared spectrometers are usually equipped with globar sources that have high stability and low brightness. While in synchrotron radiation facilities, the infrared beamline stations usually equipped the synchrotron source, in which the brightness is 100 ~ 1000 times higher than that of a globar source. So, the spatial resolution of the synchrotron FTIR imaging system can reach the infrared diffraction limitation.

2. Certain FPA detectors contain 64×64 pixels, and each pixel can generate one separate FTIR spectrum at the same time. So, imaging efficiency can be significantly improved.

3. The sizes of IR transmission windows are 20–30 mm in diameter and 0.9–1 mm in thickness, in order to match the sample holder.

4. FTIR gas generator should be equipped to remove most of the CO_2 in dry air. Therefore, the influence of CO_2 absorption bands to sample IR spectroscopy can be minimized. When purging with nitrogen, the oxygen concentration in the room should be concerned especially in a closed room.

5. The crystal of ATR accessory can be ZnSe, germanium, silicon, and diamond. Generally, these crystals have different chemical properties. Therefore, the cleaning procedures should be done according to the product manuals.

6. Micro ATR can be divided into two kinds, slide-on ATR, and crystal-changeable ATR objective. Slide-on ATR can be used in the Nicolet Continuμm microscope, so take it for example.

7. IR polarizer can be divided into two kinds, wire grid polarizer, and electrolyte polarizer. Here take the wire gird to polarize, for example. The material of the wire grid polarizer can be ZnSe or ZnS film, and 1200 parallel gold or aluminum wires per millimeter are plated on the film surface. So, one should be careful not to touch the polarizer surface.

8. Sample filling should be operated under a microscope. In the test of a solid protein sample, H_2O and D_2O are often used as mixture solution owing to that liquid can transfer pressure evenly. Since water and heavy water can become ice under high pressure, the absorption bands of ice will not affect the absorption bands of interests.

9. One crucial step in a high-pressure infrared experiment is to measure the pressure accurately. Generally, the absorbance peak of crystal quartz at 800 cm^{-1} is used as the internal standard to calculate the pressure.

10. OCT is a water-soluble mixture of polyethylene glycol (PEG) and polyvinyl alcohol (PVA), and it has an absorption band at 1500–1200 cm^{-1}, which is overlapped with the absorption of amide III bands of proteins, so it is should be careful not to contaminate the sample with the OCT.

11. Generally, a square aperture is chosen for imaging SF film, and a rectangle aperture is selected from for SF fiber. For example, a $5\mu m \times 10\mu m$ size aperture is used in spider silk measurement. In addition, the smaller size of the aperture, the worse the signal-to-noise ratio.

12. The scan times are often set to 64; the more scan times, the better the spectra signal-to-noise ratio can be obtained. However, the imaging time may be multiplied, since one mapping picture contains hundreds of pixels. Generally, the scan resolution is set as 4 cm^{-1}.

Acknowledgements

We thank the BL01B beamline of National Center for Protein Science Shanghai (NCPSS) at Shanghai Synchrotron Radiation Facility.

References

1. Dicko C, Knight D, Kenney J, Vollrath F (2004) Structural conformation of spidroin in solution: a synchrotron radiation circular dichroism study. Biomacromolecules 5:758–767

2. Li G, Shao Z, Xie X, Chen X, Wang H, Chunyu L, Yu T (2001) The natural silk spinning process. Eur J Biochem 268:6600–6606

3. Shao Z, Vollrath F, Sirichaisit J, Young RJ (1999) Analysis of spider silk in native and supercontracted states using Raman spectroscopy. Polymer 40:2493–2500

4. Monti P, Taddei P, Freddi G, Asakura T, Tsukada M (2001) Raman spectroscopic characterization of Bombyx mori silk fibroin: Raman spectrum of Silk I. J Raman Spectrosc 32:103–107

5. Rousseau M-E, Lefèvre T, Beaulieu L, Asakura T, Pézolet M (2004) Study of protein conformation and orientation in silkworm and spider silk fibers using Raman microspectroscopy. Biomacromolecules 5:2247–2257

6. Fang G, Huang Y, Tang Y, Qi Z, Yao J, Shao Z, Chen X (2016) Insights into silk formation process: correlation of mechanical properties and structural evolution during artificial spinning of silk fibers. ACS Biomater Sci Eng 2:1992–2000

7. Ling S, Qi Z, Knight DP, Shao Z, Chen X (2011) Synchrotron FTIR microspectroscopy of single natural silk fibers. Biomacromolecules 12:3344–3349

8. Ling S, Qi Z, Shao Z, Chen X (2015) Determination of phase behaviour in all protein blend materials with multivariate FTIR imaging technique. J Mater Chem B 3:834–839. https://doi.org/10.1039/C4TB01808G

9. Ling S, Qi Z, Watts B, Shao Z, Chen X (2014) Structural determination of protein-based polymer blends with a promising tool: combination of FTIR and STXM spectroscopic imaging. Phys Chem Chem Phys 16:7741–7748

10. Ling S, Qi Z, Knight D, Shao Z, Chen X (2013) FTIR imaging, a useful method for studying the compatibility of silk fibroin-based polymer blends. Polym Chem 4:5401–5406. https://doi.org/10.1039/C3PY00508A

11. Kazarian S, Chan KLA (2010) Micro and macro-attenuated total reflection fourier transform infrared spectroscopic imaging. Appl Spectrosc 64:135–152

12. Chan KLA, Kazarian S (2003) New opportunities in micro- and macro-attenuated total reflection infrared spectroscopic imaging: spatial resolution and sampling versatility. Appl Spectrosc 57:381–389

13. Zhong J, Liu Y, Ren J, Tang Y, Qi Z, Zhou X, Chen X, Shao Z, Chen M, Kaplan DL, Ling S (2019) Understanding secondary structures of silk materials via micro- and nano-infrared spectroscopies. ACS Biomater Sci Eng 5:3161–3183

14. Baker M, Trevisan J, Bassan P, Bhargava R, Butler H, Dorling K, Fielden P, Fogarty S, Fullwood N, Heys K, Hughes C, Lasch P, Martin-Hirsch P, Effiong B, Sockalingum G, Sulé-Suso J, Strong R, Walsh M, Wood B, Martin F (2014) Using Fourier transform IR spectroscopy to analyze biological materials. Nat Protoc 9:1771–1791

15. Bassan P, Byrne HJ, Bonnier F, Lee J, Dumas P, Gardner P (2009) Resonant Mie scattering in infrared spectroscopy of biological materials – understanding the 'dispersion artefact'. Analyst 134:1586–1593

16. Bassan P, Kohler A, Martens H, Lee J, Byrne H, Dumas P, Gazi E, Brown M, Clarke N (2010) Resonant Mie Scattering (RMieS) correction of infrared spectra from highly scattering biological samples. Analyst 135:268–277

17. Bassan P, Kohler A, Martens H, Lee J, Jackson E, Lockyer N, Dumas P, Brown M, Clarke N (2010) RMieS-EMSC correction for infrared spectra of biological cells: extension using full Mie theory and GPU computing. J Biophotonics 3:609–620

18. Ling S, Qi Z, Knight DP, Huang Y, Huang L, Zhou H, Shao Z, Chen X (2013) Insight into the structure of single Antheraea pernyi silkworm fibers using synchrotron FTIR microspectroscopy. Biomacromolecules 14:1885–1892

19. Lasch P, Haensch W, Naumann D, Diem M (2004) Imaging of colorectal adenocarcinoma using FT-IR microspectroscopy and cluster analysis. Biochim Biophys Acta 1688:176–186

20. Lasch P, Diem M, Hänsch W, Naumann D (2006) Artificial neural networks as supervised techniques for FT-IR microspectroscopic imaging. J Chemom 20:209–220

Chapter 19

Secondary Structure Analysis of Single Silk Nanofibril through Infrared Nanospectroscopy

Yawen Liu, Hongchong Guo, and Shengjie Ling

Abstract

Infrared nanospectroscopy (NanoIR) is a new experimental technique to research the secondary structure of protein-based nanoarchitectures in recent years. Compared with the conventional IR, NanoIR reveals to be an exquisite, sensitive, and accurate tool to analyze and image the single molecule secondary structure, which can reach up to high spatial resolution (10 nm). Here we present a detailed protocol to introduce how to study single silk nanofibril (SNF) and process the results by this routine. This protocol provides a useful method to demonstrate the microstructure of nanomaterials by NanoIR, displaying the potential application in analytical chemistry, biomaterials, and nanotechnologies.

Key words Infrared nanospectroscopy, Silk nanofibril, Second structure, Nanomaterials

1 Introduction

Silk nanofibrils, as the important components of natural materials [1], exhibit the great mechanical properties such as high tensile strength and toughness due to the unique hierarchical structure [2]. The unique structure–property–function relationship has inspired the design and applications of synthetic fibers [3–5].

Exploring the structure of a single protein nanofibril is critical to reveal the structure–property–function relationship of natural materials. The technologies such as synchrotron Fourier transform infrared microspectroscopy (S-FTIR, diffraction limit, ~2.5–75μm [6–8]), Raman spectroscopy (~2μm [7]), and synchrotron small-angle X-ray scattering (spot size, $25 \times 10\mu m^2$) [9] has been applied in detecting the structure of single animal fiber at the micrometer level. However, it remains a significant challenge to examine the structure of a single nanofibril at the nanometer level.

NanoIR, a new technology that combines the chemical analysis capability of infrared spectroscopy (IR) with the spatial resolution of atomic force microscopy (AFM), focuses the tunable infrared laser onto the sample by a probe tip [6, 10, 11] (Fig. 1). The tip will

Shengjie Ling (ed.), *Fibrous Proteins: Design, Synthesis, and Assembly*, Methods in Molecular Biology, vol. 2347,
https://doi.org/10.1007/978-1-0716-1574-4_19, © Springer Science+Business Media, LLC, part of Springer Nature 2021

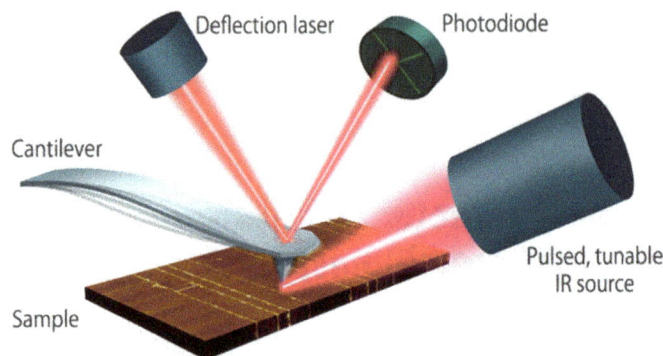

Fig. 1 Schematic diagram of nanoIR. (Adapted by permission from American Chemical Society [6], Copyright 2017)

detect the absorbed radiation when the IR laser is changed to a wavelength which is accorded with the absorption wavelength of the sample by the thermal expansion. And the thermal expansion of the sample causes the force impulse on the tip and the cantilever, inducing the oscillation of the AFM probe. Thus, nanoIR technology can overcome the spatial resolution limits of S-FTIR [6], and provide the structural information and chemical analysis with the specific spatial resolution on length scales of 10 nm.

In our recent study, we first proposed the nanoIR technology to analyze the secondary structure of the single silk nanofibril. The spectra showed amide I and amide II band, which are two of the most important bands of silk protein (Fig. 2). As one of the most important bands to investigate the conformation of the protein, amide I was similar to the conventional FTIR spectra after deconvolution calculation. And the result claimed that the nanoIR technique is suitable for semiquantitative analysis of secondary structure in single silk nanofibril. In addition, the molecular orientation of silk fibroin is also important for applications, and we first investigated the molecular orientation of a single silk nanofibril by polarized nanoIR. The difference of the spectra in the parallel and perpendicular direction is consistent with the silk fiber tested by S-FTIR. Hence, nanoIR is a promising technology to reveal the structure and chemical conformation of the protein nanofibrils.

Figure 2a, b shows the AFM and related nanoIR spectra collected from the silk nanofibrils, respectively. These sites are located on both single SNF (red points and plots), SNFs (black plots) and substrate (blue plots) [12]. The region includes amide I and amide II, which are the most important bands to investigate the conformation of proteins. To testify the reliability of the nanoIR in testing protein nanofibrils, we tested the SNF films assembled by vacuum filtration and the FTIR spectrum shows the same bands of amide I and amide II at the wavenumber between 1800 and 1400 cm^{-1} (Fig. 2c). The deconvolution of amide I can obtain the relative

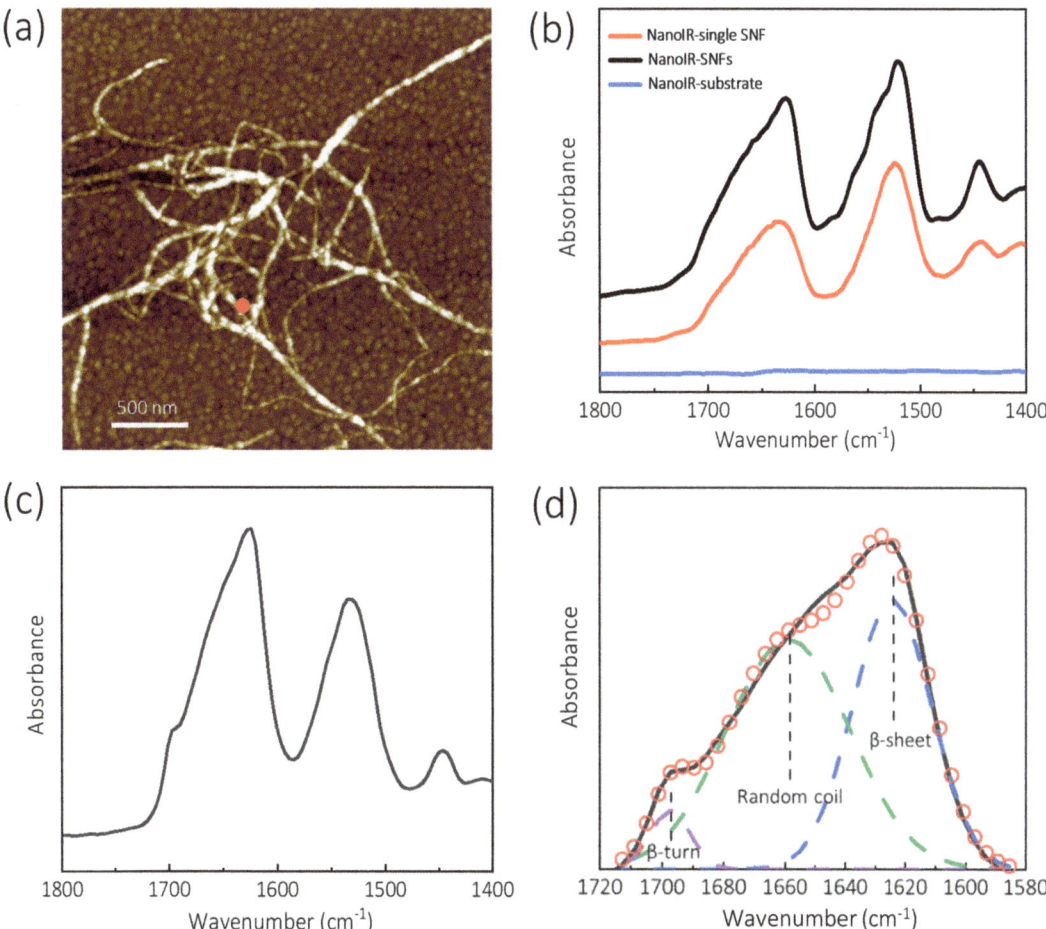

Fig. 2 (**a**) AFM image of SNFs collected by nanoIR system. (**b**) NanoIR spectra of single nanofibril, nanofibrils and substrate. NanoIR spectra of single SNF were collected from the red cross-shaped points shown in Fig. 2a. (**c**) FTIR spectrum of SNF film obtained by transmission mode testing. (**d**) the deconvolution results of the amide I band in Fig. 2c. Original spectrum (solid curve); Deconvoluted peaks (dashed curve); Simulated spectrum from summed peaks (circles). (Adapted by permission from John Wiley & Sons [12], Copyright 2020)

content of different secondary structures. As shown in the Fig. 2c, the FTIR spectrum of SNF film has a specific peak at 1630 cm^{-1} and along with a shoulder peak at 1700 cm^{-1}. The assignment of the peaks in the amide I are as follows: the peak from 1620–1630 cm^{-1} belonged to β-sheet; the peak from 1650–1660 cm^{-1} assigned to the random coil and/or helix; and the small peak from 1690–1700 cm^{-1} attributed to β-turn of the hairpin-folded antiparallel β-sheet structure [12–15]. The deconvolution of the amide I band of the FTIR spectra of SNF films provides the content of β-sheet, random coil/helix, and β-turn as 41 ± 1%, 54 ± 2%, and 5 ± 1%, respectively [15, 16] (Fig. 2d). Using the same methods to processing the nanoIR spectra of single SNF, indicating 40 ± 1% of β-sheet, 57 ± 2% of random coil/helix, and $3 \pm 2\%$ of β-turn is consistent with the conformation content obtained by deconvolution of transmission-mode spectra of FTIR. The result reveals that the nanoIR spectra can be used to evaluate the relative content of the secondary structure of a single protein nanofibril.

However, due to the limit of signal-to-noise ratio in nanoIR spectra, the prerequisite of semiquantitative analysis is that the peak position used for the deconvolution of nanoIR spectrum is determined by the transmission spectrum. Therefore, in principle, semi-quantitative analysis of nanoIR spectra is only suitable for the conformation-wavenumber relations that have been strictly verified. For an unknown protein, or a protein with unclear conformation assignments, the accuracy of the deconvolution method needs to be further studied.

2   Materials

1. Silk cocoon: *B. mori* silkworm cocoon bought from Zhejiang Province.

2. Solvents: sodium bicarbonate ($NaHCO_3$), Milli-Q water (ultrapure water), lithium bromide (LiBr).

3. Dialysis cassette (3500 MWCO), the clean substrate of mica plate with 12 mm diameter.

4. Fourier-Transform Infrared (FTIR): Data collected by infrared spectrometer of Bruker VERTEX-70 and data processed by the OPUS software.

5. Atomic force microscopy (AFM): Instrument: Dimension Fastscan. Mode: Tapping Mode.

6. AFM tips: RTESPA-300 with the resonant frequency of ~300 kHz and force constant 40 N/m.

7. AFM-IR substrate: Clean silicon wafer 10 mm × 10 mm with coated gold layer 100 nm and 600μm thickness.

8. Instrument: NanoIR3-FS AFM (Anasys Instrument). Mode: Tapping IR Mode.

9. AFM-IR tips: PR-EX-TnIR-A-10 with the resonant frequency of 75 ± 15 and Spring constant 1–7 N/m.

3   Methods

3.1 Preparation of Silk Aqueous Solution

Dissolving the *B. mori* silkworm cocoon to the silk aqueous solution is the first step to produce SNFs by bottom-up method. The concentration can influence the velocity of the silk protein assembly.

3.1.1 Degumming of Silk Cocoon	1. Weigh 10 g *B. mori* silkworm cocoon and cut into slices.
	2. Weigh 10 g NaHCO$_3$.
	3. *B. mori* silkworm slices degummed by boiling in two 30 min changes of 0.5% (w/w) NaHCO$_3$ solution with stirring, and wash with distilled water about 3 times to remove residual sericin (*see* **Note 1**).
	4. Move the degummed *B. mori* silkworm slices to air-dry at room temperature for 3 days.

3.1.2 Dissolving of Silk Fibroin Protein

1. Prepare the 9.3 mol/L LiBr solution.
2. 10 g degummed silk fibroin added to 100 mL 9.3 mol/L LiBr solution then put it to the water bath at 60 °C for 1 h with continuously stirring (*see* **Note 2**).

3.1.3 Dialyzing of the Silk Fibroin Protein

1. Dissolved silk fibroin protein is dialyzed against DI water for 3 days with exchanging water for about 3 h.
2. Centrifugation treatment of the silk fibroin protein for 30 min at 5–10 °C (9000 rpm, 1040 × *g*).

3.1.4 Calculate the Concentration of Silk Fibroin Protein

Calculate the concentration of silk fibroin protein by weighting method, % = (W$_3$ − W$_1$/W$_2$ − W$_1$) × 100.

3.2 Assembly of SNFs

1. Adjust the concentration of the silk fibroin protein to 0.1 wt% by adding DI water.
2. Heat-induced self-assembly. Incubate silk fibroin protein at 60 °C for 1 week (*see* **Note 3**).
3. Ethanol treatment self-assembly. Incubate silk fibroin protein with 7 vol% ethanol at pH 9.5 and room temperature for over 3 days.

3.3 Spectroscopy

3.3.1 FTIR Processing

1. Prepare films for the FTIR test. Films of SNF are fabricated by vacuum-filtrating the silk nanofibrils solution through filtration membranes and the pore size is 0.2μm (*see* **Note 4**). Then put the films in the vacuum drying oven for 2 days to remove water (*see* **Note 5**).
2. FTIR measurements. Open the switch and check the status of the instrument. Load the recording parameters of transmission mode. Check the signal if it reaches the detector. Set the parameters of the spectrum with a resolution of 8 cm^{-1} and 128 scans. Collect the spectra of background and films in the range of 4000 to 800 cm^{-1}.

3.4 Microscopy

3.4.1 AFM

1. Deposit 10 μL 0.001 wt% SNF solution on the clean mica plate and dry at room temperature.

2. Record height, amplitude, and phase images at tapping mode (*see* **Note 6**).

3. Process the images using the Analysis Studio v3.15 software and the images are flattened with flattening order of 2nd.

3.4.2 AFM-IR

1. Deposit 10 μL 0.001 wt% SNF solution on the clean silicon wafer with a coated gold layer and dry at room temperature (*see* **Note 7**).

2. Open the N_2 valve, circulating water, and instrument switch. After initializing the instrument and installing the tip, check the performance of the tip (*see* **Note 8**).

3. Record height, amplitude, and phase images at the AFM plate of Tapping-IR mode (*see* **Note 9**).

4. Change into contact mode then adjust the IR light under the AFM tip (*see* **Note 10**). Check and optimize the IR signal (*see* **Note 11**). Set the parameters of the spectrum with a resolution of 1 cm^{-1} and 3 scans. Record the background spectrum before recording the sample spectrum (*see* **Note 12**). Collecting the full spectrum and optimizing the signal at the wavenumber corresponding to the remarkable IR peak. Collect an optimized IR spectrum at the sample location.

5. Change into the tapping-IR mode and then set the interesting wavenumber where the infrared amplitude is strong. Select IR Imaging Enabled mode and then select the mapping image at the localized wavenumber.

6. After the test, deinitialized the instrument and close the N_2 value.

7. Data process: use the software of Analysis Studio v3.15. Flatten before output the AFM image. Select Savitzky–Golay or smooth to process the spectrum. Select the filter of Step Discontinuity to delete the shape gap of the spectrum (*see* **Note 13**).

4 Notes

1. When degumming the silk cocoon, please keep stirring to remove the sericin efficiently. Be careful of the hot water.

2. If the silk fibroin does not dissolve completely, please elongate the time for dissolving.

3. Incubate silk fibroin protein until it shows bright iridescent colors under polarized light. If not, elongate the time for incubating.

4. 40 mL 0.1% SNF solution is used to filtrate the membrane. The volume of the SNF solution should be well controlled because the quality of the spectrum is bad if the membrane is too thick.

5. The residual water of SNF can influence the quality of the FTIR spectrum. Keep the SNF membrane dry before testing.

6. The height of a single SNF is about 2–6 nm and the SNF is viscous. Tapping mode can avoid the SNF sticking to the tips to obtaining a high-quality spectrum.

7. If the sample is thin, the intensity of the IR will be weak, then using a silicon wafer with a coated gold layer as the substrate can improve the signal intensity efficiently.

8. Remove the water under the soft N_2 and ensure the sample bin is sealed.

9. Tapping IR mode or tapping mode is suitable to test the protein silk nanofibrils.

10. The contact mode is more suitable to collect the full spectrum because the tips can absorb more IR signals in this mode.

11. The optimal steps need to be done repeatedly until the strongest IR signal at the set wavenumber is acquired.

12. Collect a new background spectrum when you change a new sample. The software can perform the IR signal of the background automatically when save or load the background spectrum.

13. The shape gap may form at the junction point between the two-laser segment. The filter of Step Discontinuity can eliminate the gap effectively.

Acknowledgments

This work was supported by the National Natural Science Foundation of China [grant numbers. 51973116, U1832109, 21935002], the Users with Excellence Program of Hefei Science Center CAS [grant number 2019HSC-UE003], the starting grant of ShanghaiTech University, and State Key Laboratory for Modification of Chemical Fibers and Polymer Materials.

References

1. Ling S, Kaplan DL, Buehler MJ (2018) Nanofibrils in nature and materials engineering. Nat Rev Mater 3:18016

2. Keten S, Xu Z, Ihle B, Buehler MJ (2010) Nanoconfinement controls stiffness, strength and mechanical toughness of β-sheet crystals in silk. Nat Mater 9:359–367

3. Keten S, Xu Z, Ihle B, Buehler M (2010) Nanoconfinement controls stiffness, strength and mechanical toughness of beta-sheet crystals in silk. Nat Mater 9:359–367

4. Meyers MA, Hodge AM, Roeder RK (2008) Biological materials science and engineering:

biological materials, biomaterials, and biomimetics. JOM 60:21–22

5. Naleway SE, Porter MM, McKittrick J, Meyers MA (2015) Structural design elements in biological materials: application to bioinspiration. Adv Mater 27:5455–5476

6. Dazzi A, Prater CB (2017) AFM-IR: technology and applications in nanoscale infrared spectroscopy and chemical imaging. Chem Rev 117:5146–5173

7. Ling S, Qi Z, Knight DP, Shao Z, Chen X (2011) Synchrotron FTIR microspectroscopy of single natural silk fibers. Biomacromolecules 12:3344–3349

8. Ling S, Qi Z, Knight DP, Huang Y, Huang L, Zhou H, Shao Z, Chen X (2013) Insight into the structure of single Antheraea pernyi silkworm fibers using synchrotron FTIR microspectroscopy. Biomacromolecules 14:1885–1892

9. Van der Veen J, Pfeiffer F (2005) Coherent x-ray scattering. J Phys Condens Matter 16:5003–5030

10. Ruggeri FS, Longo G, Faggiano S, Lipiec E, Pastore A, Dietler G (2015) Infrared nanospectroscopy characterization of oligomeric and fibrillar aggregates during amyloid formation. Nat Commun 6:7831

11. Ruggeri FS, Vieweg S, Cendrowska U, Longo G, Chiki A, Lashuel HA, Dietler G (2016) Nanoscale studies link amyloid maturity with polyglutamine diseases onset. Sci Rep 6:31155

12. Xiao Y, Liu Y, Zhang W, Qi P, Ren J, Pei Y, Ling S (2020) Formation, structure, and mechanical performance of silk nanofibrils produced by heat-induced self-assembly. Macromol Rapid Commun 42(3):e2000435

13. Ling S, Li C, Adamcik J, Shao Z, Chen X, Mezzenga R (2014) Modulating materials by orthogonally oriented β-strands: composites of amyloid and silk fibroin fibrils. Adv Mater 26:4569–4574

14. Ling S, Qin Z, Huang W, Cao S, Kaplan DL, Buehler MJ (2017) Design and function of biomimetic multilayer water purification membranes. Sci Adv 3:e1601939

15. Ling S, Qi Z, Knight DP, Shao Z, Chen X (2013) FTIR imaging, a useful method for studying the compatibility of silk fibroin-based polymer blends. Polym Chem 4:5401–5406

16. Ling S, Qi Z, Watts B, Shao Z, Chen X (2014) Structural determination of protein-based polymer blends with a promising tool: combination of FTIR and STXM spectroscopic imaging. Phys Chem Chem Phys 16:7741–7748

Chapter 20

Method of Using Raman Spectroscopy to Understand the Conformation of Fibrous Proteins

Wenli Gao, Ting Shu, Qiang Liu, Hongchong Guo, Shengjie Ling, and Liang Zhou

Abstract

Raman spectroscopy has been widely used in the research of fibrous proteins because of the insensitivity to moisture, less amount of sample, and better signal-to-noise ratio. In recent years, Raman spectroscopy is adopted to investigate the secondary structures of solid or aqueous protein, the conformation transition under different conditions (concentration, temperature, pressure, pH, chemical modification, external force, etc.), the orientation of the molecular chains, and some important chemical bonds. Here, we will introduce the methods for using Raman spectroscopy to analyze the conformation and orientation of samples, which would be an efficient method to get the "structure–property" relationship.

Key words Raman Spectroscopy, Fibrous Proteins, Conformation

1 Introduction

Raman spectroscopy was discovered by the Indian scientist Sir C.V. Raman in 1928. It is an important method to investigate molecular vibration and torsion. In the case of proteins, the Raman spectra contain a lot of important information, especially amide I and amide III, which are very sensitive to conformation changes. Raman spectroscopy can be used to monitor the dynamic change of structural information of natural fibrous protein and regenerated fibrous proteins under different treatment conditions. Besides, the qualitative, semiquantitative, and quantitative analysis of the secondary structure of the fibrous proteins (α-helix, β-sheet, β-turn, random coil) can also be obtained by characteristic peak fitting. Since the 1970s, Pezolet, Lippert, Williams, Thomas, Berjot, Copeland, and others have established different methods to determine the secondary structure of proteins [1–7]. The orientation of the molecular chain is one of the important features of fibrous proteins. Due to the dichroism of 1667 cm^{-1} in the

Shengjie Ling (ed.), *Fibrous Proteins: Design, Synthesis, and Assembly*, Methods in Molecular Biology, vol. 2347, https://doi.org/10.1007/978-1-0716-1574-4_20, © Springer Science+Business Media, LLC, part of Springer Nature 2021

amide I band, the orientation of the molecular chain can be obtained by collecting Raman data in the direction perpendicular to and parallel to the laser polarization of the long axis of fibers.

The disulfide bond is a kind of chemical bond that connects the sulfhydryl groups of two different cysteine residues in the different or same peptide chain, which plays an important role in forming the three-dimensional structure of protein molecule. The conformations of gauche–gauche–gauche, gauche–gauche–trans, and trans–gauche–trans in disulfide bond occurred at 510, 520, and 540 cm^{-1}, respectively [8]. Besides, many scholars also focus on the Raman spectroscopy of protein side chains, among which there is more research on tyrosine, tryptophan, phenylalanine, and so on [9].

Compared with infrared spectroscopy, Raman spectroscopy has many distinctive advantages. Firstly, Raman spectroscopy is not sensitive to water and is more suitable for aqueous samples. Therefore, it is a benefit for studying the solution or sol samples. Next, the low brightness of the conventional infrared light source causes a poor signal-to-noise ratio and spectrum deformation. Besides, the diameter of the Raman laser beam in its focus position only 0.2–2 mm, conventional Raman spectra with just a few samples of silk or silk protein can get better Raman spectroscopy, laser beam, and Raman microscope can be further will focus to 20μm, so Raman characterization has obvious advantages over animal monofilaments or samples of low weight [10]. Moreover, Raman spectroscopy can be used to study the orientation of silk protein chains due to its light source characteristics and the polarization of Raman scattering light. However, due to the fluorescence interference, Raman spectroscopy has limited application in the chromophore of protein, but the appearance of surface-enhanced Raman (SERS) and Fourier transform Raman spectroscopy has also overcome this problem. SERS can not only quench the fluorescence quickly but also enhance the spectral lines [11]. Fourier transform Raman spectroscopy also avoid the fluorescence interference caused by the chromophoric groups [12].

In this chapter, we review a class of characterization applications of common proteins using Raman speIn this chapter, we review a class of characterization applications of common proteins using Raman spectroscopy. The research methods of characterization of animal silk (e.g., *Bombyx mori* (*B. mori*) silk, *Antheraea pernyi* (*A. pernyi*) silk, *Nephila edulis* (*N. edulis*) spider silk) were introduced. The preparation process of animal silk material sample and Raman spectrum were briefly introduced. At last, the research method of using the Raman spectral analysis test and the related matters needing attention in spectral processing are discussed emphatically.

2 Materials

1. Animal silk: *B. mori* silk, *A. pernyi* silk, *N. edulis* spider silk. Fibrous fibroin: silk fibroin, elastin, collagen, and keratin.

2. Raman spectrometer.

3. Raman microscope should be well connected to the Raman spectrometer.

4. Light source: laser device (common laser wavelength: 488.0 nm, 514.5 nm, 632.8 nm, 785 nm, 1064 nm).

5. Detectors, such as charge-coupled device (CCD) detector, Ge detector, or InGaAs detector.

6. Light filter: filtering the scattering light of laser for avoiding disturbing Raman signals and preventing the sample from being irradiated by external radiation sources.

7. Accessories, such as Raman polarizer.

3 Methods

3.1 Preparation of Samples

1. Single silk fiber: The single silk fiber was directly mounted on the glass slide and fixed with silver adhesive.

2. Silk aqueous solution: The solid silk material was dissolved into an aqueous solution using $LiBr$–H_2O, $CaCl_2$–ETOH–H_2O system, then depositing one drop on glass slide or put the solution into cuvette (*see* **Note 1**).

3. Slim film: The silk aqueous solution was cast on a culture dish and allowed to dry at air conditions, then transfer the film to a glass slide (*see* **Note 2**).

3.2 Setup of Raman Spectrometer and Experimental Parameters

1. Check the status of the Raman spectrometer and microscope.

2. Select appropriate gratings and laser with needed wavelength and intensity. (*see* **Note 3**).

3. Calibration: Put silicon slice on the stage and focus the facula on the silicon slice followed by recording its spectrum, then the obtained Raman peak was calibrated to 520 cm^{-1} through adjusting Rayleigh wavelength (*see* **Note 4**).

4. Place the sample on the stage and focus the facula on the sample.

5. Load your experimental parameters, such as wavenumber range, times of scan, time, and laser power before the sample recording. (*see* **Note 5**).

6. For polarized Raman of single silk fiber, a polarizer was carefully placed in the light path for turning the laser light into linearly polarized light. Sample spectra at different polarization angles relative to fiber axial were recorded (*see* **Note 6**).

3.3 Spectral Preprocessing

To improve the quality of the spectrum, spectral preprocessing should be made.

1. Remove outliers: After obtaining the spectrum, it is necessary to evaluate the quality of the data and remove obvious outliers from the data carefully.

2. Improve the signal-to-noise ratio: Raman spectrum is prone to generate noise, and it may be necessary to reduce part of the data to improve the quality of the spectrum. The methods to improve the quality of the spectra include preprocessing of the sample (preconcentration, freezing samples, and so on), changing the collection settings (increasing laser power and collection time), and calculation processing after collecting spectrum (such as principal component analysis, fast Fourier transform filtering, Loess, Gaussian convolution, and Savitzky–Golay) (*see* **Note 7**).

3. Fluorescence background subtraction: The fluorescence of sample and background, and thermal fluctuation of CCD have an obvious influence on the spectral baseline. Therefore, the spectral baseline needs to be corrected.

4. Spectrum normalization: In some cases, the spectrum needs to be normalized to correct the variables in the experiment after correcting the baseline. Vector normalization and min–max normalization are two common methods with a wide range of applications. The 1615 cm^{-1} in silk protein (assigned to C–C vibration in the aromatic ring) and the 1450 cm^{-1} in keratin (bending vibration of CH_2 and CH_3) are commonly used as internal standards for Raman spectrum normalization.

3.4 Spectral Analysis

3.4.1 Peak Assignment of Protein Samples

As shown in Table 1, characteristic bands of the amide group in the Raman spectrum are summarized. And Table 2 exhibits the characteristic peaks of amino acids, S–S, C–N, and other chemical bond vibrations of fibrous protein in Raman spectroscopy. Besides, the assignment of Raman characteristic bands of animal silk protein chain conformation is also summarized in Table 3.

Take regenerated silk fibroin (RSF) as an example here. The Raman spectra of RSF solution and micro hydrogel with different treatments are shown in Fig. 1. Compared with RSF micro hydrogel and RSF/HPMC sol, the characteristic band of amide I in the Raman spectra of ripened naturally and ripened with ethanol is sharper and the peak width is narrower, and the amide III band change from 1252 cm^{-1} to 1229 cm^{-1}, which suggest that the

Table 1
Characteristic bands of amide group [10]

Amide vibration type	Frequency range/cm^{-1}	Attribution
Amide A	3300	N–H
Amide B	3100	NH *str* (at Fermi resonance)
Amide I	1597 ~ 1680	C=O *str*, C–N *str*, N–H *def*
Amide II	1480 ~ 1575	N-H *def*, C–N *str*
Amide III	1229 ~ 1305	C–N *str*, N–H in-plane bending
Amide IV	625 ~ 767	O=C–N in-plane bending
Amide V	640 ~ 800	N–H out-of-plane bending
Amide VI	537 ~ 606	C=O out-of-plane bending
Amide VII	200	C–N *tor*

str stretching vibration, *def* deformation vibration, *tor* torsion

transformation transit from random coil to β-sheet. Besides, RSF hydrogels appear new peaks around 974 and 1082 cm^{-1}, while peaks in 938 and 1415 cm^{-1} disappear, indicating that the content of β-sheet in matured RSF micro hydrogel is significantly increased.

3.4.2 Content of Secondary Structure Calculating by Peak Fitting

Knowing the content of the secondary structure of the protein sample is very important for understanding the relationship between structure and property, and the Raman spectrum can help to achieve this goal through deconvolving the amide I band or amide III band.

Take keratin in wool fiber as an example here. The corresponding Raman spectra of amide I in keratin appears at 1650–1680 cm^{-1}, and the strong band at 1655 cm^{-1} can be attributed to α-helix, and the weak shoulder peak at 1675 cm^{-1} is ascribed to β-sheet. Therefore, the amide I band can be divided into Gaussian and Lorentzian bands, corresponding to the α-helix (1654 cm^{-1}), β-sheet (1677 cm^{-1}), and the v(C–C) (1619 cm^{-1}). The results of peak fitting are shown in Fig. 2. As the treatment time of oxygen plasma gradually increases, the area of the characteristic peak corresponding to α-helix (1654 cm^{-1}) gradually decreases, while the area of the peak corresponding to β-sheet (1677 cm^{-1}) gradually increased. According to the peak fitting results of amide I, the content of different conformation of keratin can be calculated by calculating the area proportion of each conformation. The specific result is shown in Table 4 (*see* **Note 8**).

Table 2
Characteristic peaks of amino acids, S–S, C–N, and other chemical bond vibrations of fibrous protein in Raman spectroscopy [9, 13, 14]

Frequency (cm^{-1})	Assignment
510	S–S *str*, **gauche-gauche**, m
520	S–S *str*, **gauche-trans**, m
540	S–S *str*, **trans-trans**, m
621	Phenylalanine, w
643	Tyrosine, w
661	C–S *str* of cystine, vw
750	Tryptophan, m
830/850	Doublet bands of tyrosine, m
890–945	C–C *str* of α-helix, vw
1004	Ring vibrations of tryptophan and phenylalanine
1031	Vibration of phenylalanine, w
1045	S–O str of cysteine, w
1079	C–N *str*, w
1126	C–N *str*, w
1176	C–N *str*, w
1248	Random coil in amide III
1318	C_α–H bending, s
1450	CH_2 and CH_3 *tor*, vs
1535	N–H *tor* of amide II
1553	Tryptophan bending, w
1617	Vibration of tyrosine, w
1653	α-Helical of amide I
1660	β-Sheet of amide I
1670	Random coil of amide I
2565	S–H *str* of cysteine

str stretching vibration, *tor* torsion
s strong, *m* medium, *w* weak, *vw* very weak, *vs* very strong

3.4.3 Analyzing Qualitatively the Orientation of Molecular Structure

The orientation of specific moieties can be obtained from the polarizing angular dependence (the angle between the polarized light and the fiber axis) of the intensity at a certain wavenumber, which corresponds to a certain vibration of the molecular group.

Table 3
Characteristic peaks of silk fibroin protein secondary structures [15, 16]

Secondary Structures	Amide I (cm^{-1})	Amide III (cm^{-1})
α-Helix	1645 ~ 1660	1260 ~ 1300
β-Sheet	1665 ~ 1680	1229 ~ 1240
β-Turn	1640 ~ 1645 1680 ~ 1690	1305
Random coil	1660 ~ 1665	1240 ~ 1250

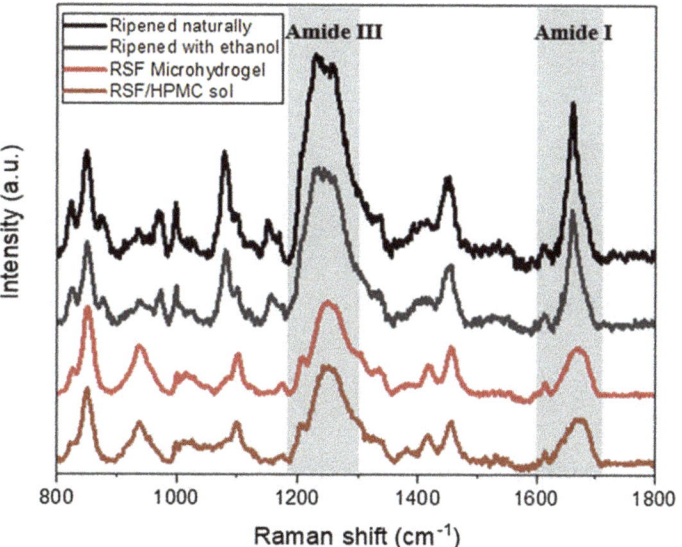

Fig. 1 Raman spectra of RSF micro hydrogel and solution: gelation and ripening process of 30%-micro hydrogel [17]

Take *B. mori* cocoon silk and *B. mori* RSF fiber as an example here. After focusing the facula on a single fiber with a microscope, the Raman scattering data are collected on the silk fiber axis with vertical and parallel to the laser polarization direction. The obtained spectra are shown in Fig. 3 (*see* **Note 9**). *B. mori* silk has obvious dichroism at the amide I band (1667 cm^{-1}) (other kinds of silk fiber is at 907 cm^{-1}). Therefore, $I_{(1667)v}/I_{(1667)p}$ (vertical/parallel)) was used to evaluate the degree of molecular chain orientation in silk fibers. The larger the value of $I_{(1667)v}/I_{(1667)p}$, the greater the orientation of molecular chains. As shown in Fig. 3, from the RSF-1x to the RSF-6x, the value of $I_{(1667)v}/I_{(1667)p}$ increase, which indicated the orientation of RSF fiber is gradually increase, and the orientation of RSF-6x is closed to that of natural cocoon silk (*see* **Note 10**).

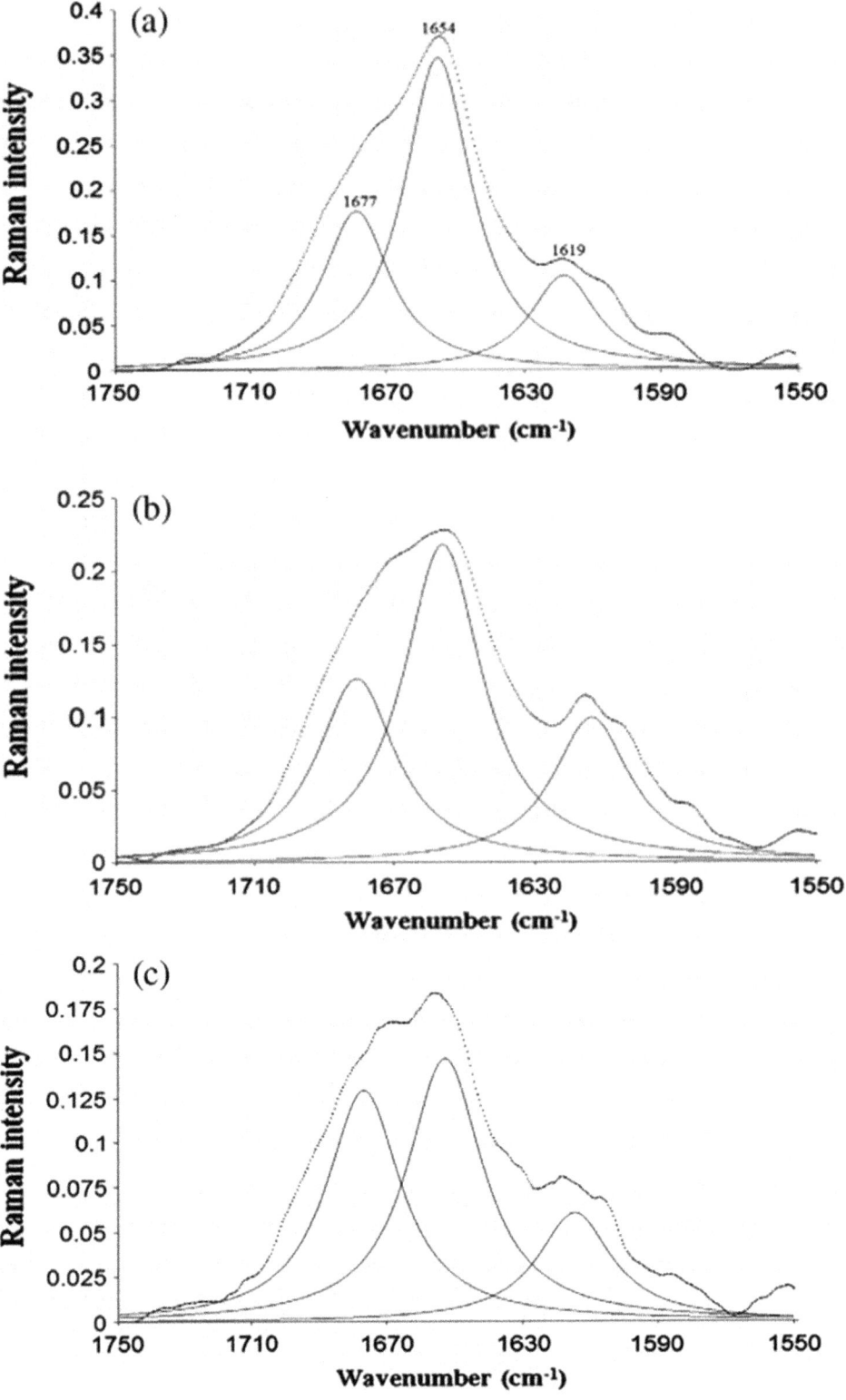

Fig. 2 Deconvolution of the amid I region of the Raman spectra of the wool fiber: (**a**) untreated wool and oxygen plasma-treated wool sample for (**b**) 2.5 min and (**c**) 5 min [18]

Table 4
Characteristic conformation of α-helix (1654 cm^{-1}) and β-structure (1677 cm^{-1}) in Raman spectra of wool fiber [18]

Sample	α-helix (%)	β-pleated sheet (%)
Untreated wool	68	32
2.5 min plasma treatment	65	35
5 min plasma treatment	52	48

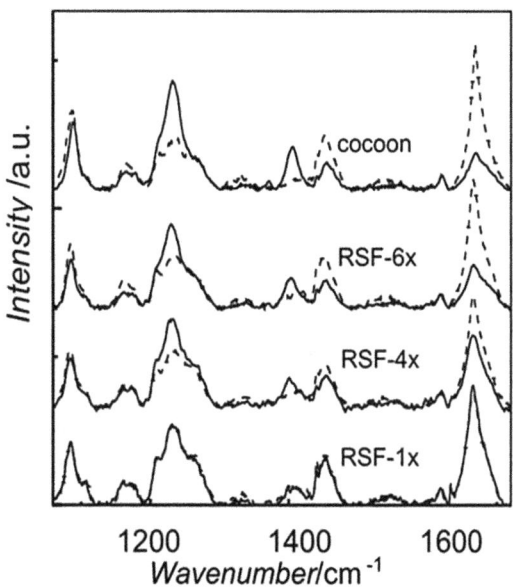

Fig. 3 Raman dichroism spectra of RSF fibers and natural cocoon silk. Single fiber aligned perpendicular (dashed line) or parallel (solid line) to the direction of polarization of the laser beam [8]

4 Notes

1. Other kinds of protein solutions using the same method to obtain the Raman spectrum.

2. Other kinds of films (except silk film) were obtained using the same method described in Subheading 3.1.

3. Gratings were classified by the number of lines per millimeter. The resolution of the spectrum is positively associated with the number of lines, but the testing range is negative correlated with the number of the line. Therefore, the choice of gratings requires careful consideration. Besides, the laser power also needs attention. The sample can be damaged while the power is too high, such as scorch sample or single silk fibers break.

4. The calibration process should be repeated after changing the wavelength of the laser or grating.

5. To obtain a more accurate spectrum, each spectrum is collected 3 times or more.

6. It should be noted that the angle displayed on the polarizer does not represent the real angle between the polarized light and the fiber axis, which depends on the direction of samples placed on the stage. The suitable method to ensure the polarized angle is to place the sample in one specific direction, then collecting Raman spectra while the analyzer changes direction from 0 to180. Such, the real direction of 0 and 90 angles can be determined according to the principle of orthogonal extinction.

7. The methods described in Subheading 3.3 do significantly improve the signal-to-noise ratio, but they also can distort the spectrum. Therefore, the smoothing step needs to be used carefully.

8. The procedure of peak fitting in the Raman spectrum is the same as that in the FTIR spectrum as described in chapter 17.

9. The Raman spectra at the direction perpendicular and parallel to the direction of polarization of the laser beam must be collected in the same position on the sample.

10. The two Raman spectra at the direction perpendicular and parallel to the direction of polarization of the laser beam need to be normalized at 1615 cm^{-1} before calculating the value of $I_{(1667)v}/I_{(1667)p}$.

Acknowledgments

This work was supported by the National Natural Science Foundation of China [grant numbers. 51973116, U1832109, 21935002], the Users with Excellence Program of Hefei Science Center CAS [grant number 2019HSC-UE003], the starting grant of ShanghaiTech University, and State Key Laboratory for Modification of Chemical Fibers and Polymer Materials.

References

1. Pezolet M, Pigeon-Gosselin M, Coulombe L (1976) Laser Raman investigation of the conformation of human immunoglobulin G. Biochim Biophys Acta 453:502–512

2. Ishizaki H, Balaram P, Nagaraj R, Venkatachalapathi YV, Tu AT (1981) Determination of beta-turn conformation by laser Raman spectroscopy. Biophys J 36(3):509–517

3. Lippert JL, Tyminski D, Desmeules PJ (1977) Determination of the secondary structure of proteins by laser Raman spectroscopy. ChemInform 8:7075–7080

4. Williams RW, Dunker AK (1981) Determination of the secondary structure of proteins from the amide I band of the laser Raman spectrum. J Mol Biol 152(4):783–813

5. Thomas GJ, Agard DA (1984) Quantitative analysis of nucleic acids, proteins, and viruses by Raman band deconvolution. Biophys J 46:763–768

6. Berjot M, Marx J, Alix AJP (1987) Determination of the secondary structure of proteins from the Raman amide I band: the reference intensity profiles method. J Raman Spectrosc 18:289–300

7. Dousseau F, Pezolet M (1990) Determination of the secondary structure content of proteins in aqueous solutions from their amide I and amide II infrared bands. Comparison between classical and partial least-squares methods. Biochemistry 29:8771–8779

8. Zhou G, Shao Z, Knight DP, Yan J, Chen X (2009) Silk fibers extruded artificially from aqueous solutions of regenerated Bombyx mori silk fibroin are tougher than their natural counterparts. Adv Mater 21:366–370

9. Thomas GJ, Kyogoku Y (1977) Biological infrared and Raman spectroscopy part AC. Marcel Decker, New York, pp 717–872

10. Monti P, Taddei P, Freddi G, Asakura T, Tsukada M (2001) Raman spectroscopic characterization of Bombyx mori silk fibroin: Raman spectrum of silk I. J Raman Spectrosc 32:103–107

11. Tague TJ Jr, Leona M (2008) Surface enhanced Raman (SERS) nanoparticle delivery device for in-situ sample analysis. Microsc Microanal 14:1114–1115

12. Edwards WAGM (1997) Fourier-transform Raman spectroscopy of mammalian and avian keratotic biopolymers. Spectrochimica Acta A Mol Biomol Spectrosc 53A(1):81–90

13. Jones DC, Carr CM, Cooke WD, Lewis DM (1998) Investigating the photo-oxidation of wool using FT-Raman and FT-IR spectroscopies. Text Res J 68:739–748

14. Gniadecka M, Nielsen OF, Christensen DH, Wulf HC (1998) Structure of water, proteins, and lipids in intact human skin, hair, and nail. J Investig Dermatol 110:393–398

15. Carey PR (1982) Biochemical applications of Raman and resonance Raman spectroscopies. Academic Press, London

16. Bandekar J, Krimm S (1980) Vibrational analysis of peptides, polypeptides, and proteins. VI. Assignment of β-turn modes in insulin and other proteins. Biopolymers 19(1):31–36

17. Dong T, Mi R, Wu M, Zhong N, Shao Z (2019) The regenerated silk fibroin hydrogel with designed architecture bioprinted by its microhydrogel. J Mater Chem B 7:4328–4337

18. Barani H, Haji A (2015) Analysis of structural transformation in wool fiber resulting from oxygen plasma treatment using vibrational spectroscopy. J Mol Struct 1079:35–40

Chapter 21

Structural Characterization of Silk Fibers by Wide-Angle X-Ray Scattering

Zhuochen Lv, Ping Qi, Leitao Cao, and Shengjie Ling

Abstract

As one of the most advanced techniques to gain insight into the structure of the materials, wide-angle X-ray scattering (WAXS) records the scattering information at wide angles which typically larger than 5° (2θ), where contains abundant and detailed atomic-scale structure information of the matter. To improve the intensity and time-resolution, the WAXS can be further coupled with a synchrotron light source. The resultant technique, that is, Synchrotron WAXS can reach and even surpass the spatial and time resolution of 0.1 nm and microsecond scale, respectively, thus is very suitable for characterization of animal silks both statically and quasi-dynamically. This chapter would show methods to understand the structure–property relationship of animal silks by WAXS.

Key words Wide-angle X-ray scattering, Animal silks, Structure–property relationship

1 Introduction

Nowadays, the rapid development of tissue engineering techniques and the increasing demands of safer and efficient biomedical materials requires the exploration and deep investigation of novel biocompatible materials [1]. Unlike the massive kinds of synthetic polymers, animal silks, as a distinguished natural material, are endowed with unique structure–property relations, nearly perfect biocompatibility, and abundant natural resources. The utilization of the most high-end techniques like WAXS and SAXS (small-angle X-ray scattering) allows people to unveil the structure–property mystery spread out by the animal silks.

X-ray diffraction occurs when X-ray travels through the matter, the resulted scattering patterns are closely linked to the microscale structures of the sample (by which way, WAXS can distinguish the crystalline and amorphous polymers). WAXS records the scattering information at wide scattering angles which typically larger than 5°,

Shengjie Ling (ed.), *Fibrous Proteins: Design, Synthesis, and Assembly*, Methods in Molecular Biology, vol. 2347, https://doi.org/10.1007/978-1-0716-1574-4_21, © Springer Science+Business Media, LLC, part of Springer Nature 2021

where stores the nanoscale structure information. As silk based fibrous proteins are generally regarded as semicrystal polymers, it is undoubtedly that WAXS can investigate their microstructural features such as crystallinity and orientation degree with high efficiency. WAXS is used to determine the crystalline structure not only for silks but also for other kinds of materials. For instance, Ye et al. [2] reported an anisotropic nanofiber-structured cellulose film, the variation of the orientation degree of which was investigated by WAXS and SAXS, the results show that the orientation degree will increase with prestretching strain. Jurkiewicz et al. [3] used WAXS to study the atomic structure of carbon materials prepared from natural tannin by high-temperature pyrolysis and low-temperature hydrothermal carbonization. Sebastiani et al. [4] determined the melting transition of DNA through WAXS, a significant crystalline phase in DNA was detected. Lin et al. [5] investigated the secondary structure differences between force-reeled silks and commonly used cocoon silk fibers through WAXS, stronger diffraction patterns were found for force-reeled silks indicating higher crystallinity.

Synchrotron radiation wide-angle X-ray scattering (SR-WAXS) is endowed with advantages such as high intensity and brighter collimation light, which can obtain more detailed and precise 2D diffraction patterns at a higher efficiency compared with WAXS, yet the principle of it is still the same as WAXS (in the following text, the term WAXS is used to represent both SR-WAXS and WAXS).

In all, WAXS can reflect microscale structural information such as crystallinity, orientation degree of molecules, and crystal size; furthermore, the evolution of the corresponding structure–property relations of animal silk during specific dynamic treatment can be systematically investigated. This chapter is a practical guide to WAXS which starts with sample preparation, then followed by the mathematical methodology of analyzing WAXS pattern.

2   Materials

1. *Bombyx mori* (*B. mori*) silk fiber.

2. *Antheraea pernyi* (*A. pernyi*) silk fiber.

3. Take one fiber at a time, stick the fiber to a paperboard which is hollowed in the center.

4. Continue sticking, the prepared closely packed silk fiber bundle should be endowed with a length and width larger than those of the X-ray light spot. The sticking procedure should ensure each fiber is well attached to the paperboard (Fig. 1).

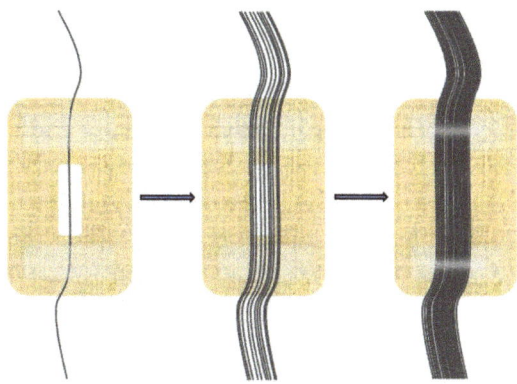

Fig. 1 Schematic diagram of sample preparation procedure

3 Methods

3.1 Basic Operational Procedure of WAXS

1. Put the sample (attached to the paperboard) onto the sample stage.

2. Adjust the relative position between the X-ray beam center and the sample and make the light beam focused on the sample.

3. Conduct quick scanning to figure out the exact position and the spatial range of the sample (during this procedure, one may conduct several trials to find the exact range where the signals emerge).

4. Set parameters such as exposure time. The exposure time of each sector (If the detector of the WAXS instrument cannot acquire the whole scattering pattern at once, it will conduct the measurement by dividing the whole target region into several different yet adjacent sectors and then carry out acquisition for each sector) depends on the intensity of the X-ray, the proper duration time should result in images with high resolution. Then, start pattern acquisition.

5. Move detector to other sectors and start pattern acquisitions (this step will be automatically carried out, the results of which are separate partial scattering patterns. The total scattering pattern is obtained by splicing patterns from all the sectors.

3.2 Data Analysis

3.2.1 Data Acquisition Based on the Operational Procedure of Fit2D Software

One may use Fit2D (a software specialized in WAXS/SAXS data analysis) to acquire information (e.g., WAXD pattern/Azimuth pattern) from scattering patterns. The basic operational procedure of Fit2D is provided below.

1. Base on experimental conditions, input the initial parameters like the dimension of the scattering pattern.

2. Load scattering pattern, locate the center of the light beam and mask off the beam stops by using the embedded mask tool.

3. Revise the experiment parameter such as the size of pixels, the S-D distance, and the wavelength of X-ray.

4. Determine the integral range of target peak (starting and end azimuth angle, inner and outer radial limits).

5. Choose the scan type (Q-space, 2-Theta, etc.) and input the corresponding number of azimuthal bins and radial/2-theta bins. For instance, if choose "Q-space" as scan type, then input "360" for azimuthal bins and "1" for radial/2-theta bins, one can gain the azimuthal integration graph; if choose "2-Theta" as scan type, then input "1" for azimuthal bins and number "360" for radial/2-theta bins, 2θ-Intensity diagram (WAXD pattern) can be acquired.

3.2.2 Calculation of Crystallinity

1. Acquire WAXD pattern (*see* **Note 1**), that is, 2θ-Intensity diagram, from WAXS pattern based on Fit2D.

2. Deconvolute each spectral peak into specific characteristic peaks standing for secondary structures like β-sheet and α-helix, which may involve the utilization of PeakFit4.0 software. PeakFit4.0 deconvolutes the spectral band into different Gaussian peaks automatically, during which one should select the number and the specific locations of Gaussian peaks and make the positions locked. Thus, only the magnitude and the width of each Gaussian peak will fluctuate during the fitting process. What is worth noting is that the position selection for each Gaussian peak should refer to the position of characteristic peaks of each conformation of the animal silk (*see* **Note 2**).

3. Conduct integration over all the peaks standing for crystalline conformation, the result of which is A_C (mathematically speaking, A_C equals to the total area of these peaks). The same procedure is done to gain A_t, which represents the total area enclosed by the 2θ-Intensity curve.

4. Crystallinity X_c then can be determined by Eq. 1.

$$X_c = \frac{A_a}{A_t} \times 100\% \tag{1}$$

3.2.3 Determination of Orientation Degree

f_C, that is, Herman's orientation parameter, which ranges from −0.5 to 1, is a reference parameter indicating the extent of chain orientation. $f_C = 1$ indicates completely axial oriented; $f_C = 0$ indicates randomly oriented; while, $f_C = -0.5$ implies the perpendicular orientation of molecules. It can be calculated through

$$f_c = \frac{3\cos^2\varphi - 1}{2} \tag{2}$$

$$\cos^2\varphi = \frac{\int_0^{\frac{\pi}{2}} I(\varphi) \cos^2\varphi \sin\varphi d\varphi}{\int_0^{\frac{\pi}{2}} I(\varphi) \sin\varphi d\varphi} \tag{3}$$

where φ is the azimuthal angle of specific crystal planes, (020) and (210) are commonly chosen for the calculation of silk fibers.

1. Determine the interval of the azimuthal angle of the chosen crystal plane (take (210) for example, the center of such an interval is 0° or 180°). Such an interval should contain specific starting and end azimuth angle as well as inner and outer radial limits.

2. Conduct azimuth integral over the determined interval.

3. Export azimuth-intensity data and use PeakFit4.0 to conduct Gaussian fit, after which the deconvoluted Gaussian peaks are gained.

4. Find the Gaussian peak standing for the plane (210) and export the corresponding azimuth (φ)-intensity ($I(\varphi)$) data.

5. Calculate the integral $\int_0^{\frac{\pi}{2}} I(\varphi) \cos^2\varphi \sin\varphi d\varphi$ and $\int_0^{\frac{\pi}{2}} I(\varphi) \sin\varphi d\varphi$ via *Origin mathematics*. Mathematically, those integrals can be determined by the area enclosed by the corresponding curves.

6. $\cos^2\varphi$ is the ratio of $\int_0^{\frac{\pi}{2}} I(\varphi) \cos^2\varphi \sin\varphi d\varphi$ to $\int_0^{\frac{\pi}{2}} I(\varphi) \sin\varphi d\varphi$, which is expressed as Eq. 3. Then f_c can be calculated through Eq. 2.

3.2.4 Calculation of Crystal Size

Since X-ray diffraction (XRD) pattern can be also acquired through WAXS, XRD methodology still works. It is worth noting that WAXS can reflect the crystal structure at the nanoscale. Start with the classic Debye–Scherrer equation.

$$d = \frac{0.9\lambda}{B\cos\theta} \tag{4}$$

where d is the crystal size along a specific direction defined by the Debye-Scherrer equation, B is termed as the peak full width at half maximum, λ is the wavelength of X-ray, and θ is the scattering angle. The procedures of acquiring the crystal size are listed below.

1. Acquire WAXD pattern (2θ-intensity diagram) from Fit2D in the manner provided in Subheading 3.2.1.

2. To fit the peaks with the Gaussian model, that is, use Peak-Fit4.0 to deconvolutes the original peak into position-fixed Gaussian peaks (the position of which should accord to crystal planes) and assign the crystal planes to each Gaussian peak correctly.

3. Then, based on each Gaussian peak (for example, peak (hkl) at some certain 2θ), B (the peak full width at half maximum) can be gained. Substitute B and θ into Eq. 4, crystal size, that is, d, along (hkl), is found accordingly.

3.2.5 Estimation of Lattice Parameters

The amino acid sequence of the crystalline region of animal silk can be regarded as poly(GA) segment with a monoclinic space group and the β-sheet crystallites can be perceived as monoclinic crystals [6–9]. For monoclinic crystals, the following formula is well developed to determine the distance between specific crystal plane, which expressed as

$$d_{hkl} = \frac{1}{\sqrt{\frac{h^2}{a^2 \sin^2 \beta} + \frac{l^2}{c^2 \sin^2 \beta} + \frac{2hk \cos \beta}{ac \sin^2 \beta} + \frac{k^2}{b^2}}} \tag{5}$$

and the Bragg equation is expressed as

$$d_{hkl} = \frac{n\lambda}{2 \sin \theta} \tag{6}$$

where n is normally 1 in Eq. 6 indicating the first-order diffraction, d_{hkl} is the distance between the (hkl) crystal planes.

1. What is worth noting is that WAXS only provides raw data patterns after measurements (*see* **Note 3**), one needs to acquire WAXD pattern (2θ-Intensity diagram) from Fit2D in the manner provided in Subheading 3.2.1.

2. To fit the peaks with Gaussian model, that is, use PeakFit4.0 to deconvolutes the original peak into position-fixed Gaussian peaks (the position of which should accord to crystal planes) and assign the crystal planes to each Gaussian peak correctly.

3. Then, based on each Gaussian peak (peak (hkl) at some certain 2θ value), all the possible d_{hkl} can be determined via Eq. 6 (for 2θ value at each (hkl) crystal plane is known).

4. List all the possible d_{hkl} in a manner, for example, the results of the calculation are

$$d_{010} = 1.43 \text{ nm}; d_{200} = 0.75 \text{ nm};$$
$$d_{020} = 0.65 \text{ nm}; d_{002} = 0.54 \text{ nm};$$
$$d_{021} = 0.54 \text{ nm}; d_{030} = 0.49 \text{ nm}$$

and substitute those calculation results into Eq. 5.

5. After substitution, equation set composed of unknowns {a, b, c, β} can be derived.

6. Solve the equations, estimation can be conducted a : b : c ≈ 1.50 : (1.25, 1.47) : 1.08 (where an assumption that sinβ ≈ 1 is introduced, the monoclinic crystal is now close to orthorhombic system), all the estimation is conducted at nanoscale.

---

4 Notes

1. When tensile load is applied on the sample to proceed in situ WAXS experiment, it is necessary to guarantee that no relative slippage between the silk fiber bundle and paperboard happens.

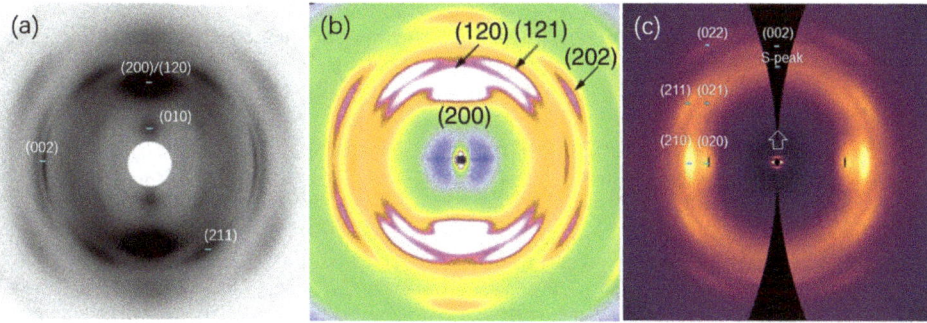

Fig. 2 Typical WAXS patterns of animal silks. (**a**) *Bombyx mori* silk [7], the crystal plane is marked abased on the work of Fang et al. [10]; (**b**) *Antheraea pernyi* silk [11]; (**c**) *Argiope bruennichi* spider silk [12]

2. The principle of WAXS is almost the same as that of X-ray diffraction (XRD). XRD-pattern like WAXD pattern (i.e., the 2θ-Intensity diagram derived from WAXS pattern) can be gained by WAXS. When deconvoluting the original data into different Gaussian peaks, the position/location of each Gaussian peak as well as the number of Gaussian peaks should accords to those of the probable conformations of animal silk, that is, each Gaussian peak should locate at some almost fixed 2θ/ azimuthal angle/q-value.

3. Typical WAXS patterns of animal silks are shown in Fig. 2 [7, 10–12].

Acknowledgments

Zhuochen Lv and Ping Qi contributed equally to this chapter. The protocol is based on the Beamline station BL16B1 of Shanghai Synchrotron Radiation Facility (SSRF). This work was supported by the National Natural Science Foundation of China [grant numbers. 51973116, U1832109, 21935002], the Users with Excellence Program of Hefei Science Center CAS [grant number 2019HSC-UE003], the starting grant of ShanghaiTech University, and State Key Laboratory for Modification of Chemical Fibers and Polymer Materials.

References

1. Yigit S, Dinjaski N, Kaplan DL (2016) Fibrous proteins: at the crossroads of genetic engineering and biotechnological applications. Biotechnol Bioeng 113:913–929

2. Ye D, Lei X, Li T, Cheng Q, Chang C, Hu L, Zhang L (2019) Ultrahigh tough, super clear, and highly anisotropic nanofiber-structured regenerated cellulose Fims. ACS Nano 13:4843–4853

3. Jurkiewicz K, Hawełek Ł, Balin K, Szade J, Braghiroli FL, Fierro V, Celzard A, Burian A (2015) Coversion of natural tannin to hydrothermal and graphene-like carbons studied by wide-angle X-ray scattering. J Phys Chem A 119:8692–8701

4. Sebastiani F, Pietrini A, Longo M, Comez L, Petrillo C, Sacchetti F, Paciaroni A (2014) Melting of DNA nonoriented fibers: a wide-angle X-ray diffraction study. J Phys Chem B 118:3785–3792

5. Lin S, Wang Z, Chen X, Ren J, Ling S (2020) Ultrastrong and highly sensitive fiber microactuators constructed by force-reeled silks. Adv Sci 7:1902743

6. Warwicker JO (1954) The crystal structure of silk fibroin. Acta Cryst 7:565–573

7. Drummy LF, Farmer BL, Naik RR (2007) Correlation of the beta-sheet crystal size in silk fibers with the protein amino acid sequence. Soft Matter 3:877–882

8. Hayashi CY, Shipley NH, Lewis RV (1999) Hypothses that correlate the sequence, structure, and mechanical properties of spider silk proteins. Int J Biol Macromol 24:271–275

9. Keten S, Xu Z, Ihle B, Buehler MJ (2010) Nanoconfinement controls stiffness, strength and mechanical toughness of beta-sheet crystals in silk. Nat Mater 9:359–367

10. Fang G, Sapru S, Behera S, Yao J, Shao Z, Kundu SC, Chen X (2016) Exploration of the tight structural-mechanical relationship in mulberry and non-mulberry silkworm silks. J Mater Chem B 4:4337–4347

11. Yang K, Guan J, Numata K, Wu C, Wu S, Shao Z, Ritchie RO (2019) Integrating tough *Antheraea pernyi* silk and strong carbon fibres for impact-critical structural composites. Nat Commun 10:3786–3797

12. Riekel C, Burghammer M, Dane TG, Ferrero C, Rosenthal M (2017) Nanoscale structural features in major Ampullate spider silk. Biomacromolecules 18:231–241

Chapter 22

Structural Characterization of Silk Fibers by Small-Angle X-Ray Scattering

Zhuochen Lv, Ping Qi, Leitao Cao, and Shengjie Ling

Abstract

Nature is rich in all kinds of unbelievably designed microstructures, which endows the natural materials with various fantastic performances. Unveiling the mystery of the sophisticated configurations and the relationship between those microstructures and the corresponding functions is helpful for the manufacture of artificial functional materials. Small-angle X-ray scattering (SAXS) is an advanced tool to gain the microstructural features of materials within a spatial scale much larger than the atomic scale (1000 nm), which can be carried out along with WAXS to conduct more systematic investigation over different kinds of materials. With the help of SAXS/WAXS, one may generate an insightful understanding of the mechanisms of structure–property evolution (which is efficient guidance for the artificial material designs). This chapter will introduce the mathematics and the methodologies used by SAXS when investigating the microstructure of natural materials.

Key words Natural materials, Small-angle X-ray scattering, Structure–property evolution

1 Introduction

Fibrous proteins (such as fibroin from silk fibers) are commonly used in biomedical applications due to their excellent intrinsic biocompatibility, abundant resources, and excellent performance. To reveal the relationship between the microstructure and the performance of natural materials, BioSAXS (biological small-angle X-ray scattering) was once adopted for the analysis of the biological macromolecules and tissues. Characterized through SAXS, abundant information, such as the structural features of both ordered and disordered proteins in solution, the size and shapes of proteins, and the orientation degree of bioparticles/voids (e.g., the orientation degree of cartilage in lesions can be obtained through SAXS) can be determined [1, 2].

When X-ray travels through the matter, scattered X-ray will emerge within a small angle range around the orientation of the incident X-ray (i.e., small-angle scattering) due to the ununiformly

Shengjie Ling (ed.), *Fibrous Proteins: Design, Synthesis, and Assembly*, Methods in Molecular Biology, vol. 2347,
https://doi.org/10.1007/978-1-0716-1574-4_22, © Springer Science+Business Media, LLC, part of Springer Nature 2021

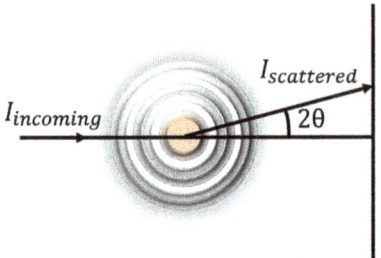

Fig. 1 Schematic graph of scattering

distributed electron density (the scale of which usually larger than 1 nm) inside the material. Distinguished from WAXS, SAXS merely records the scattering information at small angles which typically ranges from $0.1°$ to $5°$ (2θ) and can reflect the structure information at a relatively larger scale (1–1000 nm) than WAXS, the schematic diagram of scattering is presented in Fig. 1. Since the studied objects (the target scale) for SAXS are larger than those of WAXS, the signals of crystal information will be no longer significant, but the signals of macromolecular (such as the packed crystallites and biomolecules) information will dominate the SAXS pattern. SAXS is thus widely used to thoroughly investigate the morphology of polymer materials.

Based on the abovementioned principle, the shape and the size, the distribution of the particles (including voids) and their orientation can be studied by analyzing the SAXS patterns. For instance, the structure information of varied kinds of animal silks was studied by Londono et al. [3–5] through synchrotron SAXS, during which the evolution of morphological parameters such as the molecular size of silk protein, the invariant value, and distribution of the particle size, were investigated thoroughly. Grubb et al. [6] studied the microstructure characteristics of spider silk under different strains and varied relative humidity conditions through the same method (i.e. in situ SAXS). When Yang et al. [7] conducted SAXS test over dried and wet spider dragline silk, different patterns emerged due to different sample conditions. Yang et al. conducted their conclusion based on the observation and the analysis over SAXS pattern that the minimum width of the equatorial streak in the fiber axis direction implies the existence of 0.1 μm long diffracting objects along fiber direction and the aspect ratio of these objects is over 10:1. Microvoids inside polymer fiber was also studied, Th u nemann et al. [8] determined length and the diameter as well as the approximate content of microvoid inside the pristine fibers, they also elaborated the principle of their used methodology very concisely (Fig. 2).

SAXS instrument nowadays (e.g., Xeuss-3.0) can characterize particles and interface scaled larger than 500 nm without signal

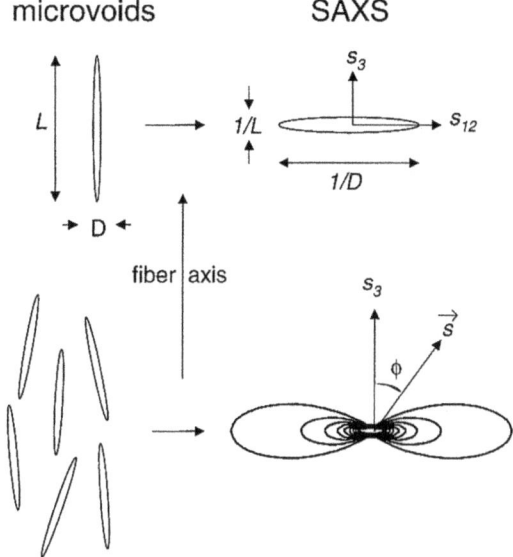

Fig. 2 The schematic representation of the methodology of microvoids characterization via SAXS [8]

absence caused by beamstops. In addition, the cosmic microwave background is also considered during SAXS trials. Both features talked above result in a more precise output of scattering pattern and an easier data processing procedure. Above all, SAXS is of great importance while comprehending the material structure at the microscale level. This chapter will introduce the methods for concise semiquantitative analysis over small-angle X-ray scattering patterns of animal silks.

2   Materials

Samples are prepared in the same manner as WAXS. All the listed preparation process accords to that of WAXS.

1. *Antheraea pernyi* (*A. pernyi*) silk fiber.
2. Take one fiber at a time and stick the fiber to a paperboard which is hollowed at the center.
3. Continue sticking; the prepared closely packed silk fiber bundle should be endowed with a length and width larger than those of the X-ray light spot. The sticking procedure should ensure that each fiber is well attached to the paperboard to avoid slippages (*see* **Note 1**).

3 Methods

3.1 Basic Operational Procedure of SAXS

1. Move the detector to the right position (*see* **Note 2**).

2. Same as WAXS, AgBeH (silver behenate powder) powder is chosen to be the standard sample used for calibration.

3. Put the sample (attached to the paperboard) onto the sample stage and get prepared for testing.

4. Adjust the position of the beam center by conduct quick scanning to figure out the exact position and the spatial range of the sample (during this procedure, one may conduct several trials to find the exact range where the signals emerge).

5. Set exposure time for each sector (this condition emerges if the instrument cannot acquire the whole pattern at once) which depends on the intensity of the X-ray. The proper duration time should result in patterns with high resolution.

6. When the acquisition is done for one sector, auto adjustment would be done by the instrument by moving the detector to other sectors. The total scattering pattern is obtained by splicing patterns from all the sectors.

3.2 Data Analysis

3.2.1 Data Acquisition Based on Operational Procedure of Fit2D Software

Same as the data acquisition and processing procedure of WAXS, Fit2D is widely used to manipulate the original data gained by SAXS.

1. Based on experimental conditions, input the initial parameters like the dimension of the scattering pattern.

2. Load scattering pattern, locate the center of the light beam, and mask off the beamstops by using the embedded mask tool.

3. Revise the experiment parameter such as size of pixels, the S-D distance, the wavelength of X-ray, etc.

4. Determine the integral range which including starting and end azimuth angle, inner and outer radial limits.

5. Choose the scan type (Q-space, 2-Theta, etc.) and input the according number of azimuthal bins and radial/2-theta bins. For instance, if choose "Q-space" as scan type, then input "360" for azimuthal bins and "1" for radial/2-theta bins, one can gain the azimuthal integration graph (if inverse the input number, i.e., 1 for azimuthal bins and 360 for radial/2-theta bins, one can get q-value graph); if choose "2-Theta" as scan type, then input "1" for azimuthal bins and number "360" for radial/2-theta bins, 2θ-Intensity diagram can be acquired.

3.2.2
Determination of B_{obs}

B_{obs} is the full width at the half maximum of the azimuthal profile from the equatorial streak [9], which is related to the orientation and the size of the particles (voids). The related theory was introduced years ago by Perret and Ruland [10]. Since the pattern acquisition procedures is basically the same as those for WAXS, here straightly starts with semiquantitative analysis and the related discussion over different parameters.

1. The basic relation between q and s can be presented as follows.

$$s = \frac{2 \sin \theta}{\lambda} \tag{1}$$

$$q = \frac{4\pi \sin \theta}{\lambda} \tag{2}$$

thus,

$$s = q/2\pi \tag{3}$$

where q is the diffraction vector and s is the reciprocal space vector.

2. What is of great importance is that the selection of integral interval should precisely cover the whole area exhibiting valid/effective signals. That is, one should find both the accurate minimum and maximum value of the "outer radial limit." The enclosed integral area is then equally divided into many small intervals (e.g., 20–40 equally spaced intervals) as shown in Fig. 3.

3. Based on Subheading 3.2.1, for each small integral interval (of which the azimuthal angle ranged from 0° to 180°),

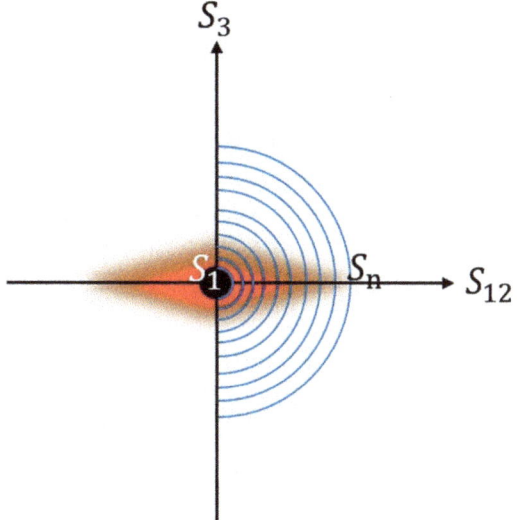

Fig. 3 Equally divided integral interval with S_1 to be the inner limit and S_n to be the outer. Each s is determined within its own interval

conduct 2θ integral within Q-space scan mode. As a result, Q-value graph will be presented by Fit2D. Read the maximum Q value in the graph (i.e., the maximum abscissa) which is set to be the q value for the corresponding integral interval. Then based on Eq. 3, the corresponding s value is calculated.

4. For each integral interval, $I(s, \varphi)$ is recorded, where s is a constant and φ is ranged from $0°$ to $180°$ (π).

5. B_{obs} is defined by

$$B_{obs} = \frac{\int I(s, \varphi) d\varphi}{I\left(s, \frac{\pi}{2}\right)} \quad (4)$$

Based on Eq. 4, each s should accord to merely one B_{obs} value. One may calculate all the B_{obs} values from the divided integral interval through Eq. 4.

6. Calculate sB_{obs}, and then use Origin to conduct linear fit over $sB_{obs} - s$ and draw the corresponding curve which is endowed with an approximate slope and an intercept. If Gaussian distribution can be used to approximate the orientation distribution of the scatterers (i.e., the particles or the voids which result in the scattering of X-ray), then the relationship can be presented as follows [8, 9].

$$sB_{obs}(s) = sB_\Psi + \frac{1}{L} \quad (5)$$

where B_Ψ reflects the orientation distribution and L represents the mean length of the scatterers.

7. Based on Eq. 5 and the linear fit result, one may find the expression or value of both B_Ψ and L.

3.2.3 Another Representation of the Orientation Degree

1. Porod's law describes the asymptote of the scattering intensity $I(q)$ for large scattering wavenumbers which yields $I(q) \sim q^{-4}$; the different power law of q indicates the different expected shapes of scatterers [11, 12]. One may thus get the insight of understanding the shape of the sample.

2. Use Fit2D to integral the pattern (*see* **Note 3**) along the vertical direction and along the horizontal direction (which is the angle equals to 0 or $90°$) to gain the one-dimensional curve. One may perceive this as to draw the Kratky plot, that is, $I(q)q^2$ vs q plot.

3. Porod's Invariant (Q_{inv}) is determined by the area enclosed by the axis and the Kratky curve.

4. The orientation degree can be easily estimated, which is as follows.

$$\text{Orientation} = \frac{Q_{inv} \text{ at } 0°}{Q_{inv} \text{ at } 90°} \quad (6)$$

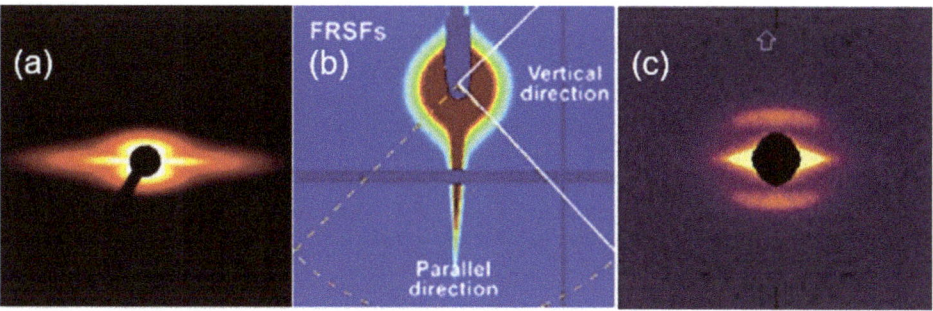

Fig. 4 SAXS of different samples, (**a**) Anisotropic nanofibers-structured cellulose films (ACFs) at 160% prestretching strain [13], (**b**) *Antheraea pernyi* silkworm silk [14] (**c**) *Argiope bruennichi* spider silk [15]

4   Notes

1. Slippages will influence the authenticity of acquired data. Especially when conducting in situ tensile SAXS/WAXS test, slippages of silk fibers should be completely avoided.

2. Since SAXS records scattering information at small angle, so detector should move away from the sample, that is, the distance between sample and detector is longer than that of WAXS.

3. Typical SAXS patterns gained from different samples (animal silks) are presented in Fig. 4.

Acknowledgments

Zhuochen Lv and Ping Qi contributed equally to this chapter. The protocol is based on the Beamline station BL16B1 of Shanghai Synchrotron Radiation Facility (SSRF). This work was supported by the National Natural Science Foundation of China [grant numbers. 51973116, U1832109, 21935002], the Users with Excellence Program of Hefei Science Center CAS [grant number 2019HSC-UE003], the starting grant of ShanghaiTech University, and State Key Laboratory for Modification of Chemical Fibers and Polymer Materials.

References

1. Kikhney AG, Svergun DI (2015) A practical guide to small angle X-ray scattering (SAXS) of flexible and intrinsically disordered proteins. FEBS Lett 589:2570–2577

2. Moger CJ, Barrett R, Bleuet P, Bradley DA, Ellis RE, Green EM, Knapp KM, Muthuvelu P, Winlove CP (2007) Regional variations of collagen orientation in normal and diseased articular cartilage and subchondral bone determined using small angle X-ray scattering (SAXS). Osteoarthr Cartil 15:682–687

3. Londono JD, Annadurai V, Urs RG, Rudrappa S (2002) Morphological parameters in varieties of silk fibers determined by small-angle X-ray scattering. J Appl Polym Sci 85:2382–2388

4. Londono JD, Annadurai V, Urs RG (2000) Small angle X-ray scattering patterns in silk fibres. Curr Sci 79:563–564

5. Londono JD, Annadurai V, Urs RG, Okuyama K, Rudrappa S (2002) Small angle X-ray scattering studies of silk fibers. J Polym Mater 19:7–12

6. Grubb DT, Jelinski LW (1996) Fiber morphology of spider silk: the effects of tensile deformation. Macromolecules 30:2860–2867

7. Yang Z, Grubb DT, Jelinski LW (1997) Small-angle X-ray scattering of spider dragline silk. Macromolecules 30:8254–8261

8. Thünemann AF, Ruland W (2000) Microvoids in Polyacrylonitrile fibers: a small-angle X-ray scattering study. Macromolecules 33:1848–1852

9. Li X, Tian F, Gao X, Bian F, Li X, Wang J (2017) WAXD/SAXS study and 2D fitting (SAXS) of the microstructural evolution of PAN-based carbon fibers during the pre-oxidation and carbonization process. New Carbon Mater 32:130–136

10. Perret R, Ruland W (1969) Single and multiple X-ray small-angle scattering of carbon Fibres. J Appl Crystallogr 2:209–218

11. Beaucage G (1996) Small-angle scattering from polymeric mass fractals of arbitrary mass-fractal dimension. J Appl Crystallogr 29:134–146

12. Beaucage G (1995) Approximations leading to a unified exponential/power-law approach to small-angle scattering. J Appl Crystallogr 28:717–728

13. Ye D, Lei X, Li T, Cheng Q, Chang C, Hu L, Zhang L (2019) Ultrahigh tough, super clear, and highly anisotropic nanofiber-structured regenerated cellulose Fims. ACS Nano 13:4843–4853

14. Lin S, Wang Z, Chen X, Ren J, Ling S (2020) Ultrastrong and highly sensitive fiber microactuators constructed by force-reeled silks. Adv Sci 7:1902743

15. Riekel C, Burghammer M, Dane TG, Ferrero C, Rosenthal M (2017) Nanoscale structural features in major ampullate spider silk. Biomacromolecules 18:231–241

INDEX

Lightning Source UK Ltd.
Milton Keynes UK
UKHW051434080922
408416UK00002B/5